PARASITES OF MAN IN NIUGINI

PARASITES OF MAN IN NIUGINI

W. H. Ewers
Lecturer in Biology
University of Papua
and New Guinea

W. T. Jeffrey
Lately Officer in charge
National Malaria School
Papua-New Guinea

THE JACARANDA PRESS

First published 1971 by
JACARANDA PRESS PTY LTD
46 Douglas Street, Milton, Q.
32 Church Street, Ryde, N.S.W.
37 Little Bourke Street, Melbourne, Vic.
142 Colin Street, West Perth, W.A.
298a Grange Road, Flinders Park, S.A.
57 France Street, Auckland, N.Z.
P.O. Box 3395, Port Moresby, P.N.G.
122 Regents Park Road, London NW1

Typesetting by Queensland Type Service Pty Ltd, Brisbane
Printed in Hong Kong

National Library of Australia
Card Number and ISBN 0 7016 8127 6

This book is affectionately dedicated
to
our long suffering wives
and
to
Guns,
Bozo,
Twisty Loo,
and
Red Ned,
whom we just happen
to be
fond of.

FOREWORD

The battle against disease has brought many heartening achievements and successes, but it is still furiously on. Its sternest encounters lie ahead. Preventable scourges in many developing countries still claim hundreds of lives and as one disease is conquered another gains new importance. In spite of the great progress being made in many fields of medicine, there are still a number of riddles to be solved.

In many villages throughout Niugini, people still believe in witchcraft and sorcery. Germ theory to them is unreal. For this reason, most of their lives have been dominated by fear.

Malaria is a great problem in many parts of the world today, including Niugini. Many Niuginians believe it is due to sorcery. Only a minority believe in the relationship between the parasite, the mosquito and man. The life cycles of all parasites found in Niugini must be fully understood before firm methods of eradication can be planned. Environment is a big factor in any eradication programme. Furthermore, people within that environment must co-operate if a degree of success is to be achieved.

In many villages throughout Niugini, local government councils are taking an increasing interest in the welfare of their people and in environmental sanitation projects. School children throughout Niugini and the South Pacific are now learning about the true causes of disease. It is through this learning process that the young people, who are emerging from the capsule of traditional fear, come to know more about the importance of parasites and germs in man.

Our country is developing. An increasing number of Niuginians are now earning substantial income from cash crops. Sanitation is improving in many villages. Some people have installed septic toilets in their homes. However, we must all adopt these ideas if we want to avoid infection of all sorts.

The book which you are about to read, has most valuable information on all of the important parasites of man in Niugini. It is written in language which many Niuginians can understand. Most of the information is local; the book indicates that no parasitological disease can be conquered unless man interferes with its life cycle. This process of interference will not take place unless all readers know where they can interfere.

I have known both authors for some years. Their book has taught me many new things about the parasites of man in Niugini. Some of the most interesting parts of this book tell the history of some important parasites.

I recommend this textbook to all Niuginian students and to the people of the Pacific area.

R. TAUREKA
District Health Officer, Goroka,
Eastern Highlands District

PREFACE

We have written this book because we think Niuginians should know more about the parasites that infect man in their country. We hope that it will be used by high school and university students, and officers of the malaria service and other departments. It will be their task in the coming years to eradicate the parasites which infect so many people and cause so much poor health. We hope that others will also read this book and find it useful and interesting.

Most of the parasites of man in this country do not cause obvious disease, but they do cause continual poor health. Poor health is difficult to measure, and it is hard to associate poor health with parasites we cannot see and often do not know we have. If a person has never been free of parasites, he does not know what it feels like to be healthy.

Some of the parasites of man in Niugini, such as malaria, hookworm and filarial worms, infect large numbers of people, and many people are infected with several parasites at the one time. There cannot be rapid economic progress in this country until its people are healthy and are able to work hard. Poor health not only prevents people from working hard, but it also inhibits (or stops) their desire to succeed.

Throughout this book we use the word 'Niugini' to mean Papua and New Guinea. It is shorter. We have hesitated to use T.P.N.G. because we dislike abbreviations, especially the colonial implications of this one. One day the people who own this country will give it their own name, and when our book gets into a second edition, in a year or two, we will use it then. Meanwhile, we hope that the name that we have used will not upset anyone who reads this book.

We should like to thank the people who have helped us with this book, especially Mrs W. Jeffrey and Mrs A. Sholl for typing; Mr A. Van Prattenberg for translating various papers from German into English; Mr S. H. Christian, MBE, for letting us use information from his early surveys in the Highlands; Mrs S. O'Neill of the Papuan Medical College Library, the staff of the library of the University of Papua and New Guinea and Mr J. Claffey of the School of Public Health and Tropical Medicine, Sydney University, for library help; Professor D. P. Drover of the University of Papua and New Guinea, Professor R. H. Black of the School of Public Health and Tropical Medicine, Sydney, and Dr W. L. Nicholas of the Australian National University, for suggestions about the manuscript; Dr. J. McMahon and Dr. R. Mylius for photographs.

Of course errors and omissions must be assigned to W. H. Ewers and W. T. Jeffrey.

Pt. Moresby, 29-6-70.

CONTENTS

Chapter 1

PARASITES AND PARASITISM

1.1 Introduction

Parasitism is a way of life where one living organism, the parasite, gets food and usually shelter and sometimes other requirements, by living in or on a larger organism called the host.

Thus parasites

(i) live inside or on the outside of the body of the host

(ii) are smaller than the host

(iii) get their food, and sometimes other essentials, from the host.

Nearly all free living animals have parasites. For example, earthworms, insects and starfish all have their own parasites. Vertebrate animals are infected with many different kinds of parasites too, and many plants are infected with parasites. About 90 species of animal parasites commonly infect man in various parts of the world; fortunately, many of these do not occur in Niugini.

Most parasites cause little or no damage to their hosts, especially if only a few parasites are present in one host. However, some parasites cause serious disease and many thousands of millions of dollars are spent every year trying to control parasites which infect man and domestic animals and plant crops.

For over a hundred years parasitologists have been doing scientific research to find out about parasites. Before we can control these parasites, we must know how they get from one host to the next, how much they affect the host, how many hosts are affected and how we can kill them and stop new infections occurring.

Vast sums of money have been spent on research on parasites and methods of controlling them. We now know a great deal about the important parasites of man and we could control all of them using the knowledge that we have. There is still a need for more research to find better ways of control, but much of the money now spent on parasitological research could be more usefully spent on control. Money, organization, education and co-operation are all that is needed to control the important parasites of man. Some of these parasites will become unimportant when social and economic conditions improve. If we all had good water supplies, houses that were screened to stop insects entering, proper toilets, and wore shoes, many of our parasites would become relatively unimportant. However, because so many of our parasites cause poor health and stop people from working hard, it is difficult to see how people can improve their own social and economic conditions without help. Money must be spent to control the parasites first—then social and economic progress will follow.

This book is concerned only with the animal parasites of man. Other kinds of organisms are also parasites of man, for example, fungi (which are plants) cause skin diseases such as grilly (*Tinea imbricata*), and bacteria and viruses cause many serious diseases of man. These other parasites of man are traditionally studied by bacteriologists and virologists; not by parasitologists, who study parasitic animals.

Most parasites can only live in one host or a few related hosts. For example, many of the parasites of man will live in some other mammals, a few will live in certain monkeys which are closely related to man, and a few will not live in any other animal except man. This restriction of parasites to certain hosts is known as host-specificity. Some parasites are very host-specific, that is they can only live in, or on, one host. Others can live in a wider range of hosts and are not so host-specific. It is important to know how host-specific each parasite is, and we must also know the hosts that it can infect. If a parasite was common in pigs, dogs and man, any attempt to control the parasite by treating man only, would fail if cross infection was continually occurring. In a case such as this, pigs and dogs are called the reservoir hosts. They would supply a steady stream of infection to man, in the same way as a water reservoir provides a steady supply of water.

The Origin of Parasitism

There are many different groups of parasites, and each group contains many different species. We will deal with the most important groups in Section 1.2. Each group has come from a free-living ancestor; that is, its ancestors were not parasitic. When an organism becomes parasitic it slowly changes, so that it is better able to live as a parasite. This slow change is called evolution and it is the mechanism by which parasites become better adapted to living a parasitic way of life. We will explain the mechanism of evolution in more detail presently.

Some parasites are very like free-living forms in the appearance of their organs and we say that they are closely related to free-living forms. Both the parasites and the free-living forms have come from (or have evolved from) the same ancestor. Other parasites are not very much like any free-living organisms that exist now. They have evolved from free-living forms perhaps a thousand million years ago and during this long period, they have slowly changed, so that they are now quite unlike any free-living forms that we can find today.

We will give some examples to show how free-living animals might have become parasitic.

(i) An insect or mite which normally lives in cracks and holes in the nest of a bird may become parasitic on the bird. It may get short of its usual food and try sucking blood from the skin of the bird. It may, by accident, find that the host provides a plentiful supply of rich food in the form of blood, and it comes to use this, rather than other food which is perhaps harder to get and less nourishing. Gradually it changes its habits and structure so that it becomes a more efficient parasite.

These changes do not occur because the organism can think and thus improve its way of life. They evolve by processes of natural selection which will be explained presently

(ii) There are many free-living nematode worms which feed on decaying plant material in the soil. It is easy to imagine plant parasitic forms evolving from them in the following way:

A free-living nematode which eats decaying plant material in the soil, gives rise to a form which eats dead cells on the surface of the root of a living plant. From this form another form evolves, which eats living plant material by pushing its head into living plant cells, while its body stays outside the plant. This gives rise to a completely parasitic form which burrows into living plant tissues and eats them.

(iii) A free-living form that feeds on the meat of dead animals (called a carrion feeder) may become parasitic by feeding on the dead flesh of a living animal (for example, a sore) and then burrowing into living flesh. The maggots of some flies which live in the meat of cows, probably evolved in this way.

(iv) Some aquatic animals (that is, animals that live in water) live on the surface of larger animals. They are not parasitic because they do not eat the host. They could just as easily live on the surface of a stone. They could easily become parasitic if they started to feed on the tissues of the surface of the host. They might even enter the mouth where there is perhaps more shelter, and from there the gut, nostrils or lungs. Or they may enter the anus or the bladder and from there move to the gut or kidneys.

(v) Most large vertebrate animals accidentally eat a great variety of other organisms. For example, an animal that eats grass will also swallow all the small organisms and eggs or cysts (resting stages), that happen to be in the grass. It will also ingest (take into the gut) many tiny organisms with the water it drinks. Some of these organisms may be able to survive in the gut and thus become parasites.

We have talked about some of the ways in which a free-living animal may become a parasite. Many parasites live in two different kinds of hosts, for example, malaria has certain stages which live only in mosquitoes. These stages cannot develop further until they are injected into man, where they produce a series of different stages. The first stage is in the liver and it gives rise to others in the blood. The life cycle is completed when a mosquito sucks up certain stages which live in the blood, and becomes infected. The malaria parasite has two hosts in its life cycle. Many worm parasites have two, three or even four hosts in the life cycle.

Life cycles with two hosts must have evolved from life cycles with only one host. We can imagine a meat eating animal (carnivore) acquiring the parasites that live in the animal that it eats. In this way, a one-host life cycle could become a two-host life cycle, and a two-host life cycle could become a three-host life cycle and so on.

Evolution

Evolution is the process whereby a species slowly changes its characteristics over a very long time, so as to become more efficient in its way of life. We say that the species becomes better adapted to its environment.

The environment of an animal is the sum of all the factors which influence its life—that is, its food, shelter, and all the other organisms that live with it, and the climate in which it lives.

This process of adaptation is best illustrated by an example. Parasites that live in the gut of their host will be carried out with the stream of food that passes through the gut, unless they can hang on in some way, or actively swim against the food stream. Many parasites have overcome this problem by evolving suckers or hooks or spines, or some other method of attaching themselves to the wall of the gut. Those individuals that can hang on, are most likely to survive, and those that cannot, will not be able to become gut parasites. Slowly, over millions of years, gut parasites have evolved very efficient organs for hanging on. They have become adapted for living as parasites.

Life began on the earth perhaps 2000 million years ago. The earliest fossils (or remains of living organisms) are found in rocks about 1000 million years old and these include some quite complex organisms. Almost certainly a wide variety of parasites had already evolved at that time, and were living in a wide variety of hosts. Parasites have very soft fragile bodies and do not often become fossils, so we cannot be sure about this, but it seems very likely. Evolutionary changes occur very slowly, but a thousand million years is a long time, and a tiny change in each generation can add up to produce quite remarkable organs. For example, our arms and legs have evolved from the fins of fishes through many intermediate stages.

The way by which evolution occurs was outlined over one hundred years ago by Charles Darwin. This was the greatest single idea in biology, because it has enabled us to understand so much about living things.

Darwin's Theory of Evolution

Darwin's theory is based on three important facts which are outlined below:

(*i*) *The Individuals of each species are not identical.*

They vary in many ways. For example, some are bigger than others, some are able to survive high or low temperatures better than others, some can live with less oxygen, some lay more eggs and so on.

(*ii*) *Living things produce many young.*

For example, one female mouse can produce about 10 young every three weeks. These are ready to mate at about six weeks and can produce their first young (or litter) when they are about nine weeks old. If we let a single pair of mice reproduce at this rate, we would have several million mice within a couple of years.

(*iii*) *Despite this ability to produce many young, the numbers of each species remain about the same all the time.*

The world is not overrun by mice, or covered by coconut trees, or elephants, or any other single species.

Darwin drew two important conclusions from these facts.

(*1*) *There is a struggle for existence.*

Many young of each species must die before they are old enough to reproduce.

(*2*) *The fittest survive this struggle.*

The things that kill off the excess numbers of a species do not act randomly (by chance). Some individuals have a greater chance of survival than others, because they are more fit. The qualities that make an individual more fit vary in each species, and with the environment in which it lives. For example, a parasite that has a slightly bigger sucker or set of hooks, is more likely to be able to hang on to the wall of the gut, and is more likely to survive than other members of the same species which have only a small sucker or set of hooks.

If the possession of a particular characteristic enables an individual to survive, and reproduce, and this characteristic is inherited, **THEN,** the proportion of a species possessing this characteristic will increase with each successive generation, until all members possess it.

The Importance of Sexual Reproduction

Nearly all living organisms have sexual reproduction at some stage in their life cycle. Some protozoa are completely asexual, and they reproduce by dividing to form two or more young individuals. In some higher organisms the females lay eggs which have not been fertilized by sperm; in these species there are no males at all. But most animals and plants reproduce sexually. In these cases each living organism starts from the joining of an egg from the female, and a sperm from the male, to form a single cell which is called a zygote. This zygote has a nucleus formed from the joining of the nuclei of the egg and the sperm. The new organism grows by the repeated divisions of this cell. The original nucleus divides to produce two nuclei exactly the same as the first one, and so on, so that every cell has an identical nucleus. Each nucleus contains the information or instructions which direct the development of the new individual. It directs the way it grows, so that certain cells become specialized for certain functions. It organises thousands of different chemical reactions within the cells of the body which direct the processes of life, such as the development of hooks and suckers, the colour of hair, size, feeding and behaviour. The nucleus contains information from both parents, and thus the new individual inherits the characteristics of its ancestors.

The Origin of Variation

Evolution cannot take place unless the individuals of a species vary and the variation is inherited, that is, passed on to the next generation. For example, children with red hair usually have parents or grandparents or great grandparents who had this characteristic.

Within the nucleus of each egg and sperm there are chemicals which contain the information or instructions which control each little bit of development, and hence inherited variation. Once in a while these chemicals change slightly and this causes a change in the instructions that they control. This change is called a mutation. Thus a person with red hair has red hair because he has inherited this characteristic or because mutation has occurred.

Mutations are very rare, but they can be caused by such things as radiation (for example, X-rays) and certain drugs or chemicals. These changes are occurring naturally all the time. They provide a continual supply of small variation in the individuals of a species. Sexual reproduction enables these inherited variations to spread within the species.

Let us consider an example which shows how evolution works.

The gut of animals secretes acids and enzymes which digest food in the gut. These substances break down proteins, carbohydrates and fats, into small particles called molecules, that can pass through the cells of the wall of the gut and into the blood stream. If these substances can digest meat, they will be able to attack the body surface of parasites, unless the parasites have some mechanism (or way) to prevent this. Gut parasites do, in fact, have very good methods of stopping themselves from becoming part of the food of the host. Some of them secrete substances called anti-enzymes which neutralize the effect of the host's enzymes, so that they are not digested. This mechanism might have evolved in the following way.

Let us suppose that a free living species lives in grass and is continually being eaten by a larger animal.

Within this species some individuals have the ability to produce anti-enzymes and they are able to survive long enough in the host to reproduce. They have this characteristic because, sometime in the past, there has been a mutation in their ancestors. Other individuals which do not have this characteristic die before they can reproduce in the host. They cannot become gut parasites. The offspring of the new parasites would all possess the ability to resist the enzymes of the host for a time. Let us suppose in a later generation of this parasite another mutation occurred which enabled the individual to survive much longer in the host. This individual would produce many more eggs or sperm within its lifetime than other individuals, because it lived longer. Many individuals of the following generation of this parasite would possess this new characteristic and would survive as parasites because they were fitter than those individuals that did not possess it.

This is the way that things evolve. Continual small changes of this kind eventually add up to large changes. They account for the evolution of complex structures and mechanisms, which enable the species to live more efficiently. Adaptations for a parasitic way of life have evolved in this way, as have all the complex organs of our bodies and the complex mechanisms by which we function efficiently.

The problem of getting to another host

One of the characteristics of living things is that each individual eventually dies while its species continues. Because of this, each species of parasite must have a way of getting some of its young into new, uninfected hosts.

As parasites have evolved from free-living organisms they have become increasingly adapted to a parasitic way of life. Different parasites have evolved different mechanisms which increase the chances of infecting a new host. We will illustrate this with a few examples.

(i) Many parasites produce enormous numbers of infective stages. For example, the nematode *Ascaris*, produces about 200,000 eggs per day, and the tapeworm *Taenia saginata*, produces 50-150 million eggs per year.

(ii) Other parasites have evolved patterns of behaviour which increase the chance of contact between host and parasite. For example, the infective larvae of hookworm and many nematodes of cattle and sheep, migrate to the surface of the soil or the top of grass, where they are much more likely to come in contact with their host, or be eaten by it.

(iii) Some parasites have developed very resistant eggs or cysts which can survive in the soil for several years. The eggs of *Ascaris* are an example of this kind of adaptation.

(iv) Many species of parasites have life cycles which involve two (or more) hosts. In such cases, the free-living stages are often eliminated and the parasite is never subjected to the hazards of the external environment. The parasite never has to live outside the host in places where it is likely to be killed by extreme temperatures or dryness, or eaten by another animal. Many such life cycles use a blood sucking intermediate host (or vector) to transmit them from one host to the next. Malaria is an example of this kind. The malaria parasite is transmitted from one vertebrate host to another, by a mosquito, which is itself a parasite of the vertebrate host.

Man picks up his parasites in many different ways. We will talk about each one in detail as we talk about each of the parasites of man in Niugini. The various ways that man becomes infected are listed below:

(i) eating food which is contaminated;
(ii) drinking water which is contaminated;
(iii) by entry through the skin from a blood-sucking insect or from contaminated soil;

(iv) contact with an infected person;
(v) from dirty hands that have not been washed after going to the toilet;
(vi) by blood transfusion.

Some Useful Terms

It is now convenient to define a few useful terms.

Ectoparasites
Ectoparasites are parasites that live on the outside of the host. The term is derived from the Greek word *ektos* meaning outside. Mosquitos, ticks, lice, fleas, leeches and scabies are all ectoparasites.

Endoparasites
Endoparasites live inside the host (from the Greek word *endon*, meaning within or inside). Malaria and hookworms are endoparasites.

Different kinds of hosts
When two or more hosts are involved in the life cycle of a parasite, the following terms are used:
(i) For life cycles involving a vertebrate host and an invertebrate host, for example, an insect, we usually refer to the host as the vertebrate host and the invertebrate or insect host. This terminology is probably best for parasites like malaria and other protozoa with two-host life cycles. It can also be used for some helminth (worm) life cycles, for example, *Wuchereria*. (See Chapter 6.)
(ii) For life cycles involving three or more hosts, or two vertebrate hosts, or two invertebrate hosts, we refer to them as follows: The final host, or definitive host, is the host in which the parasite undergoes sexual reproduction. It is the host which is infected by adult parasites.

In such life cycles, the other hosts contain larval stages (juvenile or immature stages), and are known as intermediate hosts. The first one in the life cycle is known as the first intermediate host and the next as the second intermediate host and so on. This terminology is the best for helminth parasites.

The size of parasites

Size is a characteristic or property of all living things. Each species, or kind of living organism, grows to a size which is a characteristic of that species and it is usually different from related species. Of course, there are some variations from one individual to the next, but the average size of a number of specimens, and the range of size from the smallest to the largest is characteristic of the species.

We are used to measuring the sizes of things in units of miles, yards, feet and inches. In science we usually make our measurements in metric units: centimeters, meters, etc. Over the next few years, all our measurements are going to be changed to the metric system. Many parasites are very small and we must use microscopic units of measurement.

There are ten millimeters in a centimeter, and while these units are suitable for many parasitic worms, they are far too big for smaller parasites. For small measurements, we divide each millimeter into one thousand parts. These parts are called microns, and are usually referred to by the Greek letter μ (called mew). Thus

1000 microns (μ) = 1 millimeter (mm)

10,000 microns (μ) = 10 millimeters (mm) = 1 centimeter (cm)

100 centimeters (cm) = 1 meter (m)

A red blood cell of man is about 7μ (microns) in diameter, that is 7/1000 of one millimeter.

An outline of the units of measurement of length, weight, volume and temperature are given in an appendix at the back of this book.

Nomenclature (No-men-clay-ture)

Each species, or kind of living organism, has a different name, so that we can associate the features of each species with its name. We use a system of nomenclature, or names, that was first devised (or invented) by Carl von Linne (often called Linnaeus) in 1753. It is called the binomial system of nomenclature because it uses a two-word name for each species. For example, one of the species of human malaria is called *Plasmodium vivax*. The first name is called the generic name and it indicates the genus to which this species belongs. The generic name always starts with a capital letter. The second name is called the specific name, and it always starts with a small letter, not with a capital.

When we are discussing a particular organism, and we have given its full name, we can abbreviate its name when we refer to it again. We do this by giving the initial letter of the generic name, followed by a full stop, and then the specific name in full. This can only be done when the abbreviated name cannot be confused with any other species that we are discussing. For example, we could refer to *Plasmodium vivax* as *P. vivax* once we had given its full name. We would not be able to use this abbreviation if we were discussing *Prohemistomum vivax* at the same time.

A genus contains species that are all very similar. They are placed in the same genus because we believe they have evolved from the same ancestor. All the species belonging to the same genus are related, and they share certain characteristics. For example, all species of the genus *Plasmodium* have the following characteristics:

(i) All have a life cycle involving two hosts: a mosquito and a vertebrate animal.

(ii) All have stages in certain internal organs of the vertebrate host (for example, the liver) and stages in red blood cells.

(iii) All produce a black or brown pigment called haemozoin when they infect the red blood cells of vertebrates.

(iv) All have stages which develop in the wall of the gut of the mosquito host.

The genus *Plasmodium*, together with several other genera, belong to the same family. This family, together with several other families, belong to the same order, and a number of related orders are placed together in the same class, and several classes in the same phylum.

Thus we have a system of classification which is made up of a number of larger and larger units or categories. Table 1.1 gives examples of the classification of some human parasites.

TABLE 1.1: Classification of some human parasites

Common name	Tapeworm	Mosquito	Malaria	Hookworm
species	*nana*	*punctulatus*	*vivax*	*americanus*
genus	*Hymenolepis*	*Anopheles*	*Plasmodium*	*Necator*
family	Hymenolepididae	Culicidae	Haemoproteidae	Ancylostomatidae
order	Cyclophyllidea	Diptera	Haemosporidia	Rhabditida
class	Cestoda	Insecta	Sporozoa	Phasmidia
phylum	Platyhelminthes	Arthropoda	Protozoa	Nematoda

NOTE: Different text books often have different systems of classification. For example, one book may call the Sporozoa a subphylum, and put the order Haemosporidia in a class with a different name. However, the names of species and genera should almost always be the same in all books.

Sometimes additional categories are made by dividing one category into a number of sub-categories (for example, a family may be divided into several sub-families) or by lumping a number of categories into a super-category (for example, a superfamily may contain a number of families).

Each phylum consists of animals or plants which all share the same basic pattern of organisation, and this is different from the basic pattern of organisation of all other phyla.

The description of a new species

When a biologist finds a new species he can describe it and give it a name. Provided he follows a set of rules concerned with the description and naming of species, which are called 'The International Rules of Zoological Nomenclature', this name becomes the name of the species. It cannot be changed without a good reason, and these reasons are set down in the rules.

The description of a new species should give the measurements of a specimen that is typical of the species, and other specimens as well. It should include drawings of parts or organs that are distinctive of the species, together with their measurements. The specimens which are used for the description are known as 'type' specimens and should be carefully preserved in a museum where other biologists can refer to them at a later time. It is usual to say how the species differs from related species that have already been described. The description should also say where the 'type' specimen was collected. This is known as the 'type locality'. It should also say when the specimen was collected. If the new species is a parasite, the description must give the scientific name of the host from which it was collected.

Many people collect specimens of animals and pass them on to biologists for study. These specimens are almost worthless unless they are clearly labelled with the date on which they were collected, the place where they came from, and if they are parasites, the name of the host. Sometimes it is convenient to preserve the host together with the parasite. In Niugini there are many undescribed species of parasites and other invertebrate animals. One never knows if a specimen is new until it has been carefully examined by an expert. It cannot be given a new name unless it is carefully labelled with the proper information.

The description of a new species must be published in a scientific journal or book. It must show the name of the person who is describing it and his address, or the address of the place where he works.

Sometimes two different species are given the same name. The name of the one that was described first, keeps its name, and the name of the other one must be changed. In order to further identify each species, every species should be referred to by the generic and specific name that has been given to it, together with the name of the person who described it, and the year that the description was published. For example, the full scientific name of *Entamoeba histolytica* is *Entamoeba histolytica* Schaudinn, 1903. It was given this name by a man called Schaudinn, and the description was published in 1903.

If the name of a species is changed, the name of the person who described it is put in brackets, and the name of the person giving it a new name is added on to the end. For example, *Balantidium coli* (Malmsten, 1857) Stein 1862. This indicates that Stein published a change of name for this species in 1862.

There are a number of scientific books that list the names of new species and give details of the place, date, and book or journal where their description was published. This system enables a biologist to look up the original description of any species and see if it describes the specimen he has. This process is called 'identification'. It is the way biologists give a name to a specimen.

1.2 The Main Groups of Parasites

THE PROTOZOA (Proto-zo-a)

Members of the protozoa are mostly single celled organisms and they are usually too small to see without a microscope. About 35,000 specimens of protozoa have been described and there are many more that have not been described yet. About half of the species that have been described are parasitic.

The phylum contains several classes, the most important of which are:

The Flagellates (flage-ell-ates)
The Ciliates (silly-ates)
The Amoebas (ameebers)
The Sporoza (Spore-o-zo-a)

Representatives of each group are illustrated in Figs. 1.1 to 1.6.

The Flagellates

These have at least one flagellum, which is an organ like a piece of string, which projects from the body and causes the animal to move when it is waved in a regular way. It pushes or pulls the animal along, something like a propeller of an aeroplane or a boat. Flagella are usually about as long as the body of the animal. Some species have several flagella.

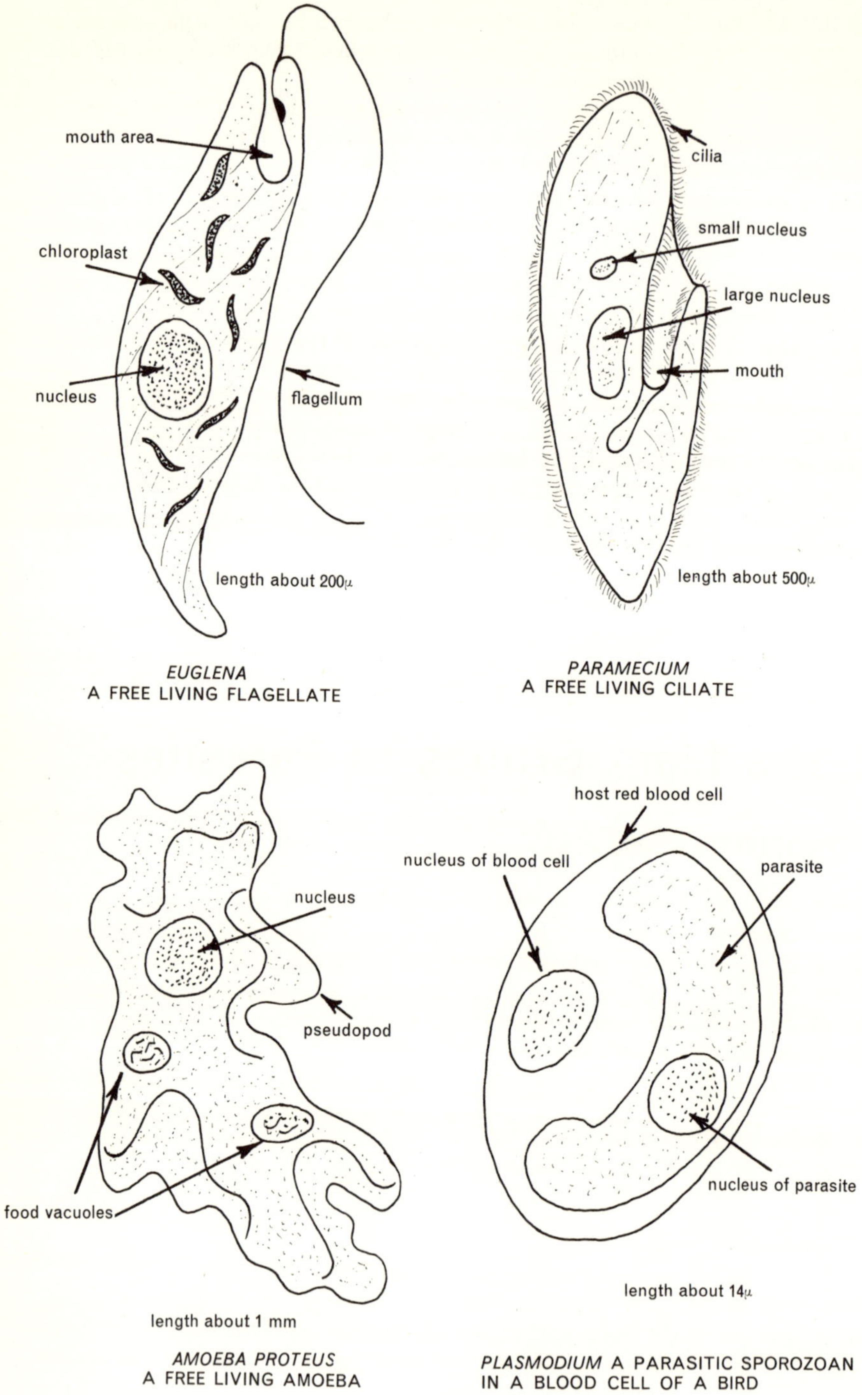

Fig. 1.1 Representative Protozoa

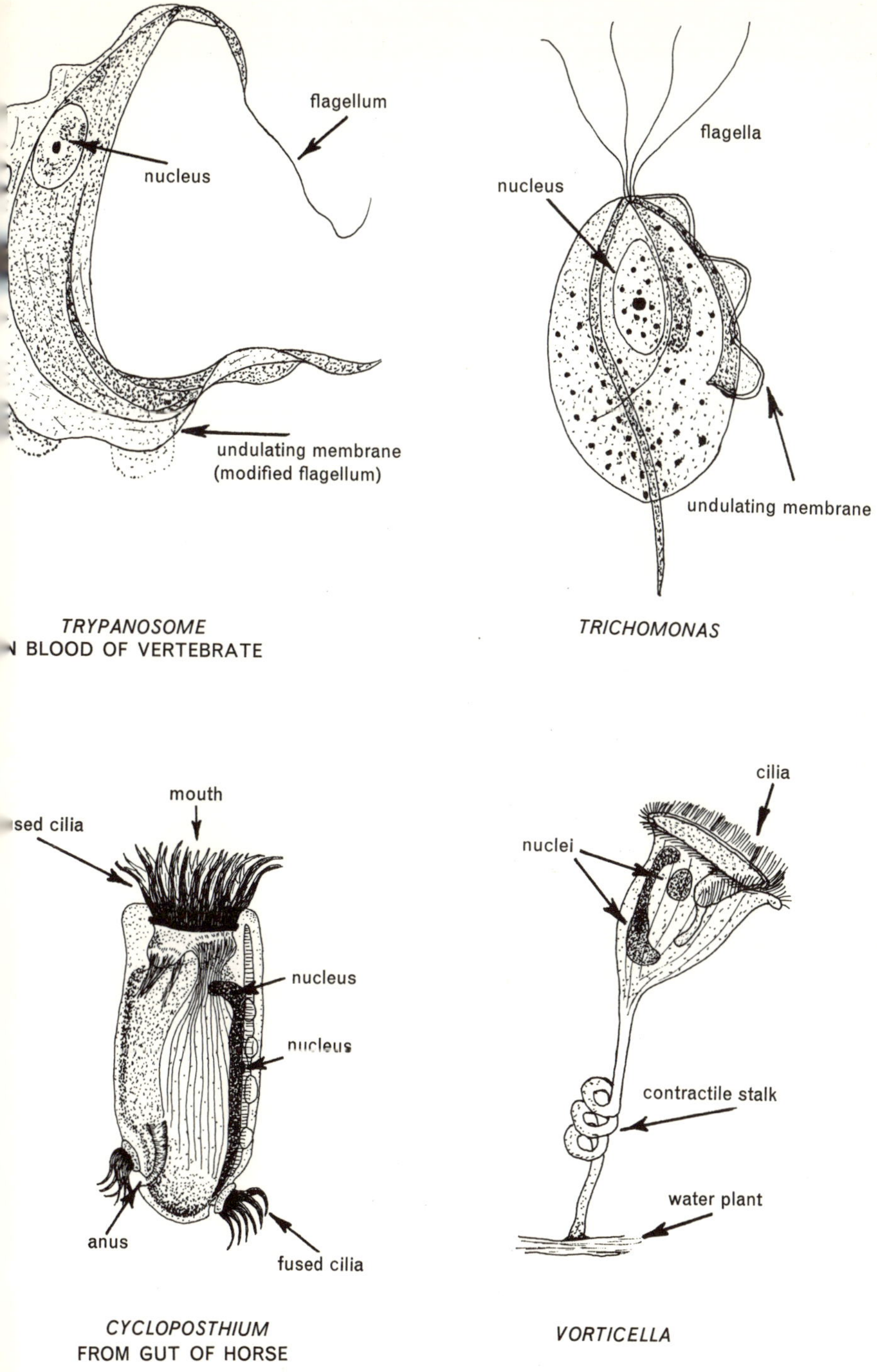

Fig. 1.2 Flagellates and ciliates

Many flagellates are free living and are found in wet soil, ponds, lakes and in the sea. Some free living species contain chlorophyll (kloro-fil) (a green pigment) and carry out photosynthesis (foto-sin-the-sis) in the same was as plants do. They are probably the ancestors of plants.

Parasitic flagellates are common in the gut of many vertebrates. They also occur in the mouth, the vagina and the bladder. One group (the Trypanosomes) is parasitic in the blood of vertebrate animals and some of these species cause serious diseases, for example, sleeping sickness in Africa. Fortunately, they do not occur in Niugini. Some flagellates are illustrated in Fig. 1.2.

The Ciliates

The surface of these protozoa is covered with many cilia which are like short hairs. As they beat back and forth, they cause the animal to move, like a canoe moves when lots of men are paddling it. In many ciliates, the cilia also make a current in the water, which carries food to the mouth. Cilia are usually very much shorter than flagella, although they have the same function, that is, they do the same work.

Many species of ciliates are free-living and like the flagellates, are found in aquatic places; many species are parasitic in the gut of larger animals. For example, they can almost always be found in the gut of frogs and white ants (termites). A wide variety of ciliates live in the stomach of cattle where they digest cellulose (the main substance in grass), and change it into other substances which can be digested by the cows. Cows cannot digest cellulose and could not live without these ciliates and other micro-organisms which do this work for them.

Most ciliates do not harm their hosts, but one species, *Balantidium*, which normally lives in pigs, sometimes causes dysentery if it infects man. Some ciliates are illustrated in Fig. 1.2.

The Amoebas

The amoebas do not have any special organs like cilia or flagella for moving, but they move in a very curious way which is called 'amoeboid movement'. Certain cells of many other animals, move in this way. Any point on the surface can flow out into a projection called a pseudopod (pseudo: false; pod: foot;) and the material in the cell slowly flows after it. In a little while another pseudopod forms somewhere else, and the old one disappears. It is exciting to watch this process going on under a microscope. (See Fig. 1.3.)

Amoebas feed by pushing out pseudopods around a particle of food. These surround the food and then fuse together so that the food is taken into the body. (See Fig. 1.4.) Certain cells of higher animals, called phagocytes, feed in the same way.

Amoebas occur in wet places such as the bottom of drains and ponds, and at the sea shore. Some free-living amoebas have a hard skeleton which may be on the inside or outside of the body. (See Fig. 1.5.)

In addition to a wide variety of free-living amoebas, there are a number of parasitic species. Several occur in man, one called *Entamoeba histolytica* causes amoebic dysentery. The other human species are harmless.

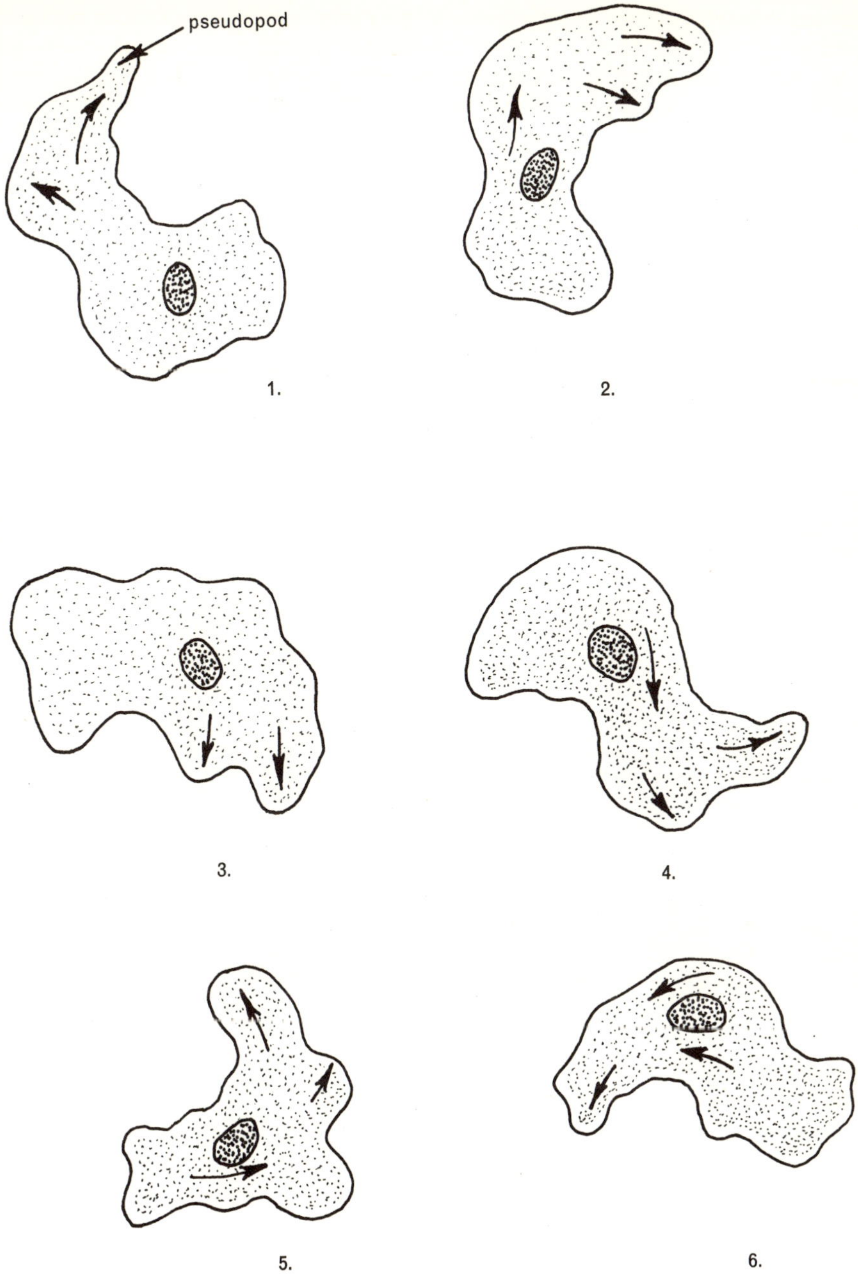

ARROWS SHOW 6 CHANGES IN DIRECTION OF THE SAME AMOEBA

Fig. 1.3 Amoeboid movement

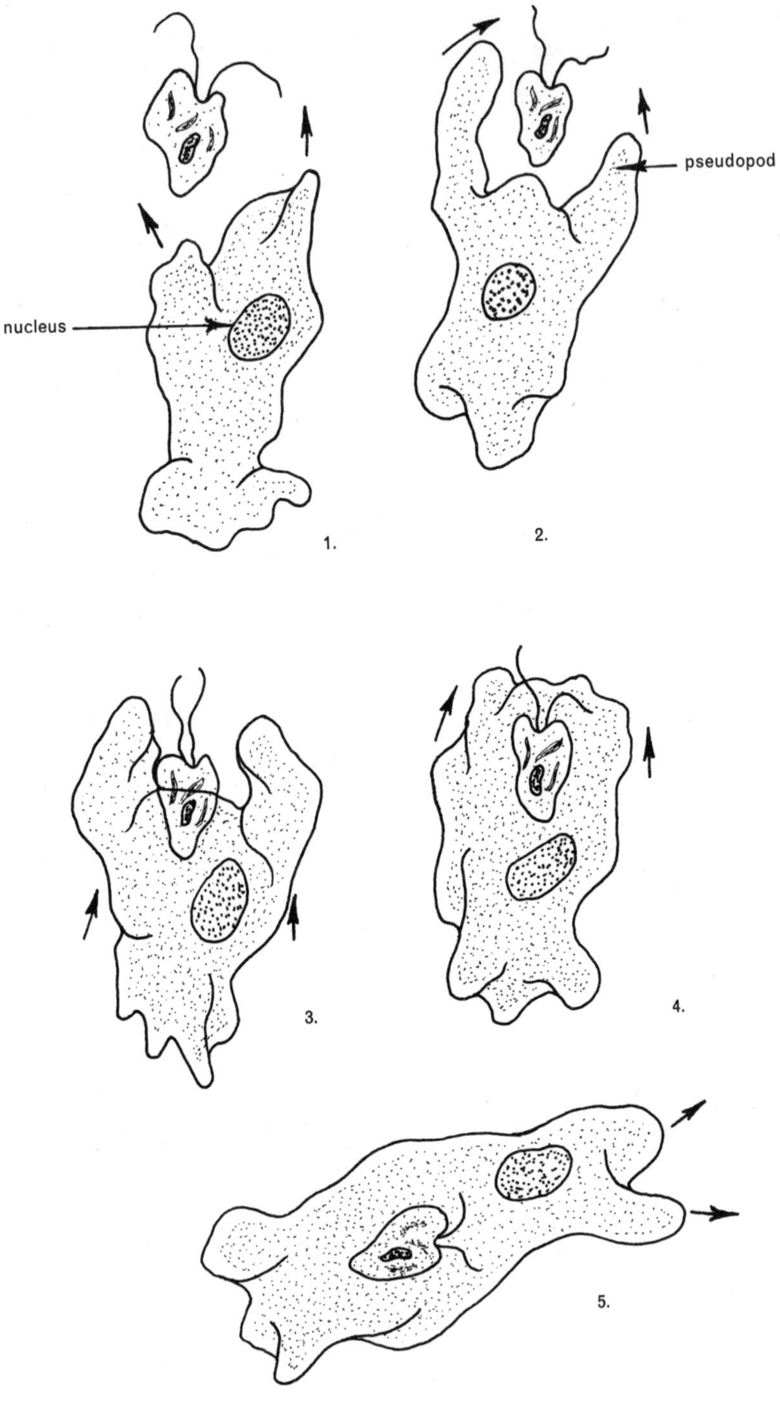

Fig. 1.4 Amoeba feeding on a small flagellate (after Buchsbaum)

Various species of *Entamoeba* have been found in the gut of reptiles, amphibians, birds, mammals, leeches and cockroaches. Other species are known to live in the mouth of vertebrates, including man. Most species of *Entamoeba* form cysts, or resting stages, which contain one, four or eight nuclei. Infection occurs when cysts that pass from the host, in the faeces, are eaten by another host.

The Sporozoa

All members of this class are parasites. Most of them live inside the cells of the blood, or the gut, or internal organs such as the liver. They are found in a wide range of vertebrate and invertebrate hosts. They do not usually have any special organs for movement, and they do not move by amoeboid movement. However, some stages of some species have flagella, and others have tiny muscle fibres just beneath the surface, which enable them to wriggle along. Most species have complicated life cycles, usually involving asexual reproduction (where one individual divides into two or more individuals), and sexual reproduction (where a male cell fuses with a female cell to form a zygote, which then reproduces to form many new individuals). Often there are two different hosts in the life cycle. Many species have a spore stage in the life cycle. Spores are resting stages that have a thick wall which protects them when they are outside the host, waiting to infect another host. Other species do not have spores, but are transmitted by blood sucking insects and ticks. Insects and related organisms which transmit disease are called vectors.

The sporozoa include the malaria parasites and many other species, which can cause serious diseases in cattle and chickens. Several species live in the sperm storage organs of earthworms. It is rare to find large earthworms that are not infected by these parasites, and they can easily be seen moving about under a microscope. Others are common in the gut of insects, such as crickets and grasshoppers. Some sporozoa are illustrated in Fig. 1.6.

THE HELMINTH PARASITES

Helminthology is the study of parasitic worms. The helminths consist of several phyla of worm-like aminals, most of which are parasitic.

THE PLATYHELMINTHES (plat-e-hell-minths)

This phylum contains one class of free-living worms (the Turbellaria), most of which live in fresh or sea water. A few species live in moist places on land.

It aslo contains several parasitic classes and we shall briefly discuss some of them. Representative platyhelminthes are illustrated in Fig. 1.7.

The Monogenea (mon-o-gen-e-a)

Most of the members of this class are ectoparasites of fish. They are commonly found attached to the gills. A few species are endoparasites of fish and amphibians (for example, frogs), and are found in the mouth, bladder, cloaca and body cavity. Most species have special organs for hanging on to the host which usually consist of suckers and hooks. There is only one host in the life cycle—hence the name monogenea, which comes form two Greek words: *monos* (single or one) and *genos* (offspring). There are about 2000 species in this class. Most of them are between 2 and 5 millimeters long.

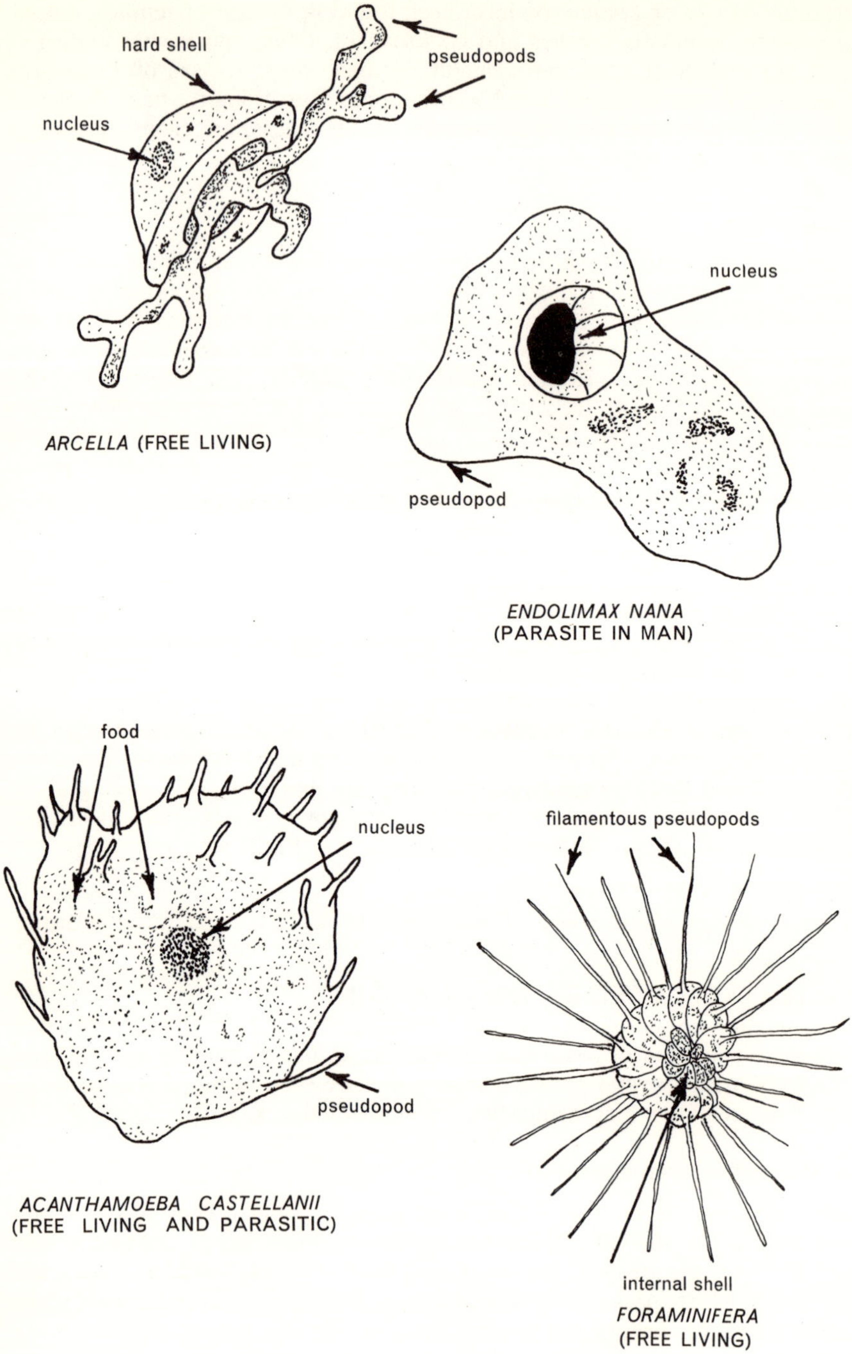

Fig. 1.5 Amoebas

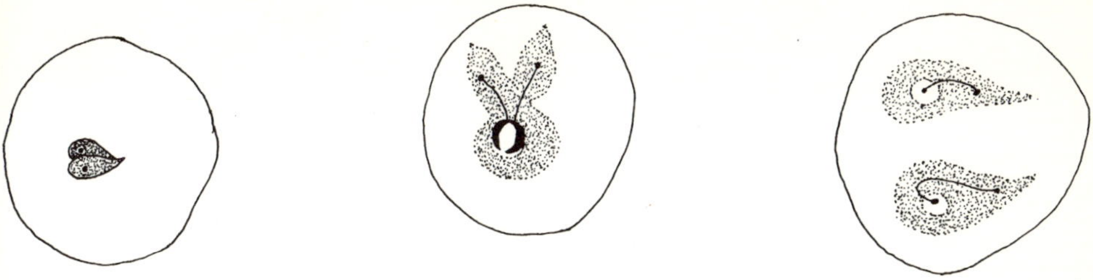

DEVELOPING FORMS OF *BABESIA* IN RED BLOOD CELLS OF CATTLE

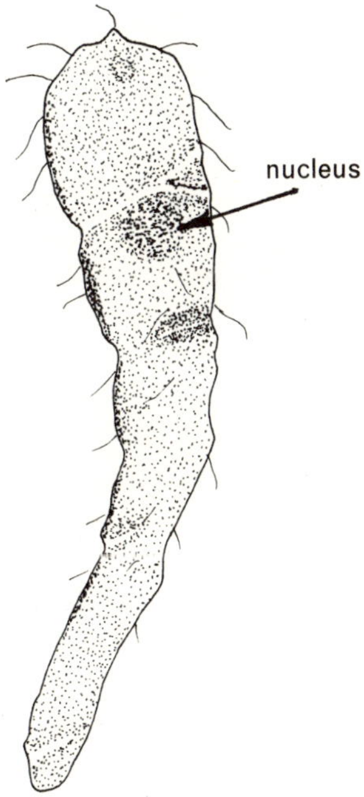

MONOCYSTIS
FROM EARTHWORM

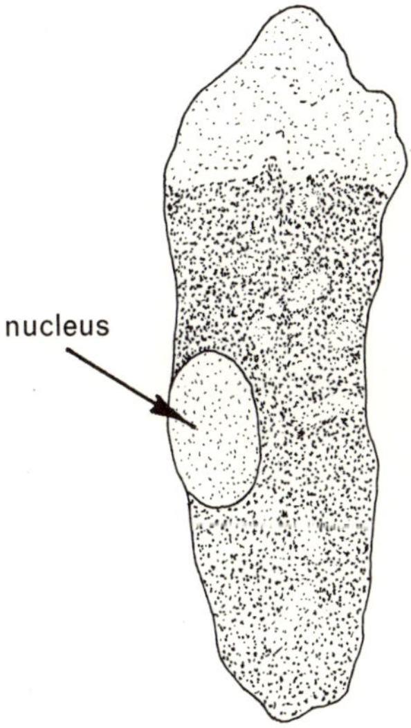

GREGARINA
FROM GUT OF CRICKET

Fig. 1.6 Sporozoa

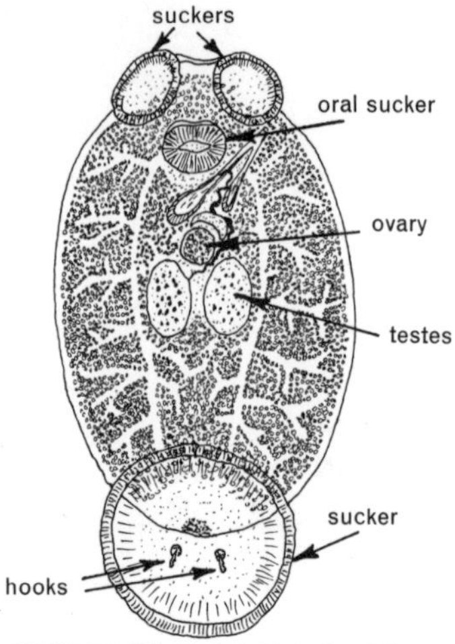

MONOGENEA FROM GILL OF FISH

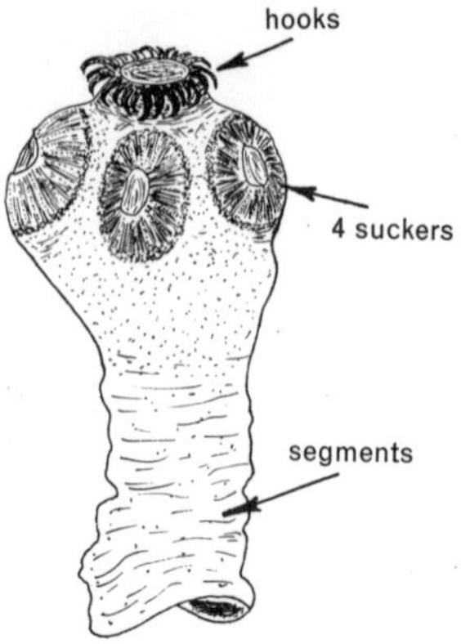

CESTODE HEAD

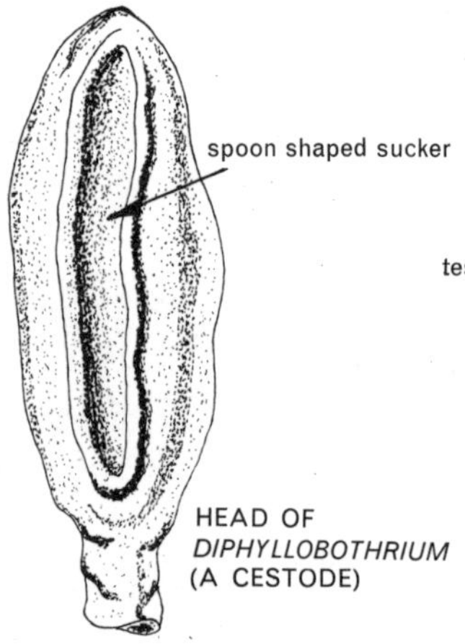

HEAD OF *DIPHYLLOBOTHRIUM* (A CESTODE)

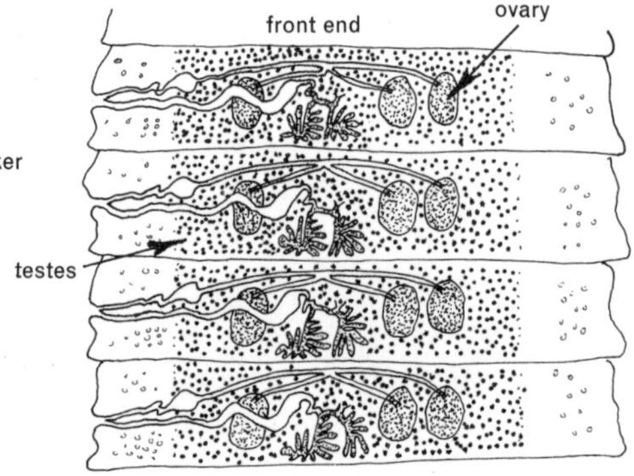

SEGMENTS OF A CESTODE

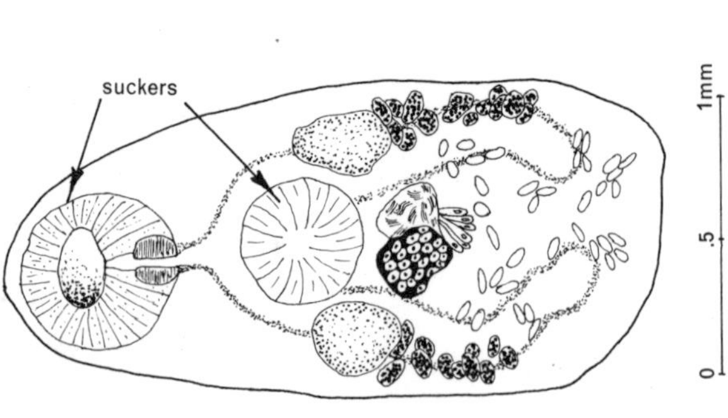

TREMATODE FROM LIZARD'S LIVER

Fig. 1.7 Platyhelminthes

The Digenea (Di-gen-ea)

About 5300 species of Digenea have been described. They are sometimes called 'flukes'. The Digenea have complicated life cycles usually with three hosts, but sometimes only two and rarely four or five. The final host is always a vertebrate, and all classes of vertebrates are infected.

The adult flukes usually live in the gut of the final host, but a few species occur in other organs such as the lung, liver and bladder. One family—the Schistosomes—lives inside the blood vessels. Certain Schistosomes are important parasites of man. They do not occur in man in Niugini.

Eggs from the adult flukes pass out in the host's faeces. These hatch in water to release a tiny ciliated stage called a miracidium, which infects the first intermediate host, which is usually a snail.

Once inside the snail host, the parasite reproduces asexually and eventually forms cercariae (sir-care-ee-ee). Many thousands of these emerge from the snail and usually burrow into a second intermediate host. This may be a crab, yabbie, snail, larval insect, fish or some other aquatic animal. The cercariae then form a cyst, or resting stage.

The life cycle is completed when the vertebrate host eats a second intermediate host. The young fluke then emerges from the cyst and starts to grow into an adult fluke.

In some families there is no second intermediate host in the life cycle, and the cercariae either burrow into the skin of the final host and then grow into adults, or, the cercariae form a cyst on vegetation and infect the final host when it eats the cyst.

There are many other variations on this basic life cycle pattern. The life cycle of the Chinese liver fluke (*Clonorchis*) is illustrated in Fig. 1.8.

Digenea vary in length from about one millimeter to several meters. But very few species are longer than one centimeter. Most species are flat like a leaf, but some are fat. We recently found one species in a fish from the Papuan Gulf, which is almost round and about two centimeters in diameter.

The digenea usually have two strong suckers. One is situated around the mouth, and the other is somewhere on the ventral surface. These are used for attachment to the host.

Most flukes are hermaphrodites, that is, each individual contains both male and female reproductive organs. Their eggs usually have a little lid (or operculum), which pops open when the miracidium is ready to hatch.

Flukes have a bifid gut, that is the gut branches into two parts.

The Cestodes (or Tapeworms)—(ses-toads)

Tapeworms usually have two hosts in their life cycle. Adult tapeworms always live in the gut of a vertebrate (the final host), and the larval stages live in a vertebrate or invertebrate, intermediate host. A few species have no intermediate host and some have two or even three intermediate hosts.

Adult tapeworms are long flat animals like a piece of tape or a tape measure, with an organ of attachment at one end, called a scolex. The structure of the scolex is different in each order of tapeworms. In one order, the scolex consists of four suckers and often a ring of hooks. In another order it consists of a flat sucker, shaped like the end of a spear. (See Fig. 1.7.) In some fish tapeworms, the scolex has four spiny tentacles which anchor the animal to the wall of the gut.

The body of the tapeworm consists of many segments. Each segment contains a complete set of male and female reproductive organs. New segments are continually being formed just behind the scolex and old segments are continually dropping off the other end. The animal has no gut, but absorbs food that has already been digested by the host, through its skin. Each segment is like a separate individual, which is attached to the one in front and the one behind. A complete tapeworm is like a machine which continually produces new individuals which slowly grow, develop reproductive organs, and finally consist of little more than a bag of eggs, which drops off the end of the machine and passes out in the faeces of the host.

Tapeworms vary considerably in size. One that lives in man grows to a length of 60 feet, but many of them are only a few inches long. About 1400 species of tapeworm have been described.

Man is the final host of several species of tapeworms. For example, there are two species of *Taenia* which infect man. Infection takes place when man eats uncooked beef or pig meat containing the larval stages. Another tapeworm that occurs in man is called *Diphyllobothrium latum*, and man becomes infected by eating raw fish. None of these species occur in Niugini.

Man can also become infected with the larval stages of some tapeworms. One of these causes hydatid disease which is a serious problem in many countries where there are lots of sheep. These larval stages form bladders which slowly grow inside organs such as the liver and lung. They sometimes grow to a foot in diameter. This species has not been recorded from man in Niugini.

The only genus of tapeworm that infects man in Niugini is *Hymenolepis*. It is described in Section 4.10.

THE NEMATODES (nema-toads)

There are many species of Nematodes. They are found in the mud at the bottom of the sea and freshwater lakes and rivers. Many species are free living in the soil and many others are parasites.

About 11,000 species have been described so far, of which about 6000 species are parasitic in vertebrate animals and about 3000 are plant parasites or live in the soil. The rest occur in the sea or are parasites of invertebrates. Zoologists have only recently become interested in the free-living and plant parasitic nematodes and new species are being found almost every day. In fact, it has been estimated that there may be 500,000 species of nematodes and only about 2% of them have been described so far.

The word 'nematode' comes from the Greek word *nema*, meaning a thread. Most nematodes look like a piece of thread, in fact one species which lives in man is commonly called 'threadworm'. They range in length from less than one millimeter to a meter or more. The parasitic species tend to be larger than the free-living ones which are mostly only a millimeter or two long. Most of the parasitic species are less than 3 centimeters long.

Most nematodes are long and thin with both ends pointed. Some female plant nematodes are almost spherical and some parasitic nematodes, such as the hookworms, have the anterior (front) end modified for attachment (Fig. 4.4). The males of many species have a copulatory bursa or copulatory spicules.

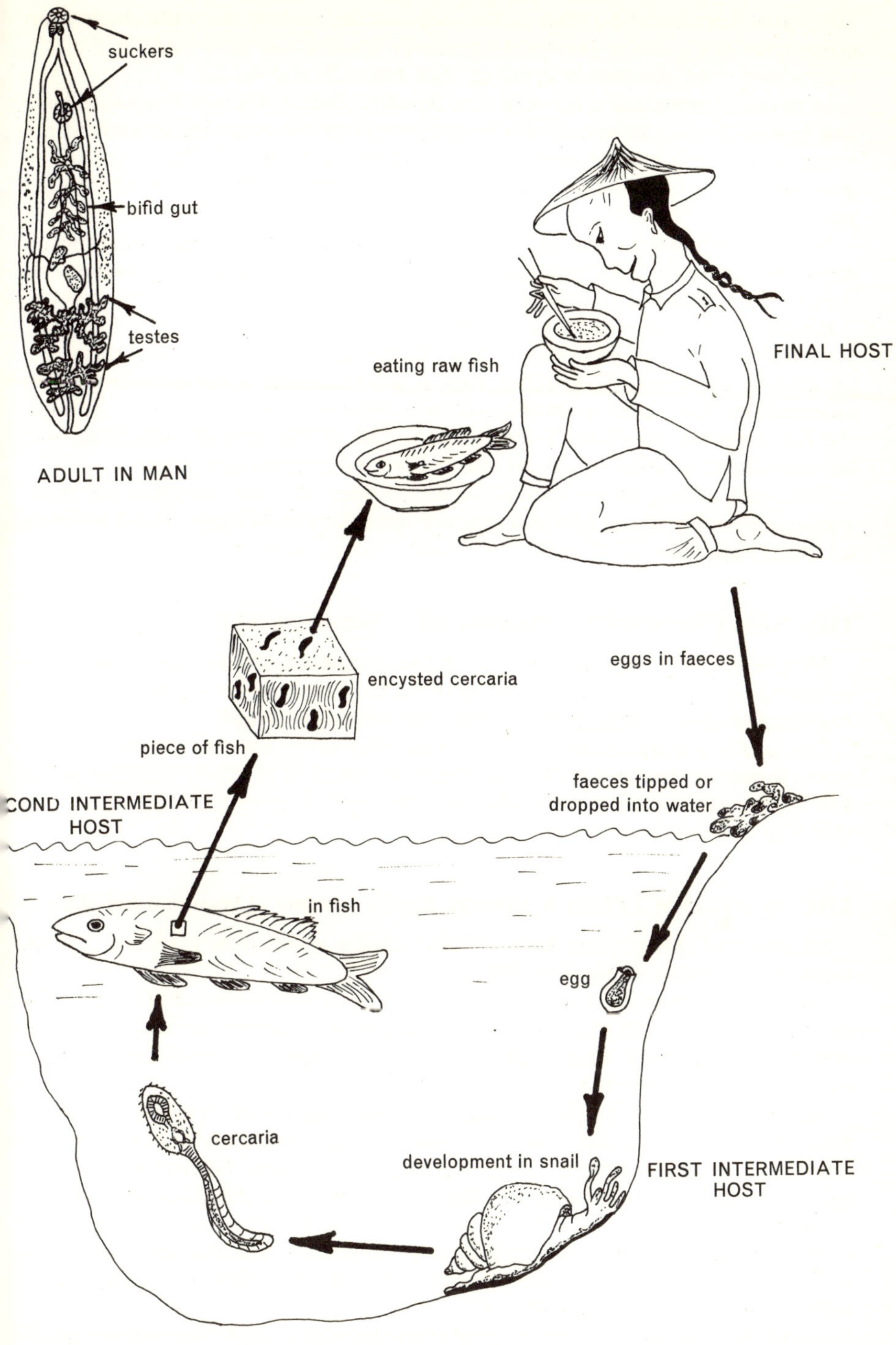

Fig. 1.8 The life cycle of the Chinese liver fluke (*Clonorchis*) (after Buchsbaum)

The copulatory bursa consists of two flaps at the end of the tail which wrap around the body of the female. Copulatory spicules are little rods in the tail, which hold open the female opening. (See Fig. 4.3. and 4.4.)

In parasitic nematodes, the sexes are separate, that is, there are separate male and female forms. All nematodes have a mouth at one end and an anus at the other, with a long straight tube (the gut) connecting them. The front part of the gut is called the oesophagus. This a muscular pump which sucks food into the gut.

Nematodes that are parasitic in animals usually occur in the gut. Some species are found in tissues, for example *Wuchereria*, which causes elephantiasis. It lives in the lymphatic vessels of man.

Parasitic nematodes may have one, two or three hosts in their life cycle. Sometimes there are larval stages in a blood sucking insect. In other cases the life cycle is based on a food chain, that is, a larval stage is parasitic in the first intermediate host, and is transferred to a second intermediate host, because the first intermediate host is the normal food of the second. The final host becomes infected when it eats an infected second intermediate host.

There are about thirty species of nematodes that are parasites of man. Fortunately, only about half a dozen are important in Niugini. Fig. 1.8 shows a typical nematode.

THE NEMATOMORPHA (Hairworms)—(Nem-at-o-morfa)

Members of this phylum are rather like nematodes. They are often found wriggling actively in ponds and pools of water. They are up to six inches long, and are thin like a piece of hair, hence the name, hairworm.

The adults are free-living, but the larval stages are parasitic in insects such as cockroaches, beetles, crickets and grasshoppers. The adults mate in water and lay eggs on aquatic plants. They are not parasites of man or any vertebrate. (See Fig. 1.9.)

THE ACANTHOCEPHALA (Thorny headed worms)—(Ac-anth-o-kef-alla)

Members of this phylum have a head which is covered with spines. It is an organ of attachment. (See Fig. 1.9.)

The Acanthocephala usually have a life cycle in which the adults live in the gut of vertebrates and there is a larval stage in an invertebrate host. The final host becomes infected when it eats an infected intermediate host. Some species have two or three intermediate hosts.

Acanthocephala do not have a gut, but absorb their food through their body wall. It is as if the gut were on the outside of their bodies instead of the inside They vary in length from a few millimeters up to about 60 centimeters. One species, with the long name of *Macracanthorhynchus hirudinaceus* (Macra-canth-or-rinkus hir-ude-in-ace-eus), is a parasite of pigs in Niugini. Each female of this species lays about 10 million eggs. Various kinds of beetle larvae act as intermediate hosts, and pigs become infected when they eat these larvae, which normally live in the soil.

About 600 species of Acanthocephala have been described. None are normally parasites of man.

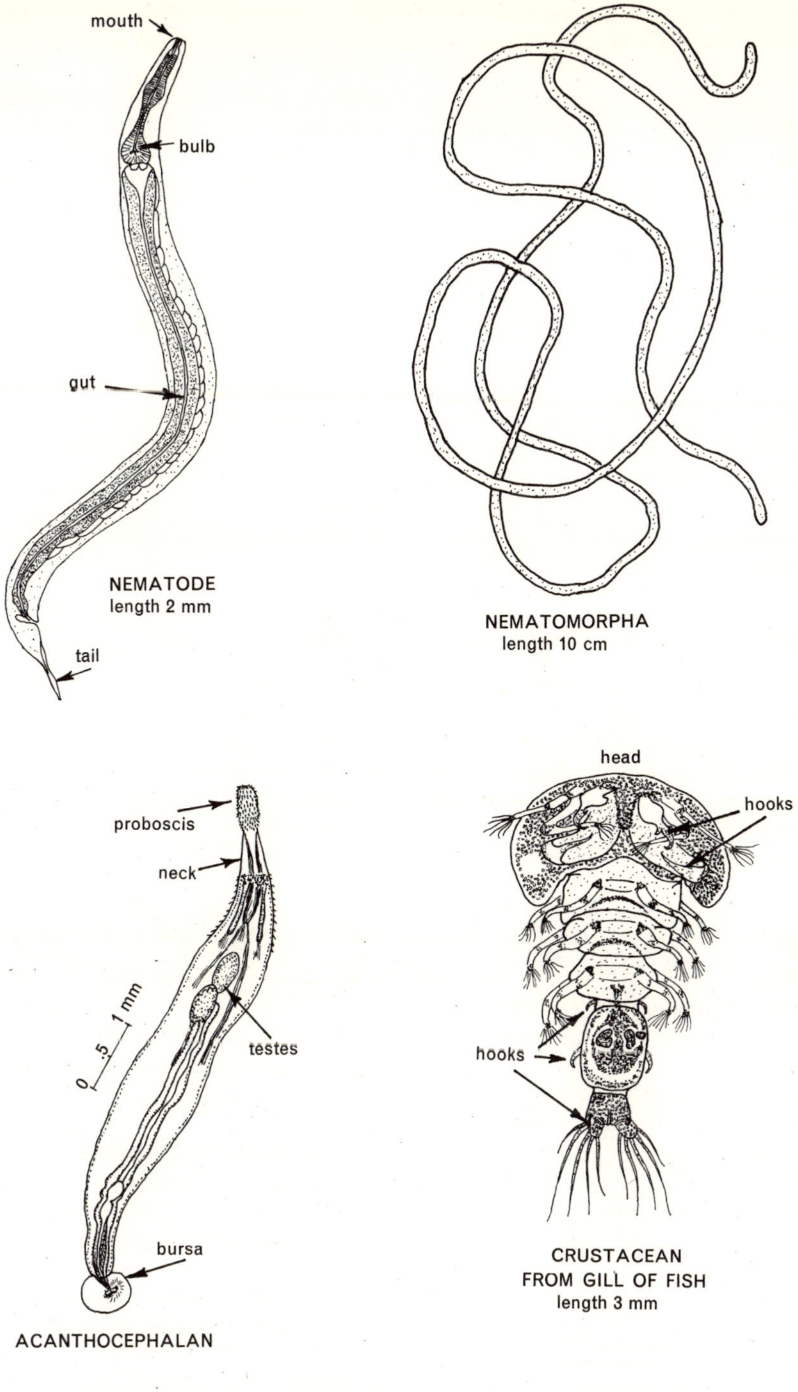

Fig. 1.9 Nematode, nematomorpha, acanthocephalan and crustacean

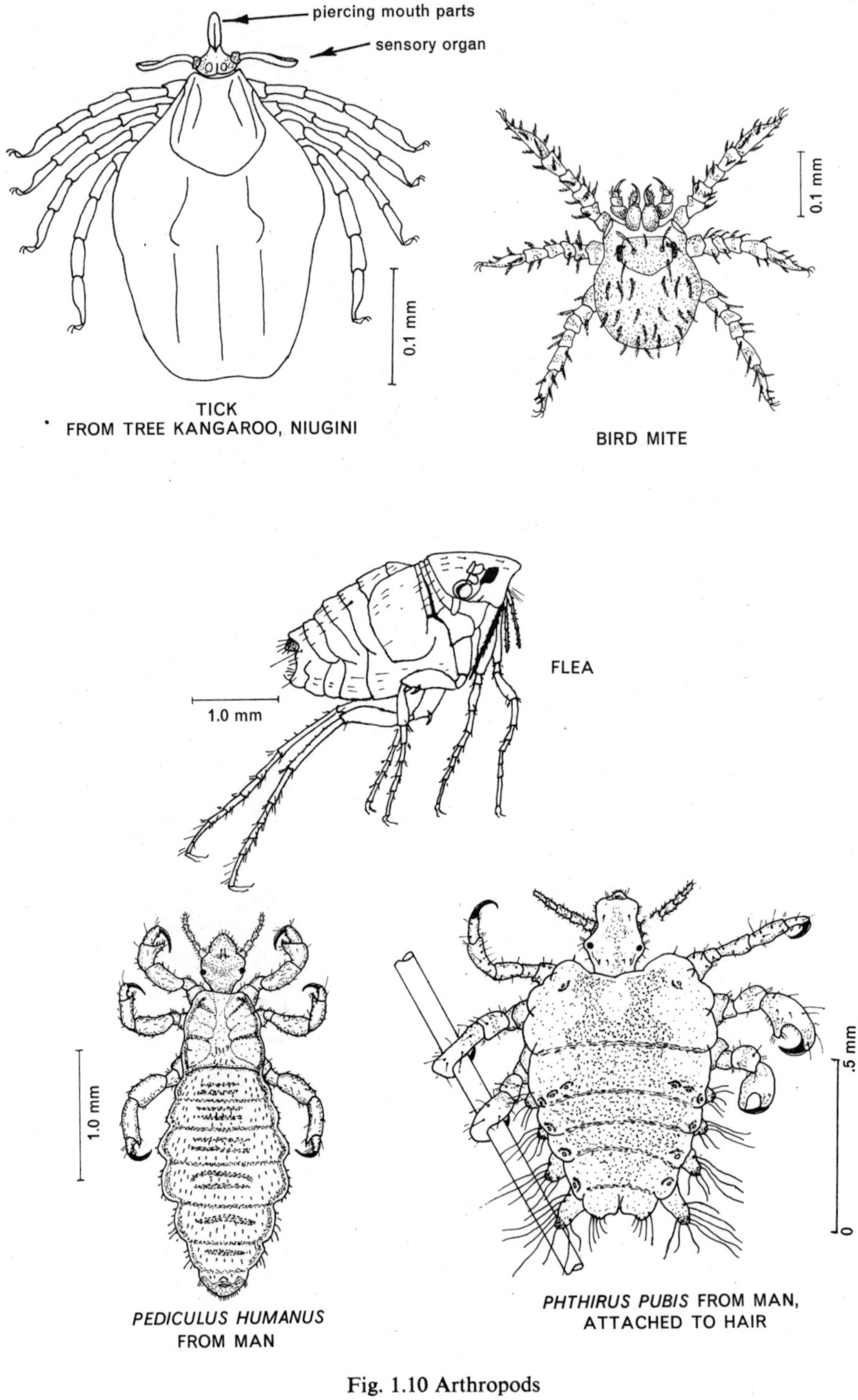

Fig. 1.10 Arthropods

THE ARTHROPODA (Arth-ro-pode-a)

This phylum contains a wide variety of invertebrates which typically have jointed legs. They include crustacea (crayfish, shrimps, crabs, water-fleas, etc.) centipedes and millipedes, spiders, scorpions, ticks and mites, and insects.

The most important parasites in this phylum are the ticks and mites and certain blood sucking insects. Most of them are important not because they cause great harm to the host, but because they carry diseases which infect man and his domestic animals.

Some crustaceans are ectoparasites of fish and other aquatic vertebrates. Many species live on the gills of fish. They often have bodies that are flattened on one side, with suckers, and lots of legs with strong claws for hanging on. (See Fig. 1.9.) The adults of a few forms are so modified for a parasitic way of life, that they do not look like crustaceans at all, but the larval stages of these forms are like the larvae of other crustacea.

Most parasitic ticks and blood sucking insects are temporary parasites. They do not live permanently on the host. They cannot lay their eggs until they have had a meal of blood.

Ticks and Mites

These belong to the class Acarina. They have a rounded body which is not divided into segments. The adults have four pairs of legs and the larval stages three pairs. (See Fig. 1.10.)

Ticks

Ticks are usually big enough to see without a microscope. Some of them grow to about one centimeter long. They have a brown leathery body which often has short hairs coming from it. About 800 species of ticks have been described and all of them are blood sucking ectoparasites of vertebrates that live on the land. The females lay large numbers of eggs (up to 30,000) after feeding. The eggs hatch to produce a six legged larva which must have a blood feed before it can moult. Moulting is a process where an animal sheds its skin, and while its new skin is still soft and able to stretch, quickly increases in size. Most ticks moult twice before they become adults. Usually the ticks remain attached to the host for only a few hours. The life cycle takes up to two years. Both males and females suck blood. Ticks cannot move very much and most of them crawl to the top of grass and wait until a host comes along and touches them. Then they hang on, feed, and drop off.

Ticks sometimes parasitise man. Some of them cause swelling and itching and a few species inject substances which cause paralysis. This can be serious and may even cause death.

Ticks serve as vectors for certain viruses, bacteria and protozoa. Some of these cause serious sickness in man and cattle. For example, *Babesia* is a blood parasite like malaria, which is carried by ticks to cattle. It is a serious problem in cattle in Niugini.

Mites

Mites are usually small, often microscopic. They have long hairs on their bodies and their mouth parts are smaller than those of ticks. Many mites are free-living and are found in the soil, in fresh and salt water, and in decaying plant material. They are common in the nests of birds and mammals.

Many species are either temporary or permanent ectoparasites of vertebrates and invertebrates. Some species that are parasitic on other animals, or which normally live a free-living existence, sometimes accidentally infect man and cause itching.

Two species of mites are common 'permanent' parasites of man. Both occur in Niugini. They are *Sarcoptes scabiei* which causes human scabies, and *Demodex folliculorum* which infects the skin of the head, but is generally harmless. These mites live in shallow burrows at the base of hairs and sweat glands. (See Section 6.4.)

Some of the temporary mites transmit disease to man. Larval mites of the family Trombiculidae are vectors of a disease called scrub typhus, which occurs from India and Japan, through most of Asia to Niugini and Northern Queensland. This disease is characterised by a sore or ulcer which develops where the mite attaches itself; fever, often a very bad cough, and damage to the heart, lungs, brain and kidneys. It can be fatal (that is, cause death).

Some mites are the intermediate hosts of certain tapeworms.

Insects

Insects have a segmented body divided into three separate parts: the head, thorax and abdomen. They usually have two pairs of wings and six legs, all of which come from the thorax. Some ectoparasitic insects do not have wings and in a few of them, the body is so modified that the three separate parts are all fused together.

Two different kinds of life cycles occur among the insects. In one, the eggs hatch to give nymphs, which are rather like the adults, except that they are smaller and lack wings. These nymphs moult several times and each time they become bigger and look more like the adults. Their wings become a little larger each time they moult. Crickets, cockroaches, grasshoppers and bugs have this kind of life cycle. In the other kind of life cycle, the eggs hatch to produce a worm-like larva which is quite unlike the adult. The maggots of flies, the caterpillars of butterflies, the larva of mosquitoes and fleas, and beetle grubs are examples of this kind. These larvae also moult several times. They then change into a resting stage called a pupa. In some insects (for example, certain moths), the pupa is protected by a cocoon which is made by the larva. The adult insect develops inside the skin of the pupa and emerges when it is mature.

Flies

Most of the insects which parasitise man are blood sucking flies that come to man just long enough to take a meal. They are very temporary parasites. The Diptera or flies, are an order of insects which only have one pair of wings; the hind pair of wings has become modified to form two small knobs called halteres, which help the fly to maintain its balance. This order includes blowflies and bush flies, mosquitoes and a lot of other flies, some of which are temporary parasites of vertebrates. They usually cause little harm to their hosts apart from an itchy bite. However, many of them carry diseases. For example, mosquitoes transmit over 50 virus diseases to man, as well as malaria and filariasis. Mosquitoes are dealt with in Section 3.6.

Lice

Man is a host to two permanent ectoparasitic insects called lice. Both occur in Niugini. They are *Pediculus humanus* (Ped-ik-u-lus) which lives in the hair of the head and body, and *Phthirus pubis* (Pa-theer-us pew-bis) which lives in the pubic hair around the external sex organs. (See Fig. 1.10). Both cause itching. The females lay small white eggs called nits, which are attached to hairs. Each female lays about two hundred eggs which hatch in about a week. They take about two weeks to become adult. The adults live for about a month. Lice transmit several fever diseases to man.

Fleas

The fleas are another order of insects. They are ectoparasites of birds and mammals. Over 1000 species of fleas have been described. They are not very host-specific, and fleas from rats, mice, pigs, cats and dogs often turn to man for a quick feed of blood. One common temporary man biter is called *Pulex irritans* (Pew-lex), because of the irritating bites it causes.

Fleas do not look much like other insects. (See Fig. 1.10.) Their head, thorax and abdomen are almost fused together and they have no wings. Their bodies are flattened from side to side; an adaptation for gliding between hairs and feathers. They have legs that are modified for jumping, so that they can reach new hosts. Fleas usually lay their eggs in the nests or burrows of their host. The females lay about 300 eggs during their lifetime. These hatch within a week or so, to produce a small larva which looks like a fly maggot. When the larva is ready to pupate it spins a silk cocoon and rests in this, while it changes into an adult flea. The complete life cycle takes about three to four weeks.

Fleas transmit several diseases to man, including plague, which in the past swept through whole countries causing many millions of deaths. Fleas also act as intermediate hosts for certain tapeworms.

Bugs

We must mention one more order of ectoparasitic insects—the bugs. They have sharp sucking mouthparts that are adapted for piercing plants and animals and sucking juices from them. Many bugs are important pests of food plants. A few are blood suckers and feed on birds and mammals. One group of blood sucking bugs are the vectors of a serious disease caused by Trypanosomes, which infect man in South America.

The bedbug is the only important bug of man in Niugini. They hide in cracks and crevices in the walls of houses and beds during the day time. At night they come out and bite people while they are sleeping. Unlike most other Arthropod parasites of man, they are not vectors of any diseases. We will discuss these further in Section 6.3.

Chapter 2

MAN—THE HOST

2.1 Introduction

In order to understand the interactions between host and parasite (that is, the host-parasite relationship) we need to know a little about the structure and function of those organs of the host in which parasites live, and those that are concerned with defending the body against infection. Most of the parasites of man live in the digestive (die-jest-ive) and circulatory (sirc-u-late-ory) systems. We will deal with these first and then outline the mechanisms by which the body resists infection.

2.2 The Digestive System

The digestive system consists of a long tube, the gut, which extends from the mouth to the anus, and some organs which are associated with it. As food passes down this tube, it is digested and absorbed. Digestion is a chemical process whereby enzymes (en-zimes) break down the large molecules of fats, carbohydrates and proteins (which make up food), into molecules which are small enough to pass through the cells of the gut so that they can be used by the body. The uptake of these small molecules is called absorption.

The gut tube is lined by cells which secrete enzymes and other substances that are essential in digestion. The wall of the tube contains three layers of muscle. These are continually contracting and relaxing along the entire length of the gut, and so push the food toward the anus. Some of the food cannot be digested and this is excreted as faeces.

The gut of man is illustrated in Fig. 2.1. It consists of:

(a) **the oesophagus** (o-sof-ag-us) This tube, about 10 inches long, extends from the mouth to the stomach. It is not concerned with digestion.

(b) **the stomach** This is like a large bag, which can stretch when it is full of food. The wall of the stomach is lined with cells which secrete hydrochloric acid and an enzyme called pepsin. Together these two substances break down proteins in the food into medium sized molecules. They are digested into smaller molecules, called amino acids, in the intestine. Other cells lining the stomach and the intestine produce a substance called mucin which sticks to the lining of the gut and prevents it from being digested.

The end of the stomach is closed off by a muscular valve which acts like a tap. It periodically lets small amounts of food flow out of the stomach into the intestine.

Digestion and absorption is completed in the intestine, which consists of two separate parts. In an adult man, the intestine is about 30 feet long.

(c) **the small intestine** The first part of the small intestine is called the duodenum (dew-o-deen-um). Ducts (small pipes) enter it from the pancreas (pan-cree-ass)

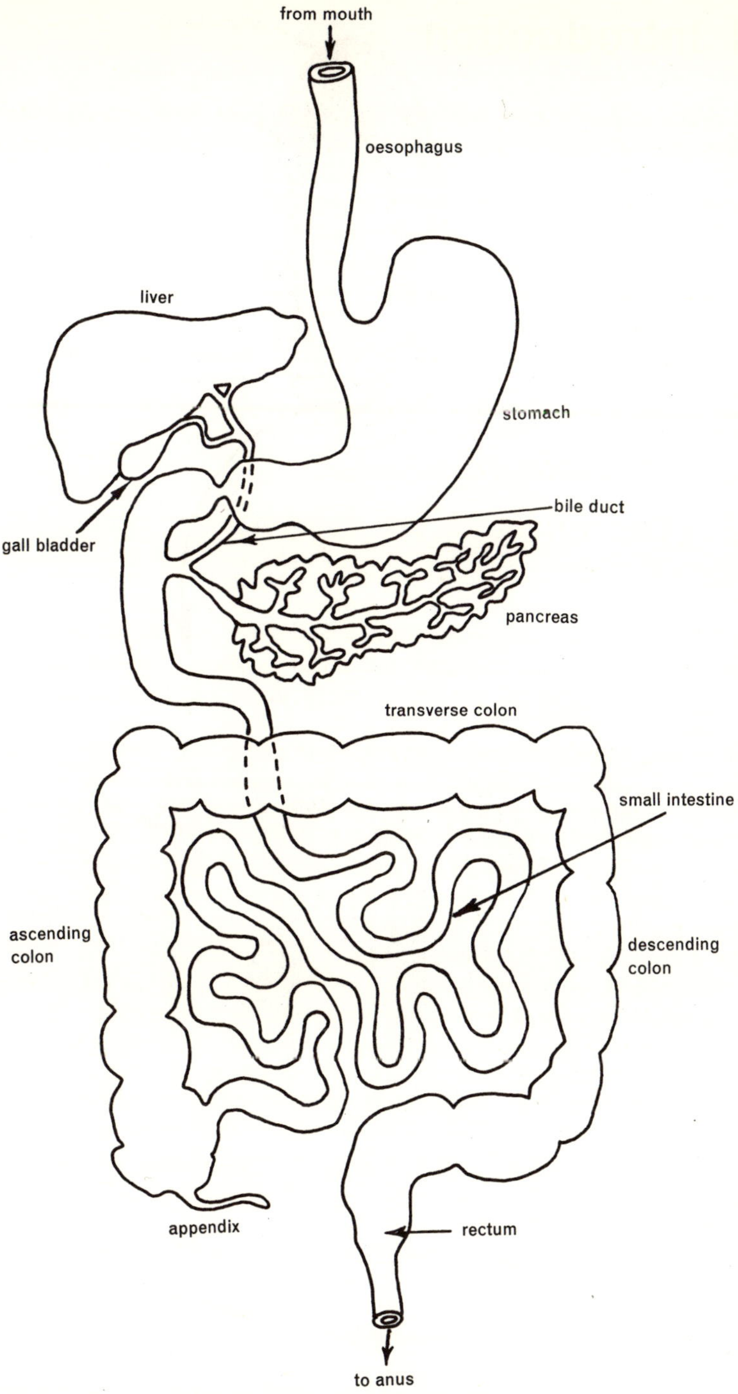

Fig. 2.1 The digestive system of man

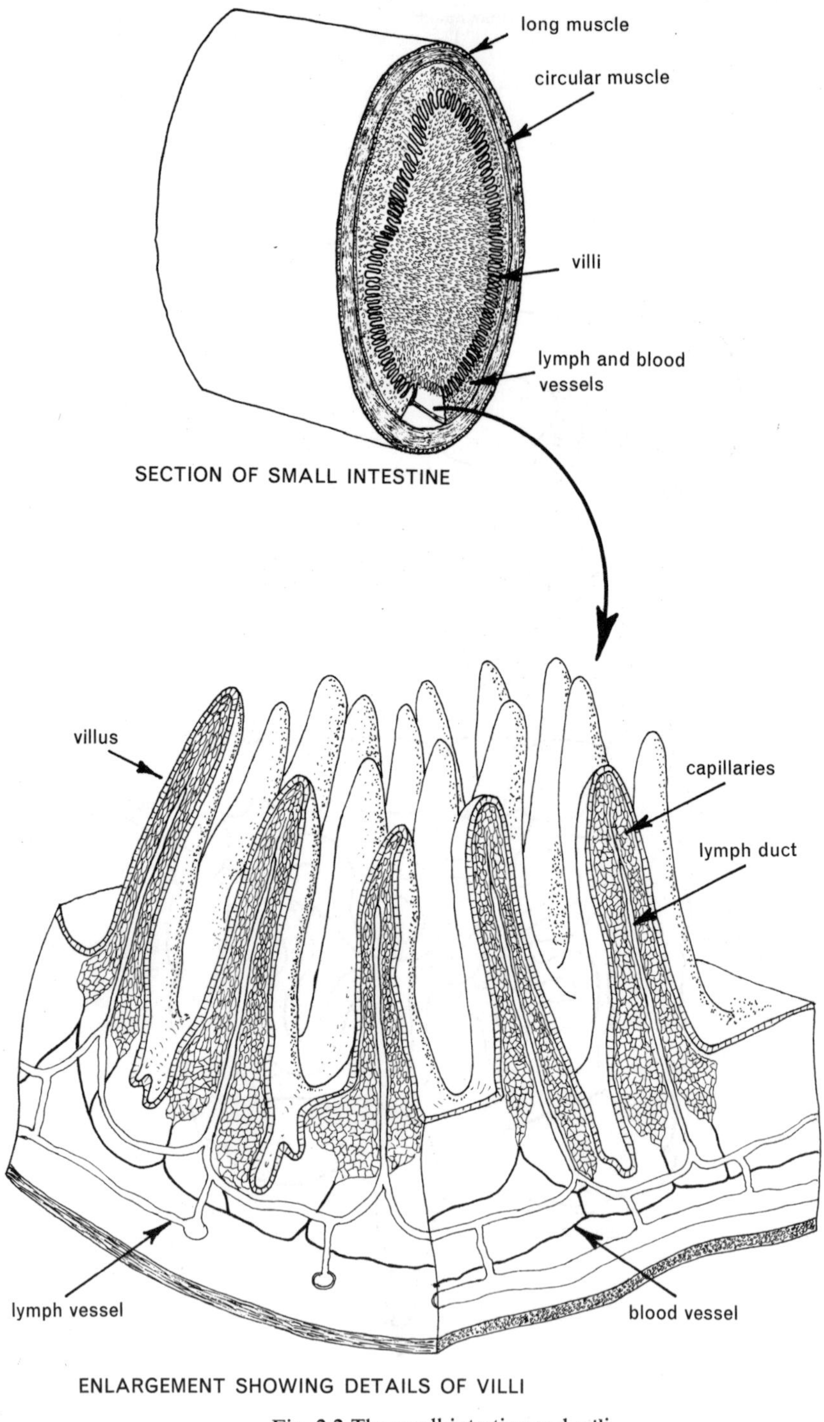

ENLARGEMENT SHOWING DETAILS OF VILLI

Fig. 2.2 The small intestine and villi

and the gall bladder. The pancreatic duct pours enzymes into the gut. The gall bladder is a storage organ attached to the liver, containing a greenish-yellow fluid called bile. Bile is made in the liver, stored in the gall bladder, and transferred to the gut by the bile duct. It breaks down large masses of fat into tiny microscopic particles which can be absorbed. This process is not due to enzymes, but is similar to the action of soap or detergents on grease or fat.

The surface of the small intestine consists of millions of tiny fingerlike projections called villi, each of which has its own blood capillaries and lymphatic vessel (we will talk more about these later). The villi greatly increase the surface area of the gut, and hence the area for absorption of small molecules (see Fig. 2.2). The blood capillaries and lymphatics are so arranged that they are very close to every part of the surface of the villi, thus making it easy for substances to be absorbed.

The blood capillaries from the villi join up to form a blood vessel which goes to the liver. The liver acts as a storage organ for food that has just been absorbed.

The lymphatic vessels also join up and eventually enter the blood vessels. They receive most of the fat that is absorbed while the blood capillaries take the amino acids and small sugar molecules.

(d) **the large intestine** (or colon) This has a much larger diameter than the small intestine and is about five feet long. The appendix is a small closed tube about the size of a baby's finger, which comes off from the large intestine near where it joins the small intestine.

The large intestine consists of three parts called the ascending colon, the transverse colon, and the descending colon, which becomes the rectum. The food must actually pass uphill as it goes through the ascending colon. The main function of the large intestine is the absorption of water. Digestion can only take place if the food is mixed with a lot of water. Much of this water is secreted into the gut. By the time the food has got to the large intestine, all the usable substances in it have been digested and absorbed. Waste matter is left over. This contains a lot of cellulose—a carbohydrate from plants—which cannot be digested. The water is re-absorbed from this, into the blood stream so that it can be used again. The waste matter is passed out as faeces.

2.3 The Circulatory System

This consists of a network of blood vessels which carry blood to all parts of the body. No living cell in the body is more than about twenty microns from a blood vessel. Blood is continually being pumped around the body by the heart. The lymphatic system is linked with the blood system in a way that will be explained later.

The circulatory system is a transport system. It carries dissolved substances around the body. Its main functions are outlined below.

(i) It carries oxygen from the lungs to all cells.

(ii) It carries carbon-di-oxide from all cells to the lungs.

(iii) It carries digested food substances from the gut to the liver where many of them are stored. The liver releases food substances into the blood so that their concentration is kept fairly constant and they are carried in the blood to all cells of the body.

(iv) It carries waste products to the kidneys where they are filtered out and excreted in the urine.

(v) It carries hormones, or chemical messengers, from the glands where they are made, around the body. These messengers cause certain organs to change in various ways. For example, when a person becomes frightened, the adrenal glands release a hormone called adrenalin into the blood. This causes the heart to pump faster and the liver to release extra sugar into the blood, and many other effects. Together these effects provide energy so that the person can fight or run away.

(vi) It also plays an important part in keeping certain things, such as body temperature and the amount of water in the tissues, constant, so that the body functions most efficiently.

(vii) It also contains substances called antibodies and white blood cells which function in fighting infection; these will be dealt with later.

The Circulation of the Blood

The heart is a single organ which consists of two pumps. One pumps blood to the lungs, where it comes in close contact with air and takes up oxygen and releases carbon-di-oxide. Oxygenated blood comes back to the second pump in the heart, which pumps it around the body.

Blood leaves the heart in large thick walled vessels or pipes, called arteries. These give off branches which go to all parts of the body. These branches become smaller and smaller until they are little bigger than the blood cells which pass through them. This network of thin walled vessels or capillaries, comes close to all cells in the body. Because the capillaries are thin walled, substances can move out of the blood and into the cells, and at the same time, substances can move from the cells into the blood. The capillaries join up with larger and larger vessels called veins, which carry the blood back to the heart. (See Fig. 2.3.) It has been estimated that there are about 60,000 miles of blood vessels within the human body. This is over twice the distance around the earth at the equator.

The human body contains about 10 pints of blood and this circulates about 1000 times every day (or every $1\frac{1}{2}$ minutes).

The heart is a strong pump. It has to be, to do all this work. All the way through the arteries, capillaries and veins, the blood is under pressure. Many of the small molecules in the blood leak out through tiny spaces in the capillaries, in the same way that air will leak from a puncture in a tyre. These particles include water, salts, fats and proteins. Red blood cells have a fixed shape and are too large to leak out. But white blood cells can squeeze out between the cells lining the capillaries. One of the functions of the lymphatic system is to collect these substances and return them to the blood, so that the tissues do not become water logged.

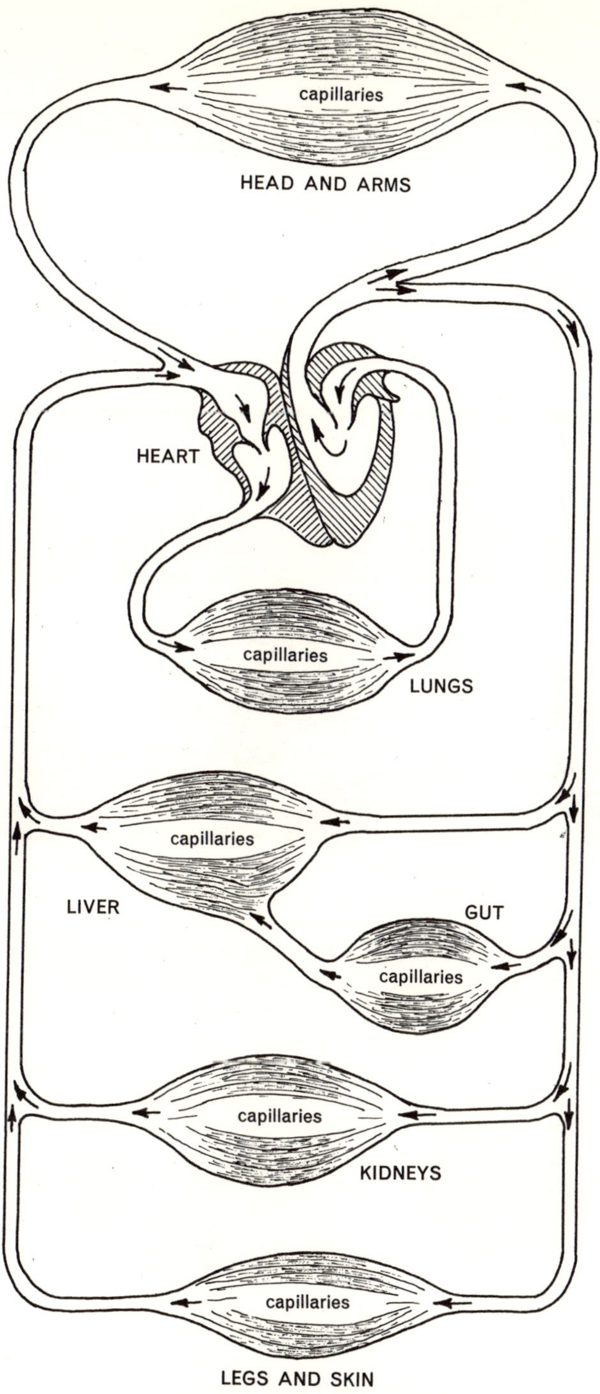

Fig. 2.3 The circulation of blood

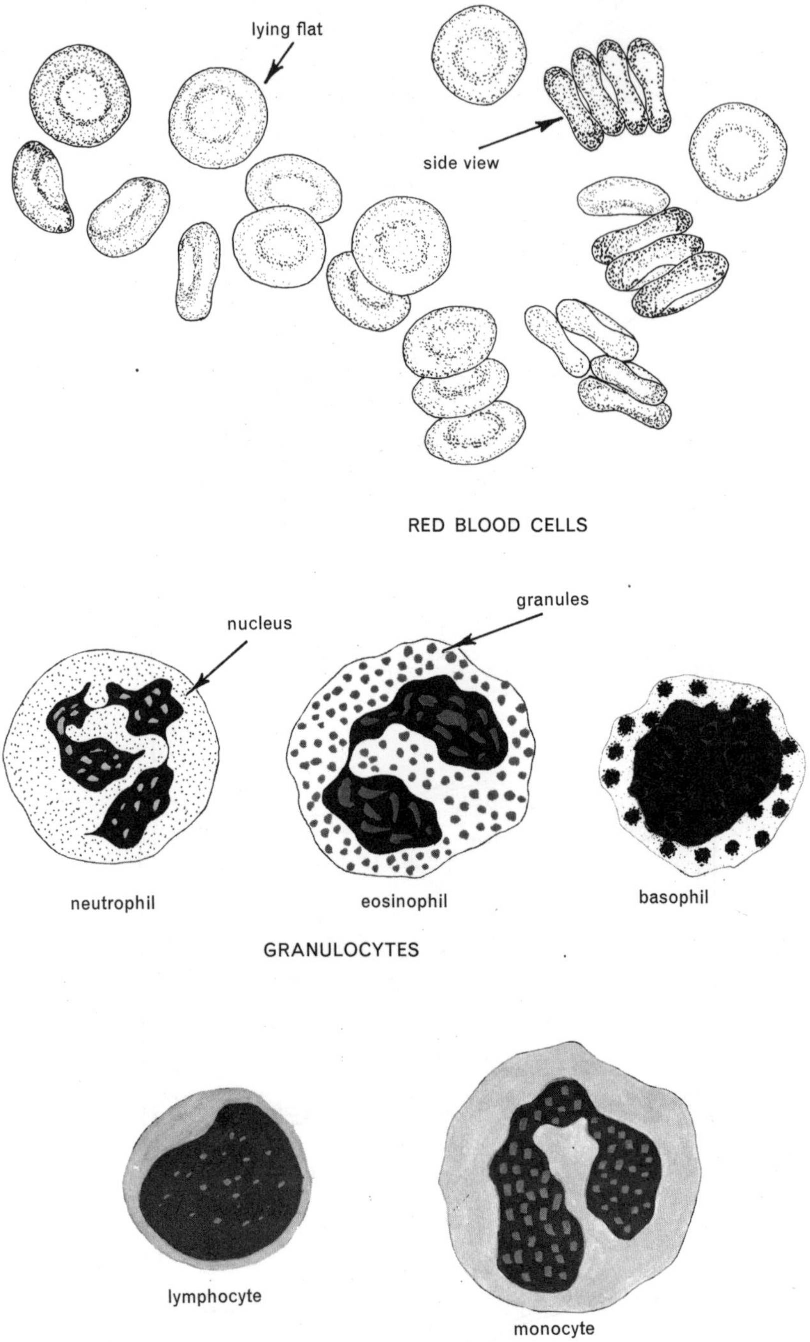

Fig. 2.4 Blood cells

The Blood

Blood is a fluid containing cells. This fluid is called plasma and it is mostly water. It contains many dissolved substances such as sugars, fats, amino acids, inorganic salts and many different kinds of proteins. Several kinds of cells are suspended in this fluid. The most common are the red blood cells or corpuscles. There are about 5 million of these in each cubic millimeter of blood, and about 35×10^{12} (35 million million) in the body.

It is the red blood cells that carry oxygen and carbon-di-oxide. They contain a chemical called haemoglobin, which is a protein containing iron. The red blood cells are flattened (see Fig. 2.4) and do not have a nucleus, but they grow from nucleated cells which lose their nuclei before they are released into the blood. New red blood cells are continually being made in the bone marrow inside certain kinds of bones, and old ones are continually being destroyed in the liver and spleen. The red blood cells live for only about 3 months. If, due to certain kinds of sickness, more red blood cells are being destroyed than are being made, then the person becomes anaemic. There is not enough haemoglobin in the blood.

The blood also contains several kinds of white blood cells. They are much less common than red blood cells. There are about 7000 white blood cells per cubic millimeter of blood. The white blood cells are larger than red blood cells and have a nucleus. They are divided into two basic types (see Fig. 2.4).

(a) Granulocytes (gran-u-low-sites)

These contain many tiny, dark staining bodies called granules. They also have a nucleus which is usually divided into two or three lobes. There are three kinds of granulocytes. They are all formed in the bone marrow.

(i) *Neutrophils* (newt-ro-fils) which make up about 65 per cent of all white cells. Their granules are very fine and the nucleus often consists of three or four lobes, joined by fine strands, like a string of sausages.

(ii) *Eosinophils* (e-o-sinofils) which make up about 1 to 4 per cent of all white cells. Their granules stain red with a dye called eosin—hence their name. Their nucleus usually has two lobes. Helminth (worm) infections usually cause an increase in the number of eosinophils in the blood, so that they may account for 10 to 80 per cent of all white cells.

(iii) *Basophils* (bays-o-fils). These are the rarest of the white cells and account for only one half of one per cent of the total. They have large granules which stain dark blue-purple with certain blood stains. Their nucleus is usually bi-lobed (bi=two).

(b) Non-Granular Leucocytes (lew-co-sites)

There are two kinds of these: Lymphocytes (lim-fo-sites) and Monocytes.

(i) *Lymphocytes* come from the lymphatic system which is part of the circulatory system. They can live for months or years. Two distinct sizes of lymphocytes are found in the blood. The small ones account for about 20 to 25 per cent of all white blood cells. They are round and have a large round nucleus. The large lymphocytes represent only about 3 per cent of the total white cells. They have a large oval nucleus which is often a little indented, and the cell is usually oval shaped.

(ii) *Monocytes* account for 3 to 7 per cent of the white cells. They are large, round, or nearly round, cells, with a large irregular nucleus.

Scientists have only recently started to find out what white blood cells do and how they do it, and there is still a lot to be learned about them. They seem to be mostly concerned with fighting infection. We will talk about their function soon, but first we will say something about the lymphatic system.

The Lymphatic System

Because the blood is under pressure from the heart, fluid is pushed out into the tissues through tiny holes in the walls of the capillaries. One of the main functions of the lymphatic system is to drain this fluid back into the blood system. Certain white blood cells are also able to squeeze out of the capillaries into the tissues.

The lymphatic system consists of an extensive network of vessels. It drains all the tissues that are served by the blood capillaries, thus the lymphatic system and the blood system together form the circulatory system. The lymph vessels arise from tiny blind-ended lymph capillaries which join together to form larger and larger lymph vessels. These eventually drain into the veins in the neck, and thus return fluids back to the blood system.

The lymph vessels contain a fluid called lymph which is very much like blood without any red cells in it.

Lymph vessels from the gut also transport digested fats to the blood system. Movement of the lymph is slow. It is not pumped along like the blood, and it probably takes several hours for protein molecules which have leaked out of the capillaries to get back to the blood by way of the lymphatic system.

If the lymphatic vessels are blocked, the tissues become waterlogged and swell. This sometimes happens when a person is infected with the filarial worm called *Wuchereria bancrofti*. This may result in elephantiasis (see Chapter 5).

Along the lymphatic vessels there are lymph nodes. They vary in size but many of them are about as big as a grain of rice. Each node has a tough wall. Inside the node there is a reservoir of lymphocytes which are held ready to attack any infection which occurs in the body.

As the lymph passes through these nodes, any bacteria, foreign particles, or bits of dead cells, are eaten by white blood cells.

THE SPLEEN

This organ is situated near the stomach and is attached to the small intestine by mesentries, which are sheets of tissue that hold it in place. Under normal conditions, the spleen is about six inches long and four inches wide and is a dark purple colour. It can become greatly enlarged if the body is infected with malaria or certain other diseases.

The spleen has lots of spaces between the tissues which are filled with blood cells, both red and white.

One of the main functions of the spleen is the destruction of old worn out red blood cells. These are eaten by certain white cells and are broken down.

The spleen, together with the lymphatic system and to a lesser extent the liver, are sites for the production of non-granular white blood cells. Together they form what is called the Lymphoid-Macrophage System which is concerned with defence against infection.

The Functions of the White Blood Cells

The granulocytes are like front line soldiers. It is their job to attack foreign organisms such as bacteria, viruses and other parasites that enter the tissues. They do this by eating small foreign particles in the same way that an amoeba eats. This process is called phagocytosis (fago-si-toe-sis), and the white blood cells which eat in this way, are often called phagocytes. Almost any kind of foreign substance is recognised as being different from self and is phagocytosed if it is small enough. Some parasites are too big to be eaten but they are attacked just the same by enzymes released by the phagocytes.

Let us look at what happens when bacteria infect a cut in the skin. Much more is known about defence mechanisms against bacteria than other parasites. However, the principles must be the same for all parasites.

The first thing that happens is that mast cells, which are present in the tissues, burst or disintegrate near the bacteria. Mast cells are always present in the tissues, but not normally in the blood. They have large granules and are very like blood granulocytes. When they burst they release substances which attract white blood cells to the infected area. The neutrophils are the first to arrive, and they start to eat the bacteria (which are foreign material). The bacteria are digested by enzymes inside the neutrophils. Within a few hours the neutrophils burst and a second wave of soldiers—the lymphocytes and monocytes—arrive and eat the rubbish that has been left on the battlefield. This includes bits of burst granulocytes and foreign material from the bacteria, some of which has been partly digested.

The lymphocytes and monocytes grow as they feed and become macrophages (*macro*: large; *phagos*: to eat). Then they become the front line soldiers and start to eat the bacteria.

Some of the protein from the bacteria is stored within the macrophages without being digested. These macrophages continue to live in the body and they divide to form new cells. The cells pass on, to other cells, the ability to make substances called antibodies which act in a special way against foreign proteins. If at a later stage the body is still infected, or becomes reinfected, with the same kind of bacteria, these antibodies will act against the infection. Antibody production takes at least four or five days to get going. Once the body has been attacked by a particular parasite it starts to develop a new weapon against it. This weapon can be brought into battle straight away if the parasite is still present, or, it can be held in case reinfection by the same parasite occurs some time later. This new weapon is called an antibody.

Antibodies

Antibodies are large protein molecules that are made by cells of the lymphoid —macrophage system and set free into the blood and lymph. There are many different kinds of antibodies and they combine with foreign substances, which are collectively called antigens, in a way that prevents the antigen from harming the host. Almost any foreign protein or carbohydrate molecule of large size, can act as an antigen. That is, it can cause the body to make an antibody which combines with it. When a macrophage eats a particular protein or carbohydrate, it passes on information which allows other cells to make antibody molecules which combine with that particular antigen and no other. For each kind of antigen molecule that enters the body, a corresponding antibody is made which is different from all other antibodies. Because of this, we say that antibodies are highly specific; they will only react with a particular antigen.

The combination of an antigen and an antibody is called an immune reaction. We will talk about snake bite to illustrate this. The venom of snakes contains large protein molecules of a certain kind which are toxic or poisonous. The venoms of different snakes have different kinds of toxic substances in them, which affect the body in various ways. For example, some cause the heart to slow down, others affect the nervous system, some cause bleeding, others make the blood clot or the red blood cells disintegrate. They are also antigens because, when they are injected into an animal, they cause the production of antibodies, which combine with them so that they cannot cause toxic effects.

Snake venom is so toxic that it usually kills the animal that is bitten before it can make any antibodies against it. But if small quanties of snake venom are injected, each day, for several days, into a large animal like a horse, it does not die, and after a few days it starts to make antibodies against the toxic proteins in the snake venom. These antibodies appear in the blood of the horse, and can be separated from it by special techniques or processes. If these antibodies are injected into the blood of a man, who has been bitten by the same kind of snake, they combine with the toxic protein molecules of the snake venom, so that they do not cause any damage. This is how snake antivenene is made and how it works. Modern antivenenes are usually mixtures of antibodies that act against the antigens of several different kinds of snakes. This is a good idea because people often do not know what kind of snake has bitten them.

Antibodies always act by combining with antigens. Various things can then happen, all of which are designed to defend the body against infection or the toxic effects of foreign substances. The following examples illustrate this.

(i) The antibody attaches itself to a particular antigen on the surface of a parasite and this acts as a label, which attracts the phagocytes and tells them to eat the parasite. They gather round in the same way that people would, in response to a large sign saying 'free beer and lollies'. Phagocytes eat any foreign particles that they happen to meet, but they come much faster, and eat them much more quickly, if they are labelled with antibody.

(ii) Many antibody molecules are banana shaped and at each end of the molecule, have areas which join with antigens. There are lots of antigen molecules on the surface of a parasite and if there are lots of antibody molecules present, each one will become attached to the surface of two parasite or antigen molecules. This combination of antigen and antibody molecules, will cause the formation of a tight mass, or precipitate, of antibody and antigen molecules, which is easily removed by phagocytes.

(iii) Many parasites are far bigger than phagocytes, so they cannot be eaten in one meal. However antibodies can kill them in two ways.

(a) They can combine with antigens on their surface so that they cannot function properly. Parasites that do not have a gut must absorb food and excrete their waste products, through their skin. If their skin is clogged up with antibodies they will die.

(b) Worm parasites are often killed by antibodies which combine with antigens and form precipitates which block their mouth and anus.

(iv) When certain antibodies combine with antigens on the surface of a parasite, they attract enzymes which are present in the blood. These make a tiny hole in the parasite which causes it to burst and die.

Immunology is the name given to the study of the defence mechanisms which the body has against infection. It is a new science and much has to be learnt about it yet. Most of the work done in this field so far, has been concerned with defence against bacteria and viruses. This book is concerned with other kinds of parasites such as helminths and protozoa. It seems that the body uses the same mechanisms against them, as it uses against bacteria and viruses, but they often do not work as efficiently. In many parasitic infections, the immune mechanisms are effective only so long as the body is already infected with parasites of the same kind. So long as parasites of a particular kind are present, immune mechanisms prevent the development of more parasites of that kind. If these parasites die, the immune mechanisms usually do not last very long.

Chapter 3

MALARIA

3.1 Introduction

Malaria is one of the most important diseases of man in the world today. About 100 million people are infected with it every year and about one million of them die from it. It is the most important disease in Niugini. Malaria is more common in warm tropical countries than in colder countries. In areas where it is common, malaria kills many babies and children and some adults too. Sometimes the sickness is severe and a person with malaria may spend several weeks in bed. In other cases the disease is mild and people suffering from it do not feel very sick. But they are not fit enough to do hard work or to concentrate properly at school.

Malaria has been eradicated, that is to say, wiped out, from some parts of the world and a program to control malaria has been started in Niugini. Eventually malaria will be completely eradicated from this country and our people will be much healthier and happier.

Malaria is caused by tiny parasites which live inside red blood cells. These parasites belong to the Phylum Protozoa. There are about 100 species of malaria which infect various vertebrate animals such as certain snakes, lizards, birds, rats, bats, monkeys and other mammals. Only four species of malaria infect man. They are called *Plasmodium falciparum*, *Plasmodium vivax*, *Plasmodium malariae* and *Plasmodium ovale*. Each one produces a slightly different disease.

Malaria parasites can be found in the blood of people suffering from the disease. When the blood is stained with certain chemicals or dyes, and examined under a microscope, the parasites can be seen more easily. The parasites of each species look different and an experienced malaria technician can easily identify the species that are present. Fig. 3.1 shows typical parasites of the various species.

All species of *Plasmodium* that infect man are transmitted from one person to others by mosquitoes. The parasites are sucked into the mosquito when it feeds on an infected person. They then reproduce within the mosquito and other people are infected when the mosquito bites them. Not all kinds of mosquitoes transmit malaria. The human malarias are only transmitted by certain species of mosquitoes belonging to the genus *Anopheles*. Only female mosquitoes suck blood. They are the vectors of malaria. They transmit it or spread it from one person to others.

Some of the methods used to control malaria aim to kill the vectors before they can infect other people. The best method for doing this at present, is to kill them with DDT.

We will now consider the life cycle of malaria in some detail. (See Figs. 3.2 and 3.3.) There are several stages in the life cycle, each of which is different in size and shape, and stains differently with special stains which make them easier to see under the microscope. Each stage grows and changes to form the next stage. Often one stage multiplies or reproduces so that one parasite produces many parasites of the next stage. In the life cycle of malaria and related organisms, there are two main kinds of reproduction, which are called schizogony (shy-zog-ony) and sporogony (spore-og-ony). Both involve the splitting of one cell to form many new parasites of the next stage. Schizogony occurs in

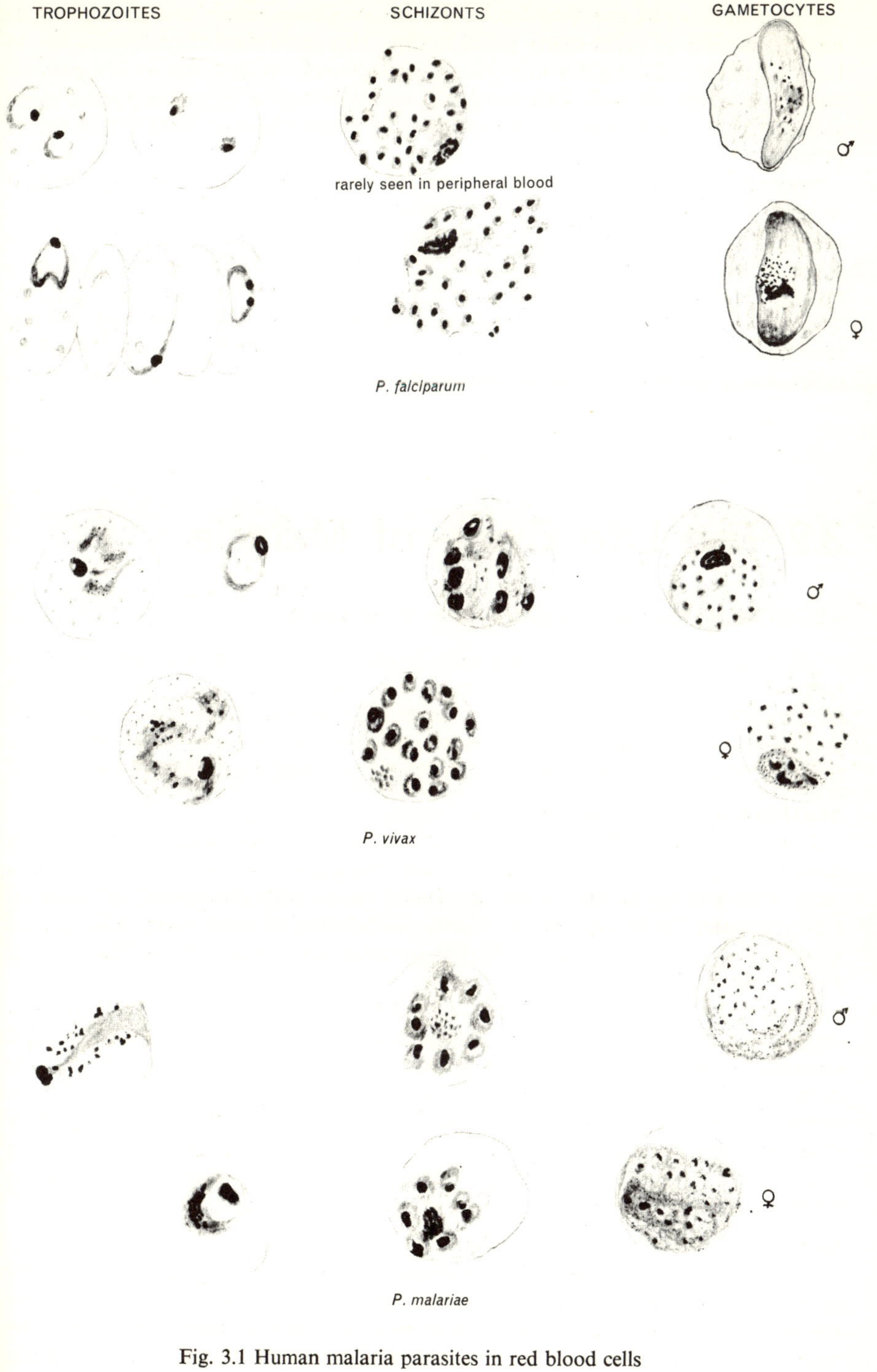

Fig. 3.1 Human malaria parasites in red blood cells

the liver and in the red blood cells. In this process a single cell grows and its nucleus divides to form many small nuclei. This individual then breaks up to form a number of daughter individuals, each one of which contains one nucleus and has its own cell wall. The word schizogony is derived from the Greek word *schizo*, to split, and refers to the splitting of one nucleus to form a number of others.

Sporogony occurs in the wall of the gut of the mosquito. It is very similar to schizogony, but it occurs immediately after fertilization or the union of a male gamete (or sperm) and a female gamete (or egg), to form a zygote. It is the stage which gives rise to sporozoites. In some parasites which are related to malaria, sporogony gives rise to spores, or stages which are covered by a protective spore wall. They can survive in an environment away from the host. The sporozoites of malaria do not have a protective wall, but they are like spores in other ways, and are formed by sporogony.

3.2 The Life Cycle of Malaria

SPOROZOITES (Spore-o-zo-ites—ites as in kites)

Man becomes infected with malaria when he is bitten by an infected mosquito. Infected mosquitoes inject many tiny stages, called sporozoites, into the blood or lymph of man. These sporozoites are carried around the body, in the blood, and when they get to the liver they move into liver cells. All sporozoites disappear from the blood within half an hour after they have been injected by the mosquito.

SCHIZOGONY IN THE LIVER

Once in the liver, each sporozoite grows into a much larger stage called a schizont (shy-zont), which produces many thousands of merozoites (mero-zo-ites). The number of merozoites produced varies with the species of malaria. For example, there are about 10,000 merozoites in each liver schizont of *Plasmodium vivax* and about 40,000 in each liver schizont of *Plasmodium falciparum.*

The production of merozoites is a kind of asexual reproduction in which a single nucleus divides many times to form many new nuclei. This process is called schizogony. When the liver schizonts are mature, they burst, and their merozoites spill into the blood. Sometimes a few of these merozoites re-enter liver cells where they may stay dormant (or inactive) for many months. Later they grow into schizonts, which may burst and start a new infection. This does not occur with falciparum malaria infections. It usually takes about 6 to 12 days from the time of infection to the first appearance of malaria parasites in the blood.

SCHIZOGONY IN THE BLOOD

The merozoites produced in the liver then enter red blood cells. Usually each merozoite enters a different red blood cell but sometimes two may enter the same cell. The parasites are tiny at first but grow so that they occupy the whole cell. Soon after they enter the red blood cell, the parasites have a shape

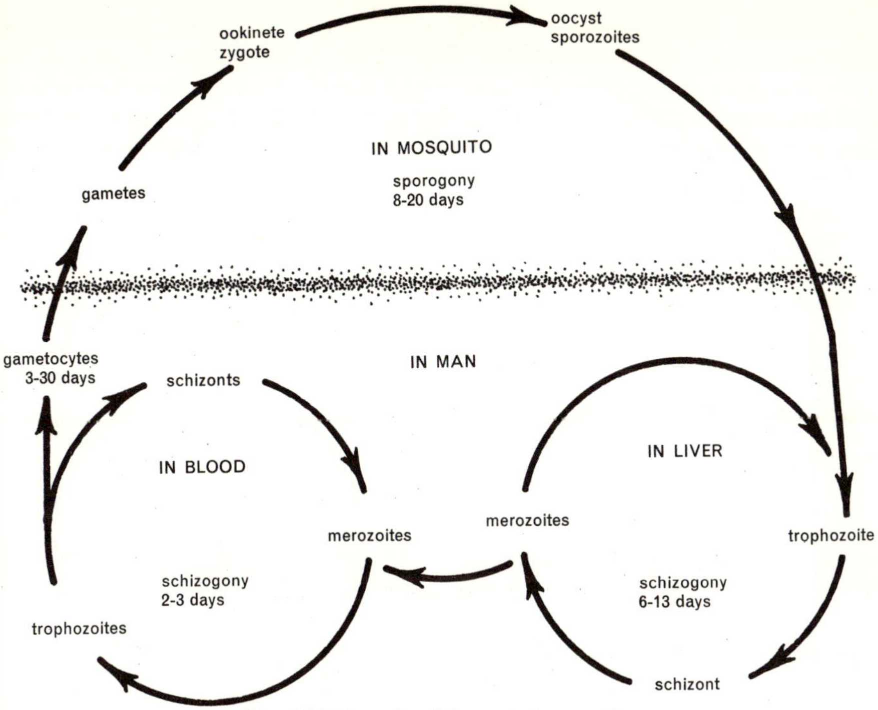

Fig. 3.2 Life cycle of the malaria parasite

like a finger ring with a diamond in it. The nucleus is like the diamond. It is a single dot which stains red and the rest of the parasite stains blue. The ring stage gradually grows into an amoeboid form which moves about inside the red blood cell, like an amoeba. It eats the haemoglobin of the red blood cell and changes it chemically to form its own body. Some of the haemoglobin is broken down into a black or brown pigment which is called haemozoin. Small granules of haemozoin can usually be seen inside malaria cells. Eventually the parasite undergoes schizogony. Its body divides to produce between 4 and 36 merozoites. The number depends on the species and strain, for example, *Plasmodium falciparum* generally produces about 16 merozoites while *Plasmodium malariae* generally produces between four and ten. There appear to be many strains or varieties of each species in Niugini, and they vary considerably in the number of merozoites in a schizont. The Niugini strains are different from those in other parts of the world.

The merozoites are freed into the blood when the schizont bursts and they then enter more red blood cells.

Schizogony in the blood usually occurs regularly at fixed intervals. All the schizonts in the blood mature and burst at about the same time, and this occurs about every 48 hours in *P. vivax*, 72 hours in *P. malariae*, and 36 to 48 hours in *P. falciparum*.

When the schizonts burst, particles of pigment and pieces of cells are released into the blood, and these are thought to cause fever, which always occurs just after schizogony in the blood.

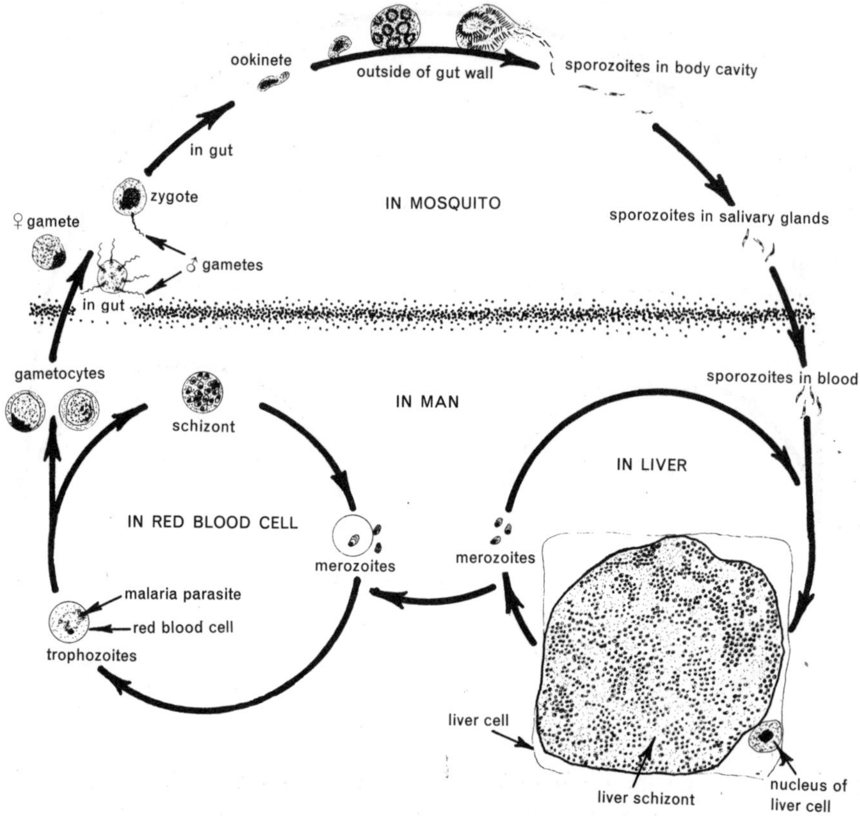

Fig. 3.3 Stages in life cycle of the malaria parasite

Schizogony in the blood always occurs about midday in human and monkey malarias. Recent experiments suggest that slight changes in the temperature of the host cause schizogony to occur in all the parasites at the same time. The temperature of the human body varies by three or four degrees throughout the day. It is highest about midday, and lowest during the night when a person is asleep. It seems that the schizogony cycle of the parasite follows the temperature cycle of the host. That is, the two cycles are synchronised (one is in step with the other).

THE FORMATION OF GAMETOCYTES (gam-eat-o-sites)

From 3 to 30 days after the parasites first appear in the blood, gametocytes start to appear. These are sex cells which grow from merozoites. No one knows what suddenly makes some merozoites change so that they produce gametocytes instead of schizonts, but it always happens. Gametocytes have a different appearance from other blood forms. In *P. falciparum* they are banana shaped, but they are usually round or oval in the other species. Once they are fully grown, the gametocytes do not develop any more unless they are eaten by a mosquito.

There are two kinds of gametocyte; male and female. Their nuclei look different and they stain different colours.

Experiments done recently with bird and monkey malarias have shown that gametocytes vary throughout the day, in their ability to exflagellate and to infect mosquitoes. They are most infective and most ready to exflagellate around midnight, and least infective around midday. This system has probably evolved because the mosquitoes usually feed on the host at night.

DEVELOPMENT IN THE MOSQUITO

When an *Anopheles* mosquito sucks blood containing gametocytes, it becomes infected. Within a few minutes of entering the gut of the mosquito, the male gametocytes undergo a process called exflagellation. Each male gametocyte produces 4 to 6 flagellate sperms or gametes, which wriggle out of the gametocyte after a great deal of struggling. They then swim to the female gametocytes, or eggs, and fuse with them to form zygotes. This process is called fertilization. One sperm fuses with one egg. The zygotes change their shape to become small worm-like organisms called ookinetes. They wriggle to the wall of the stomach of the mosquito using tiny muscles which lie just underneath their skin. They then burrow into the stomach wall where they grow into oocysts (o-o-sists). These are round bodies which poke out from the stomach of the mosquito. Sporozoites develop inside each oocyst by the process of sporogony. Mature oocysts may contain up to 10,000 sporozoites and there may be up to 50 or more oocysts in one mosquito.

The oocysts burst when they are mature and the sporozoites are liberated into the blood cavity of the mosquito. They then wriggle their way into the salivary glands of the mosquito. The salivary glands produce a substance which prevents blood from clotting. Some of this substance is injected into man, when the mosquito bites, so that the blood will remain liquid while it is being sucked into the mosquito. Up to 50,000 sporozoites may be injected into man each time he is bitten by an infected mosquito. Development in the mosquito takes about 8 to 20 days.

SUMMARY

Man becomes infected when a mosquito injects sporozoites into his blood or lymph.

The sporozoites infect liver cells and grow to become schizonts. Liver schizonts burst and liberate thousands of merozoites into the blood. These merozoites infect red blood cells and grow to become schizonts. This schizogony cycle in the blood occurs every 2 or 3 days. Thus there are two different asexual cycles or processes of schizogony; one in the liver and the other in red blood cells.

After a time, some merozoites develop into gametocytes instead of schizonts. Male gametocytes undergo exflagellation in the gut of the mosquito to produce sperms. The sperms fertilize the eggs to produce ookinetes. Thus, sexual reproduction occurs in the mosquito. It is followed by asexual reproduction when the oocyst produces sporozoites.

3.3 History of the Discovery of the Life Cycle of Malaria

Malaria has been a disease of man for a very long time.

Many monkeys have malaria parasites which are similar to those found in man, and it seems likely that as man evolved from a monkey-like ancestor, the parasites evolved with him.

Malaria was probably not a disease of major importance in nomadic hunters, but it probably became important when man started to live in settlements with an agricultural way of life.

Writings of the ancient Egyptians, Chinese and Hindus contain descriptions of fevers which were almost certainly caused by malaria. Some of the Egyptian writings date back to 1600 B.C.; about 3600 years ago.

Hippocrates, a Greek doctor who lived about 400 B.C., described the different symptoms of the three types of malaria now known as falciparum, vivax, and malariae malaria. He first thought they were caused by bile and later by gases from swamps. Most people still believed in this swamp theory until about 1900. One writer explained it as follows:

'With the commencement of the rains, the vegetation sprouts (grows) overnight in the tropics and as quickly decays in the stagnant (still) pools, the heat and moisture combine to excite the deadly principles of malaria, which are wafted in the early morning, like smoke or steam over the streets of the nearby town to infect the inhabitants.'

The name 'malaria' was derived from two Italian words: '*mala*', meaning bad, and '*aria*', meaning air. The French people call the disease 'Paludism' and this name comes from the Latin word '*palus*', meaning marsh or swamp.

From primitive times some African natives had always believed that they caught malaria by being bitten by mosquitoes. By 1850, a few doctors in Europe and America had suggested the same idea, but it was a long while before this idea was proved to be right.

In 1847, a German doctor called Meckel saw black pigment in the blood and spleen of a patient with malaria, when he examined these organs under a microscope. This pigment had been seen and described in 1716, but Meckel made a big advance by associating the pigment with the disease, malaria.

The parasites were first seen in the blood by Alphonse Laveran, who lived from 1854 to 1922. Laveran was a Frenchman, and was working in North Africa as an army doctor, when he made his discovery. He had often seen malaria pigment in the organs of people who had died of malaria. He also watched small bodies inside red blood cells and saw them grow in size, and saw malaria pigment develop within them. In 1880 he watched male gametocytes exflagellating and became convinced that the bodies he was studying were living parasites, and that they caused malaria.

Laveran went to Paris and told of his discoveries at a meeting of the Academy of Medicine, but the doctors there did not believe him.

Only a year before, a group of Italian doctors, which included a Professor of

Medicine, had convinced the world that they had discovered the cause of malaria. They had grown bacteria which they called *Bacillus malariae* from air and mud in a swamp, and also from a patient with malaria.

After hearing of Laveran's discovery, a doctor called Richards looked at blood from patients with malaria and saw the same parasites that Laveran had described. But others did not bother to look for themselves and they did not want to believe an army doctor who was not an important person.

In 1884, Laveran showed the parasites to Louis Pasteur and other doctors in Paris. Pasteur was an important scientist and once he believed Laveran's ideas, other people followed. A German called Gerhardt injected blood, containing malaria parasites, into a healthy person who then became sick with malaria. This really proved that Laveran was right. Malaria **was** caused by parasites inside the red blood cells. (Malaria is sometimes transmitted to people when they are given a blood transfusion which contains malaria parasites.)

The next big question to be answered was 'How does malaria move from one person to another?' In the period from 1850 to 1897, there were several discoveries which provided clues to the answer.

In 1878, Patrick Manson, then a missionary doctor in China, watched the development of *Wuchereria bancrofti* in the body of a mosquito, which had been fed on blood from an infected man. (The story of this parasite is given in Chapter 5 of this book). These observations established the idea that mosquitoes could transmit disease. In 1894 Manson published a series of reasons why he though that malaria was transmitted by mosquitoes.

Between 1893 and 1895, two diseases of domestic animals which are caused by blood parasites, were shown to be transmitted by insects. This further strengthened the idea that blood sucking insects might transmit other diseases.

Ronald Ross was a British army doctor who worked in India. While he was home in England on leave, he went to see Manson, who was then living in London. Manson asked him to try and prove that mosquitoes carried malaria. He also arranged for the Government to allow Ross to spend all his time working on this problem. Until then, no one employed by the government had ever been allowed to work full time on research. Ross returned to India in 1895 and started to work on his research project.

Over the next five years, he wrote 110 letters to Manson, telling of his progress and seeking further advice. Ross soon showed that exflagellation occurred in the gut of mosquitoes that had been fed on infected humans. Two years later, on the 20th August 1897, he found cysts on the stomach of an infected mosquito. These cysts contained pigment which looked like the pigment in human malaria. At this time, Ross was sent back to medical duties in an area where there was no human malaria. He decided to carry on his studies using bird malaria. By the end of 1898, he had succeeded in tracing the development of the parasites in the mosquito and had, in fact, transmitted the disease from an infected bird to uninfected birds, through mosquitoes.

Meanwhile, a group of Italians led by Grassi, had been working along the same lines as Ross. They had been unsuccessful for a time because they were using Culicine mosquitoes instead of Anophelines. By 1899 they had infected mosquitoes with human malaria and described all the stages in detail, and they had produced malaria in a human volunteer after he had been bitten by an infected mosquito.

For a time, it was thought that the story of the life cycle was complete. Many people thought that the sporozoites entered red blood cells soon after a person was bitten by an infected mosquito. One worker mixed sporozoites with blood and watched what happened under his microscope. He said that he saw the sporozoites enter red blood cells and even made drawings which showed this happening. He was a liar.

Many people have repeated this experiment, but they have never seen sporozoites enter red blood cells.

Experiments showed that when very large numbers of sporozoites, from many mosquitoes, were injected into the blood, they disappeared from the blood within half an hour. Other experiments showed that the parasites did not reappear in the blood for at least seven days. Furthermore, it was difficult to imagine how a sporozoite could change into a ring form.

In 1908, a scientist in Brazil found schizonts of a malaria-like parasite in the lungs of a bird. He found that the schizont produced merozoites which infected red blood cells.

In 1934, schizogony in tissues was described in detail for a species of bird malaria, but no similar stages could be found in tissues of people who had malaria.

Finally, in 1948, the British workers, Shortt, Garnham, Covell and Shute infected a volunteer by injecting into his arm the sporozoites from the salivary glands of 2010 infected mosquitoes. Seven days later, doctors operated on the volunteer and cut off a piece of his liver. It was found to contain liver schizonts.

This proved, finally, that sporozoites do not enter red blood cells, but first reproduce in cells in the liver.

Some interesting lessons can be learned from the story of malaria. We have seen, for example, that people make mistakes and think they see things that do not happen, and this can hold up progress for a long time. We saw that for a long time, malaria was associated with swamps because the vectors lived there, but it was a very long time before it was shown that mosquitoes carry malaria. People had to think about disease in a new way before science could progress. New observations may not be readily accepted when they are contrary to popular beliefs. We should always be asking ourselves questions about the things that we and other people believe. Are they based on good observations, or are they just ideas that seem to be true, but have never been properly tested. Ideas are very important, we cannot progress without them, but they should always be tested before we accept them as being true.

A disease is first recognised from a set of symptoms that always occur together; that can be seen and measured. For example, malaria was recognised as a disease with the following symptoms:

(i) fever that often occurs regularly every two or three days which starts with shivering and ends with sweating
(ii) the presence of black pigment in the blood and spleen
(iii) enlargement of the spleen
(iv) anaemia
(v) the presence of living organisms in the blood.

A great deal of work had to be done to prove that the organisms in the blood caused these symptoms. We have already told how people once thought that malaria was caused by a bacterium, that had been found in the air and mud from a swamp. The Italian doctors also thought they found this same organism in a patient who had malaria. This was a mistake. There must be much better proof than this, that a particular organism causes a particular disease.

Robert Koch was a German bacteriologist who lived from 1843 until 1910. He laid down a set of procedures or steps that must be followed in order to prove that a particular disease is caused by a particular organism. (He also did some important work on malaria in Niugini, as we shall see in Section 3.5). They are called Koch's postulates and are still followed today. A postulate is something that is asked for or demanded. Koch's postulates are questions that must be answered in order to show that a particular disease is caused by a particular organism. Koch's postulates are:

(i) the suspected organism must be present in all cases of the disease
(ii) the suspected organism must be isolated and its characteristics studied
(iii) it must produce the disease when a susceptible host is infected in an experiment
(iv) and it must be isolated from the experimental host and shown to have the same characteristics as the organism that was first isolated.

3.4 The Pathology of Malaria

Pathology is the study of the effects of a disease on the organism infected with it. In this case pathology is the study of the effects of malaria parasites on the human body.

Malaria affects man in several ways. The most important of them are listed below.

(i) Regular fever

Every two or three days the patient becomes hot, due to a rise in temperature of the body. Fever may occur regularly for several weeks if no treatment is given.

(ii) Anaemia (an-eem-ia)

The parasites destroy red blood cells. Red blood cells are important because they carry oxygen to all parts of the body. If too many red blood cells are destroyed, the body cannot work properly, because there is not enough haemoglobin to carry oxygen. The patient is then said to be anaemic or to have anaemia.

(iii) Blockage of the blood capillaries

The heart pumps blood to all parts of the body through large pipes called arteries. These continually branch to form smaller and smaller pipes which supply blood to all organs such as the brain, liver, stomach and intestines and tissues such as muscles and skin. The smallest of these pipes are called capillaries and are often not very much bigger in diameter than the blood cells which pass through them. The capillaries join up to form larger and larger pipes, called veins, through which the blood returns to the heart. In certain kinds of malaria, the infected red blood cells stick together and stick to the walls of the blood capillaries, and block the flow of blood. This has severe effects on some organs and tissues.

The schizonts in the liver do not cause any obvious disease. All the symptoms of malaria are due to the stages in the blood.

We will now discuss the pathology of malaria in more detail.

FEVER

Fever occurs about two weeks after infection. The temperature of the body rises from its usual value of 98°F (or 37°C) to 100 to 105°F (or 38 to 41°C). The time between infection (the bite of an infected mosquito) and the start of fever, is called the incubation period. It varies with each species of malaria as follows:

P.vivax 6 to 30 days; average 14 days
P.malariae 18 to 60 days; average 30 days
P.falciparum 7 to 27 days; average 11 days
P.ovale 8 to 16 days; average 14 days

Sometimes the incubation period may be several months.

A few days before the first real attack of malaria, there may be mild symptoms such as headache, backache, loss of appetite, slight pains in the abdomen (belly) and mild diarrhoea (the runs). The infected person feels unwell but not very sick.

A typical attack of malaria comes on suddenly. It can be divided into four stages. The patient first feels cold, then hot. This is followed by a period of sweating, after which the body temperature returns to normal.

The cold stage may last from a few minutes to a couple of hours. The patient shivers and his teeth may chatter. These symptoms continue even when the patient is covered with lots of blankets and clothing.

When the shivering stops, the patient becomes hot and takes off all the blankets and warm clothing. The body temperature may rise to about 106°F or 41°C. This fever may continue for up to eight hours, and the patient is usually very sick. During the fever stage there may be severe headache, nausea, vomiting and a great thirst.

The fever stage starts soon after the bursting of schizonts in the blood. As they burst, merozoites are released into the blood, together with pigment and other waste products from the parasites, and small pieces of red blood cells. Injections of foreign protein into the blood of small mammals, causes the same kind of fever, and it is thought that malaria fever is caused by the release of proteins into the blood when schizonts burst. (See Fig. 3.4.)

The fever stage usually stops suddenly and the patient starts to sweat so much that the bed clothes become wet.

The patient is glad that the fever is finished. Vomiting and headache stop, and the body temperature returns to normal. The sweating may last for several hours. The patient may sleep or try to work.

After the sweating is over the patient has few symptoms, but may be very weak and irritable.

If a person is not treated with antimalarial drugs, he may have attacks of malaria periodically for several weeks. The interval between attacks varies with the species of malaria. Attacks of vivax malaria usually occur every 48 hours

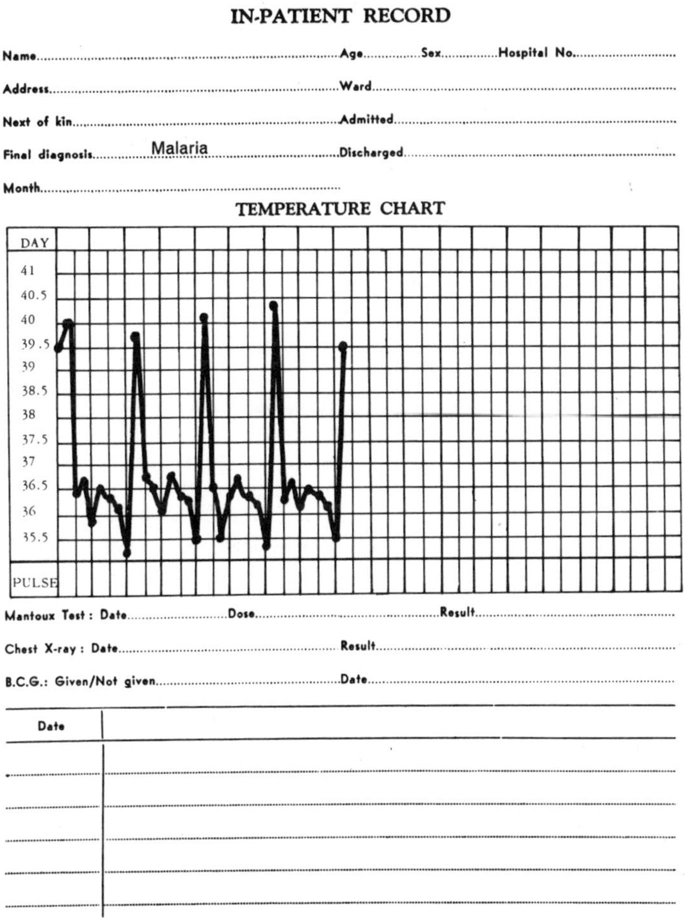

Fig. 3.4 Temperature chart of a patient with malaria

or two days, although they may be irregular during the first few days. In falciparum malaria, the attacks are sometimes irregular, but they usually occur every 36 to 48 hours. Sometimes they occur every 24 hours. Attacks of malariae malaria usually occur every 72 hours (three days).

Sometimes a person will be bitten by infected mosquitoes several times within a few days. In such cases, parasites of two or more species may be found in the blood at the same time, or there may be two separate lots of schizonts which mature several hours apart. Regular fever may not occur in such cases.

So far we have been discussing a typical attack of malaria. People who have been continually infected with malaria for many years, may not show the typical symptoms that have just been described. They may have a few parasites in their blood most of the time, but show few signs or symptoms of disease. As a result of continual infection, their bodies develop the ability to resist the disease, so that they do not become very sick. They are said to be partly immune to malaria. However, while people have malaria parasites in their blood, they are never completely healthy and may still be very sick from time to time, especially if they become infected with some other disease as well.

ANAEMIA

Each liver schizont of *P.vivax* releases about 10,000 merozoites into the blood, and those of *P.falciparum* release about 40,000. Because thousands of sporozoites may be injected into man at the one time, by a single mosquito, there might be hundreds of schizonts developing in the liver at once. Some of the sporozoites injected by a mosquito probably do not develop, because they are destroyed by white blood cells before they become established in the liver.

Once schizogony begins in the blood, each merozoite will produce between 4 and 36 new merozoites everytime a schizogony cycle occurs. As each one matures it destroys another red blood cell.

Anaemia will develop if more red blood cells are being destroyed than are being produced in the bone marrow for release into the blood.

To illustrate the way that malaria parasites can increase, let us suppose that—

(i) 100,000 merozoites are released into the blood as schizonts in the liver burst, and

(ii) each merozoite produces ten more merozoites every cycle of 48 or 72 hours.

The number of parasites in the red blood cells would increase as is shown below:

Number of parasites at—

first cycle:	100,000 x 10	=1 million
second cycle:	1,000,000 x 10	= 10 million
third cycle:		= 100 million
fourth cycle:		= 1,000 million
eighth cycle:		= 10,000,000 million
ninth cycle:		= 100 million, million

Now the body of a normal adult man contains about 35 million, million (35 x 10^{12}) red blood cells, so that if the parasites continued to reproduce at this rate, all the red blood cells would be destroyed by the ninth cycle. This does not happen. In the most severe falciparum malaria, only about 20 per cent of the red blood cells are infected. And in heavy infections of the other species, only 1 to 2 per cent of the red blood cells are infected. There are two reasons for this; firstly, soon after infection starts, defence cells start to remove parasites and infected red blood cells, so that the infection does not keep on increasing at this great rate. Secondly in vivax malaria the merozoites prefer to infect immature red blood cells (reticulocytes) and there are not very many of these in the blood.

However, many red blood cells are destroyed at each cycle and a severe anaemia develops, especially in children. The body responds to this by producing new red blood cells at a faster rate. New blood cells are continually being made in the bone marrow, to replace old worn out blood cells which are destroyed in the spleen. This process goes on all the time in uninfected people. In persons suffering from malaria, the process is speeded up, so that usually the anaemia is not severe and does not last long. However, most Niuginians have about 20 per cent less haemoglobin in their blood than people in Australia, where there is no malaria. Malaria is not the only cause of anaemia in Niugini, but it is an important cause.

BLOCKAGE OF THE BLOOD CAPILLARIES

In falciparum malaria, schizogony usually occurs in the capillaries of the internal organs. Schizonts and free merozoites are seldom found in the peripheral blood (blood of the skin). It seems likely that merozoites actually enter red blood cells in the capillaries of the internal organs, perhaps because the blood flows more slowly there. In this situation, the infected red blood cells often stick together and stick to the walls of the capillaries of internal organs. This slows down the flow of blood to these organs and may block it completely. Blockage of the blood capillaries sometimes occurs with infections of the other species of malaria, but this is not common. The symptoms that occur as a result of these blockages depend on where the blockage occurs. If a blood capillary is blocked, it may burst and blood will leak out into the surrounding tissues. Sometimes tissues and organs are starved of oxygen because oxygenated blood does not reach them, and then many cells may die. The tissues will also be affected because they are not being supplied with food substances and because toxic wastes are not being removed.

Blockages in the brain occur sometimes. They may lead to loss of consciousness (coma) and sometimes the patient may die. This form of the disease is called 'cerebral malaria'. Blockages in the tiny vessels supplying the heart muscles may cause heart failure, and blockages in the lungs may produce symptoms resembling pneumonia. If the gut is affected, the symptoms may resemble cholera, dysentery or appendicitis. The kidneys and other organs may also be affected.

ENLARGEMENT OF THE SPLEEN

The spleen is a dark red organ which lies near the stomach, on the left side of the body. Inside, it is like a sponge with all the spaces filled with blood.

The spleen has several functions; that is, it does several kinds of work. Two of the most important of these are listed below:

(i) It destroys old, worn-out red blood cells. Red blood cells are made in the bone marrow, and after about 100 days are eaten by cells called phagocytes (or macrophages) in the spleen. These phagocytes eat only the oldest red blood cells. The red blood cells are broken down into amino acids and iron and some other substances. One of these substances is used to make bile, but the iron and amino acids are released into the blood, where they are carried around the body. Cells in the bone marrow remove iron and amino acids from the blood and use them to build new red blood cells. In a healthy man about ten million old red blood cells are destroyed every second; and of course, the same number of new cells are made in the bone marrow every second.

(ii) The second main function of the spleen is to remove unwanted bodies such as bacteria and other parasites from the blood. Red blood cells infected with malaria or other parasites are also removed in the spleen. They are all eaten by phagocytes in the same way as old red blood cells.

People suffering from malaria have enlarged spleens. They are bigger than usual because the phagocytes in the spleen increase in number and are continually eating parasites and thus removing them from the blood. Children who get malaria a lot, always have very big hard spleens which can be felt when they are lying down, on the left side of the abdomen, below the bottom rib.

The spleen remains enlarged for some time (several months) after an attack of malaria if drugs have not been given.

In small children that are continually infected with malaria, the spleen gradually increases in size until they are about nine years old. It is always enlarged and hard. Immunity slowly develops and the spleen may become smaller after they reach the age of about nine. The spleen never shrinks back to its normal size if they are continually infected. If malaria is controlled in an area, the spleens become softer and gradually shrink back to normal, over a period of about three years.

Because enlargement of the spleen is so characteristic of malaria, surveys which determine the incidence of enlarged spleens provide good measurements of the amount of malaria in a group of people. We will discuss spleen surveys in more detail in Section 3.8.

BLACK WATER FEVER

This is a rare condition nowadays. In the past it was associated with falciparum malaria.

In this disease large number of red blood cells burst so that the red pigment, haemoglobin, is set free into the blood. This pigment is removed from the blood by the kidneys and excreted in the urine. The haemoglobin gradually turns black, and causes the urine to become dark. Hence the name 'black-water fever'. Malariologists think that black-water fever was associated with the use of quinine, a drug that was one commonly used for the treatment and prevention of malaria attacks.

THE DIAGNOSIS OF MALARIA

Diagnosis means identifying a disease. Each disease can be recognised by certain features which together make it different from all other diseases. Fever, especially fever which occurs regularly every 2 or 3 days, is one of the most obvious signs of malaria. However, fever is also a characteristic of many other diseases. In fact one of the most common symptoms of infection is an increase in body temperature. To be certain that a person has malaria, it is necessary to find the parasites in the blood. This is done by making a smear of blood on a microscope slide, and staining it with special stains so that the parasites can be easily seen when examined under a microscope at a magnification of about 800 times.

Several different stains can be used, but the essential chemicals are the same as those used by Romanowsky in 1891 to stain malaria parasifes. They are known, collectively, as Romanowsky stains.

Romanowsky was a Russian army doctor who happened, by accident, to find a very good stain for malaria parasites. No better stain has ever been discovered.

He was not a very careful scientist and often did not replace the stoppers in his bottles of staining solutions. One day he stained some blood from a patient with malaria, with an old solution of methylene blue in which mould was growing. He then stained it with eosin and found that the nuclei of the malaria parasites stained deep red and the cytoplasm bright blue. This method showed up details of the parasites that had not been seen before. Only when methylene blue is oxidised to a mixture of several substances does the stain work like this. The oxidation (or chemical reaction with oxygen) occurred because the solution had been left for a long time and mould had grown in it.

Today a number of blood stains are used, such as Giemsa, Wrights, Fields, Leishmans and Yagaum. All have been named after the people who first published their formula, except the Yagaum stain, which is named after the Yagaum hospital, near Madang, where it was first made. All these stains are modifications of Romanowsky's formula.

RELAPSES

Once a person is infected with malaria, he may become sick at irregular intervals, over several years, without renewed infection from mosquitoes. These recurrent attacks are called relapses and may be divided into two kinds. Very little is known about the mechanisms involved in relapses. There are several theories which explain what happens, but they have not yet been tested experimentally.

In the first kind, or true relapse, no parasites are left in the blood after the last attack of malaria. The new attack, occurs soon after a new batch of merozoites are released from a liver schizont which has remained inactive, or dormant, for months or even years. Many malariologists think that some merozoites from the last attack of malaria, enter liver cells and grow into a second lot of schizonts. They must come from liver schizonts, because infections caused by the innoculation of blood from an infected person do not relapse. An alternative theory for true relapses is that many schizonts usually start to grow in the liver, as a result of one bite from an infected mosquito, but only one bursts at a time, to start a blood infection. Perhaps when a liver schizont bursts it releases a chemical which delays the maturation of the other schizonts in the liver, so that they stay dormant. This chemical would gradually be removed from the liver, thus allowing another schizont to burst at a later time.

True relapses do not occur in falciparum malaria. But they occur in all other human species. In one study, 90 per cent of the soldiers who had malaria while in the Pacific area during World War II, had at least four relapses after they went home. Some had over thirty

In the second kind of relapse (known as recrudescence), a few parasites continue to live in the blood long after the symptoms of malaria sickness have gone. These suddenly start to reproduce in the blood and cause a second series of attacks. This kind of relapse may occur with infections of all species.

It is usually difficult to tell if a new attack of malaria is due to a new infection or a relapse. Relapses are said to appear more suddenly than new infections and the attacks are usually regular (every 48 or 72 hours) right from the start.

Very little is known about the causes of relapses. They sometimes occur in people who become very tired or hungry, or in people who have been exposed to unusually hot or cold temperatures, or a sudden change in climate. Sometimes they occur after childbirth or surgery, or sickness due to other disease, or after a person has drunk too much alcohol. Sometimes relapses are not associated with any of these things. They just happen for no apparent reason.

Public servants who come from villages in malarious areas of Niugini to work in the Highlands, where malaria is rare, often develop malaria fevers soon after they arrive. They also frequently have attacks of malaria soon after returning from leave, but they seldom have attacks of malaria while they are on leave. Many of these cases are possibly relapses brought on by a change in climate or altitude.

Malaria is common in Niugini and people should take drugs which prevent attacks of malaria. (These are discussed in Section 3.8.) It is a good idea to keep taking these drugs for about one month after leaving a malarious place. Sometimes an attack occurs soon after a person stops taking anti-malaria drugs. This may be a relapse or it may be the first attack a person has ever had. In other words, while a person has been taking drugs, no symptoms have developed. If a person leaves a malarious country and becomes sick with fever, the doctor should be told that it might be malaria. It is very easy for the doctor to tell if a person has malaria by looking for parasites in the blood. But doctors in countries where there is no malaria often don't think to look for it, or fail to ask whether their patients have been in a malarious area.

Relapses of vivax malaria may occur up to four years after infection, although it is rare for a person to have attacks after more than two years. Relapses of malariae malaria have been recorded 30 years after infection occurred. Recrudescences of falciparum malaria usually do not occur more than six months after infection occurred.

A NOTE ON MALARIA TERMINOLOGY

The incidence of an infection is the proportion (percentage) of a particular population or community that is infected with it at a particular time. We use this term throughout this book, and it is commonly used by parasitologists, in the way we have defined it.

However, malariologists and the World Health Organisation do not use the term 'incidence' in this sense. They use the terms 'prevalence' or 'parasite rate' to mean the proportion (per cent) of people that are infected with malaria at a particular time. And they use the term 'incidence' to refer to the number of cases of malaria that occur, per 1000 people, throughout a period of one year.

3.5 Immunity to Malaria

Professor Robert Koch, the famous bacteriologist, was probably the first person to study malaria in Niugini. In 1900, he published an article in the German Medical Weekly, on his work here. He suggested that if children are continually exposed to malaria, they develop an immunity to it. As far as we can find out, Koch was the first scientist to recognise the importance of immunity in malaria infections.

He studied the incidence of malaria in isolated villages. The incidence of a disease is the proportion (percentage) of a particular population or community that is infected with it at a certain time.

In the village of Bogadjim (Madang district) he found that 80 per cent of the children under two years were infected with malaria and 42 per cent of the children between two and five were infected. But, he could not find parasites in the blood of anyone over the age of five years.

In the nearby village of Bongu he found a similar situation. His data for Bongu is given below.

Children under 2 years; 6 out of 6 were infected; incidence 100%
Children 2-5 years; 6 out of 13 infected; incidence 46%
Children 5-10 years; 4 out of 17 infected; incidence 24%
People over 10 years; 0 out of 39 infected; incidence 0%

Koch collected a number of other similar examples.

He recognised that when malaria occurred in isolated villages it was restricted to children. In each of the above villages there was one adult that had malaria, and in both cases, that person had recently come from another area. Koch also found that children between three and six years old had very enlarged spleens. Older and younger children did not have very enlarged spleens. He suggested that the spleen became enlarged and became small again as each person grew older. He further suggested that the incidence of malaria decreased with age because the people were becoming immune to it. They were developing or acquiring, immunity to the disease.

He observed that people who had lived in a malarious area could go to another malarious area without becoming sick, provided the same species of malaria occurred in both places. He also noticed that previous infection with one species did not give any protection against other species. People taken from malaria free areas to malarious areas always became sick with malaria.

Koch said that, at that time, some areas of Niugini only had one or two species of malaria. Today the three common species are widespread, and occur in all areas of the country. This has probably occurred because people move about from place to place much more than they used to.

Since Koch's work was published a great amount of evidence has accumulated to support his ideas about immunity. Nowadays if there is a decrease in the incidence of any disease with increasing age most people think that it may be a sign of developing immunity; yet few people know that Koch put forward this idea in 1900.

Today, even in the most malarious areas of Niugini, parasites occur in some old people. We will never know why modern data is different from Koch's, but it is. Table 3.1 gives the incidence of malaria in different age groups in the Maprik area, before the malaria control program started in 1957. At that time Maprik was one of the most malarious areas in the world. The table shows that most children (97 per cent) had parasites in their blood before they were one year old. In fact, it is highly likely that everyone was infected by the time they were one year old.

Table 3.1 Incidence of malaria, in different age groups, at Maprik, in 1957, before the malaria control project started

Age	*Incidence %*	*Age*	*Incidence %*
0-2 months	46	3 years	95
3-4 months	83	4 years	96
5-6 months	94	5-9 years	82
7-8 months	93	10-14 years	62
9-11 months	97	15-19 years	48
1 year	95	> 20 years	34
2 years	93		

After the children reached the age of about five, the incidence of infection started to fall and it continued to fall with increasing age. Only 34 per cent of the people over twenty years old had parasites in their blood.

The most likely explanation for this is that immunity slowly develops with increasing exposure to infection. This is known as acquired immunity.

If, after a host has been infected with a parasite, and it is later reinfected with the same parasite, and it does not develop the disease, it is said to have acquired immunity to the infection. As a result of previous infection, it has acquired or developed mechanisms which prevent that infection from becoming established in its body again.

Many bacterial and viral infections of man produce an acquired immunity after one infection, which lasts for ever, or for a very long time. Single infections of malaria and most other parasitic diseases, produce only a weak, short lasting immunity. We do not know why the first group of diseases produce a permanent immunity, while most parasitic diseases do not.

In malaria infections, the host develops immune mechanisms as the infection proceeds, which reduce the effects of the parasite on the host. Almost certainly, several different mechanisms are involved, but we do not know very much about them yet.

As the macrophages, or phagocytes, of the lymphoid macrophage system eat any parasites they come across, they will start to make antibodies. Each different protein made by the parasite will cause the production of up to five different antibodies, which are made by the host. Thus the host produces numerous antibodies which can combine with the antigens of the parasite. Only some of these can act against the parasite. An antibody produced against an antigen which is always inside the parasite, will have no effect against the living parasite, because it cannot combine with it. Only antibodies which combine with antigens on the surface of the parasite can have any effect against the living parasite. Probably the most important antibodies are those which label the parasites, so that they are quickly recognised and eaten by the macrophages. As the production of these gets going, the macrophages remove the parasites at an increased rate. Eventually they remove parasites faster than the parasites are reproducing, and the number of parasites in the blood decreases.

The host also develops a tolerance to the parasites. At the time of the first attack of chills and fever, there are very few parasites in the blood. For example, there may be only 10 parasites per cubic millimeter of blood, in a *P.vivax* infection at the time of the first attack. The patient will gradually recover without taking drugs, after having attacks every two or three days, for about two or three weeks. The attacks gradually get less and less severe. But large numbers of parasites may still be present in the blood after all attacks have ceased. For example, in *P.vivax* infections as many as 500 parasites may be present in each cubic millimeter of blood. The host has developed a tolerance to the parasites. Schizogony is no longer followed by chills and fever. The parasites are still alive and will cause disease if injected into a non-immune person.

What probably happens is this; the chills and fever that follow schizogony in the blood, are thought to be due to foreign proteins which are released into the blood, as the parasites split apart. A week or so after the fevers start, large amounts of antibodies are probably being produced, which quickly combine with these foreign proteins and they would be removed by the macrophages before they could cause fever. These antibodies have no effect on the living parasites, because they can only combine with the proteins inside the parasite which are released at schizogony when the parasite bursts.

In areas where people are continually reinfected with malaria, several other things happen which suggest the gradual development of immunity. As age increases, the proportion (per cent) of people with gametocytes in their blood decreases, so does the densities (the number per unit volume of blood), of gametocytes and all blood parasites. This is shown in Table 3.2 which gives data from a survey done by Dr W. Peters in the Western District of Papua, and published in 1957.

Table 3.2 Changes in the mean density of malaria parasites (per cubic millimeter of blood) and the proportion (%) of people with gametocytes in their blood (the gametocyte rate) with age. Data from the Western District of Papua, from Peters, 1957. (The density figures include gametocytes)

	Age Groups					
	<2	2–4	5–9	10–14	15–19	>20
Mean Parasite densities	452	292	82	96	32	31
Gametocyte rates (%)	29	27	4	7	0	2

When there is a successful malaria control program, the incidence of malaria falls, and people soon lose their immunity to malaria. This is shown by the work of Metselaar, which was done in Dutch New Guinea (West Irian). He found that as the incidence of malaria fell with successful control, the proportion of people with gametocytes in their blood increased. This was especially noticeable in older children and adults who had the most immunity at the start. It was much more noticeable in falciparum infections than the other species. His data is given in Table 3.3. Before spraying started in 1953, the incidence of malaria was 46 per cent; it fell steadily to 6 per cent in 1959.

Table 3.3 Changes in the percentages of malaria positive slides which contained gametocytes, with the introduction of malaria control. Gaps have been left in the table where the number of cases was too small for accurate analysis. (After Metselaar)

	Percent of gametocyte positives, in infections of the following species											
	P.falciparum				*P.vivax*				*P.malariae*			
	1953	1955	1957	1959	1953	1955	1957	1959	1953	1955	1957	1959
Age Group in years 0–2	19	28	75	47	22	33	43	60	30	27	—	—
3–8	7	20	23	40	9	16	28	50	22	12	30	—
9–14	5	5	27	15	9	11	—	—	8	5	17	—
>15	2	12	12	20	3	—	8	—	10	3	10	—

All this evidence supports the idea of the gradual development of immunity to malaria. It was not until the 1950s that experiments were done, which led to an understanding of the mechanisms involved.

Before we go on to talk about these experiments, it is necessary to talk about antibodies again. The body makes five different kinds of antibodies. They vary in their size, structure and chemical formulae. One kind, for example, is a very large molecule with five arms like a starfish. Most antigens cause the production of all five kinds of antibody. One kind of antibody occurs mostly in tears, saliva and the mucus that lines the gut, but the other kinds are found in the blood. The different kinds of antibody are known collectively as immuno-globulins. Each kind has its own name and the most important kind, as far as malaria is concerned, is called gamma globulin. It is actually a mixture of many different antibodies. Gamma globulin can be separated from the blood by special techniques, and the amount of gamma globulin in a unit volume of blood can be accurately measured. Because the different antibodies in gamma globulin are alike, it is very difficult to separate them one by one, and to measure the concentration of each one on its own.

We can now tell the story of the research work which led to an understanding of the mechanisms of immunity to malaria, in man. Most of this work was done by two people called Cohen and Macgregor, who worked together.

In 1951, two groups of newborn babies in a highly malarious area of West Africa, were treated as follows: One lot was given an antimalarial drug, called chloroquin, each week, which prevented them from getting attacks of malaria. The other group was given pills that looked the same, but which had no effect. This kind of treatment is called a 'placebo' and it is used as a comparison. The results of the group that took antimalarials can then be compared with the other group.

After three years, the concentration of gamma globulin in the blood was measured in both groups. All the comparison group had been infected with malaria and the group that had been taking chloroquin had not had any attacks of malaria. The result showed that the comparison group had much higher concentrations of gamma globulin in their blood than the other group.

This suggested that gamma globulins are important in the development of immunity to malaria.

A second series of experiments showed that:

(i) The concentration of gamma globulin in the blood of new born African babies from malarious areas, is higher than the concentrations in the blood of European babies, who come from areas where there is no malaria.

(ii) After birth, the concentration of gamma globulin in Africans from malarious areas, decreases for 3 to 6 months and then increases again.

(iii) The gamma globulin levels in children 2 to 6 years old, who suffered from malaria, was always higher than the levels in children of the same age who took antimalarial drugs.

(iv) When adults from a malarious area took antimalarial drugs the level of gamma globulin decreased and not nearly so much was produced by the body per day. However, the concentration of another group of blood proteins, called albumins, stayed the same and their production was not affected by the taking of drugs.

Up to this stage the results show that:

(i) Higher concentrations of gamma globulin are found in people who have had malaria.

(ii) The concentrations drop when they take antimalarials.

(iii) The concentration of gamma globulin in new born babies, born to mothers who have had a lot of malaria, is higher than in other new born babies.

This also provided evidence that gamma globulin concentrations increase with malaria. Other workers had already shown that gamma globulin consists of antibodies and that antibodies are highly specific, that is, a particular antibody is produced in response to the presence of a particular foreign protein, or antigen.

The results also suggested that gamma globulin was passed from mothers who had had a lot of malaria, to their babies, before they were born.

As a result of all this information, Cohen and Macgregor made the hypothesis that gamma globulin contains antibodies to malaria. They then designed an experiment to test this.

Blood was taken from adults who had been infected with malaria many times and the gamma globulin was separated from it and stored in a refrigerator, so that it would keep until it was needed.

Children with falciparum malaria, with high densities of parasites in their blood (up to 4 per cent of all red blood cells were infected) were then given injections of the gamma globulin. They were given up to 2.5 grams of pure gamma globulin over a period of three days.

Blood slides were taken every twelve hours and the density of parasites was determined by counting. A careful record was also kept of the symptoms in the children.

It was found that:

(i) There was no immediate decrease in the density of the parasites.

(ii) After the first cycle of schizogony in the blood there was no increase in the density of parasites.

(iii) By the end of the second cycle of schizogony the density of parasites had fallen to 1 per cent of the initial value.

(iv) By the ninth day no trophozoites could be found in eight of the twelve patients.

(v) As the density of parasites decreased the symptoms went away.

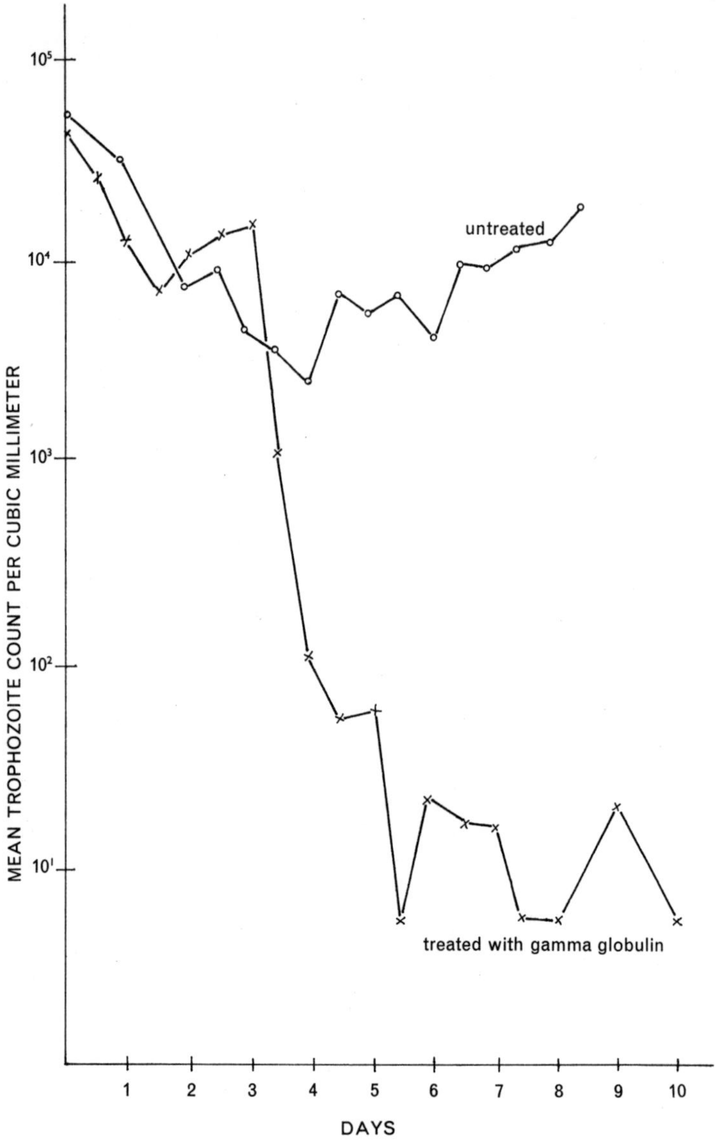

Fig. 3.5 Decrease in parasites after injection with gamma globulin containing malaria antibodies (after Cohen and McGregor)

Fig. 3.5 shows how the number of trophozoites decreased in the children who had been treated with gamma globulin. The number of trophozoites in untreated children did not change very much.

Three different lots of comparison were done in this experiment:

(i) Some children were not treated.

(ii) Some children were given injections of gamma globulin from people in England where there is no malaria.

(iii) Some children were given injections of blood serum, from which the gamma globulin had been removed.

In the comparison groups, the density of trophozoites was unaffected by the treatment.

This experiment showed that the gamma globulin reduced the number of parasites in the blood. It could have done this in two ways; by inhibiting or preventing schizogony, or by labelling the young merozoites so that they were quickly recognised and eaten by the macrophages.

The protection provided by the gamma globulin did not last long. One child was readmitted to hospital three months later with another severe infection of *P.falciparum*.

We can summarise the important points about acquired immunity to malaria as follows:

(i) It develops slowly in people who are continually exposed to infection.
(ii) It does not last very long.
(iii) Immunity to one species of malaria does not give much protection against other species of malaria (and possibly against other strains of the same species).
(iv) Several mechanisms are probably important in the development of immunity to malaria. The production of antibodies of the gamma globulin kind is probably the most important.
(v) Mothers pass on gamma globulin antibodies against malaria to their babies before they are born.

So far, we have been talking about acquired immunity to malaria. There is another kind of immunity which is called Natural Immunity.

When a host that has never been exposed to a particular infection cannot be infected, the host is said to have natural immunity to that infection. The parasite is also said to be host specific, that is, it can only develop in certain host species. Host specificity and natural immunity are two different names for the same phenomenon, or thing. Although we have two different names for this, we do not know very much about the mechanisms which prevent a parasite from becoming established in a particular host.

We know the lymphoid macrophage system may play a part in natural immunity, but other mechanisms may also be important.

The spleen plays a major part in defence against blood parasites and, when it is removed, a foreign parasite can sometimes become established. We will give an example; *Plasmodium berghei* is a malaria that lives in rats and mice. One of us recently tried to infect flying foxes by injecting blood into them from infected mice. The flying foxes did not become infected.

The spleen was then removed from several flying foxes which were then injected with parasites. After the spleen had been removed, *P.berghei* became established in the species *Dobosonia moluccensis*, but not in flying foxes of the genus *Pteropus*. In one case, the lymphoid macrophage system was eating the parasites and preventing the infection from becoming established, but, once the spleen was removed, infection occurred. In the other case, some other mechanism may have prevented the infection from becoming established after the spleen had been removed.

Natural immunity is an inherited characteristic of a species. Each species inherits a whole set of mechanisms which have evolved so that it does not become overwhelmed (sick), by every infection that it meets.

Sometimes certain individuals have a natural immunity to certain diseases. These too are inherited. For example, vivax malaria is much less common in American Negroes than it is in white people. Many American and African Negroes are naturally immune to it.

In parts of West Africa, in areas where there is a lot of malaria, up to 40 per cent of the people have an inherited characteristic, which, in some way, prevents malaria infections from being very severe. It is called the sickle cell characteristic. This characteristic is not found in Niugini.

3.6 The Vectors in Niugini

Mosquitoes are insects belonging to the Order Diptera. This order has the following characteristics:

(i) Only one pair of wings (most other insects have two pairs). The rear pair of wings are represented by the halteres, which are small knob shaped organs.
(ii) The mouth parts are adapted for piercing or sucking.
(iii) The life cycle contains larval, pupal and adult stages.

Mosquitoes belong to the Sub Order Nematocera (Nem-atoss-era), of the Order Diptera. Members of this group have at least six joints in their antennae. Members of the other sub order have only three joints in the antennae. The antennae are paired appendages on the head of insects. They have a sensory function, that is, they are used to smell and touch.

The Nematocera are divided into several families, one of which is the Culicidae (ku-liss-id-e), or mosquitoes. The mosquitoes are small insects which have the following characteristics.:

(i) A long abdomen.
(ii) Long narrow wings.
(iii) A long slender proboscis. This consists of a number of needle like mouth parts that are covered by a sheath.
(iv) Tiny scales along the edge of the wings and on the wing veins and body. They can only be seen under a microscope.
(v) The larvae and pupae are aquatic (live in water).

The Culicidae are further divided into three sub-families or tribes. They are the Culicinae (ku-liss-in-e), the Anophelinae (An-off-el-in-e), and the Toxorhynchitinae (Toxo-rin-kite-in-e). All the vectors of human malaria belong to the Anophelinae. The Culicinae contain many species that are vectors of various diseases, but they never transmit human malaria. The Toxorhynchitinae have a curved proboscis which bends under the head. The adults feed only on plants (flowers and fruit). The differences between Anophelines and Culicines are summarised in Table 3.4 and Fig 3.6.

There are two genera of Anophelines. They are *Bironella* and *Anopheles*. They have different wing patterns. (See Fig 3.7.)

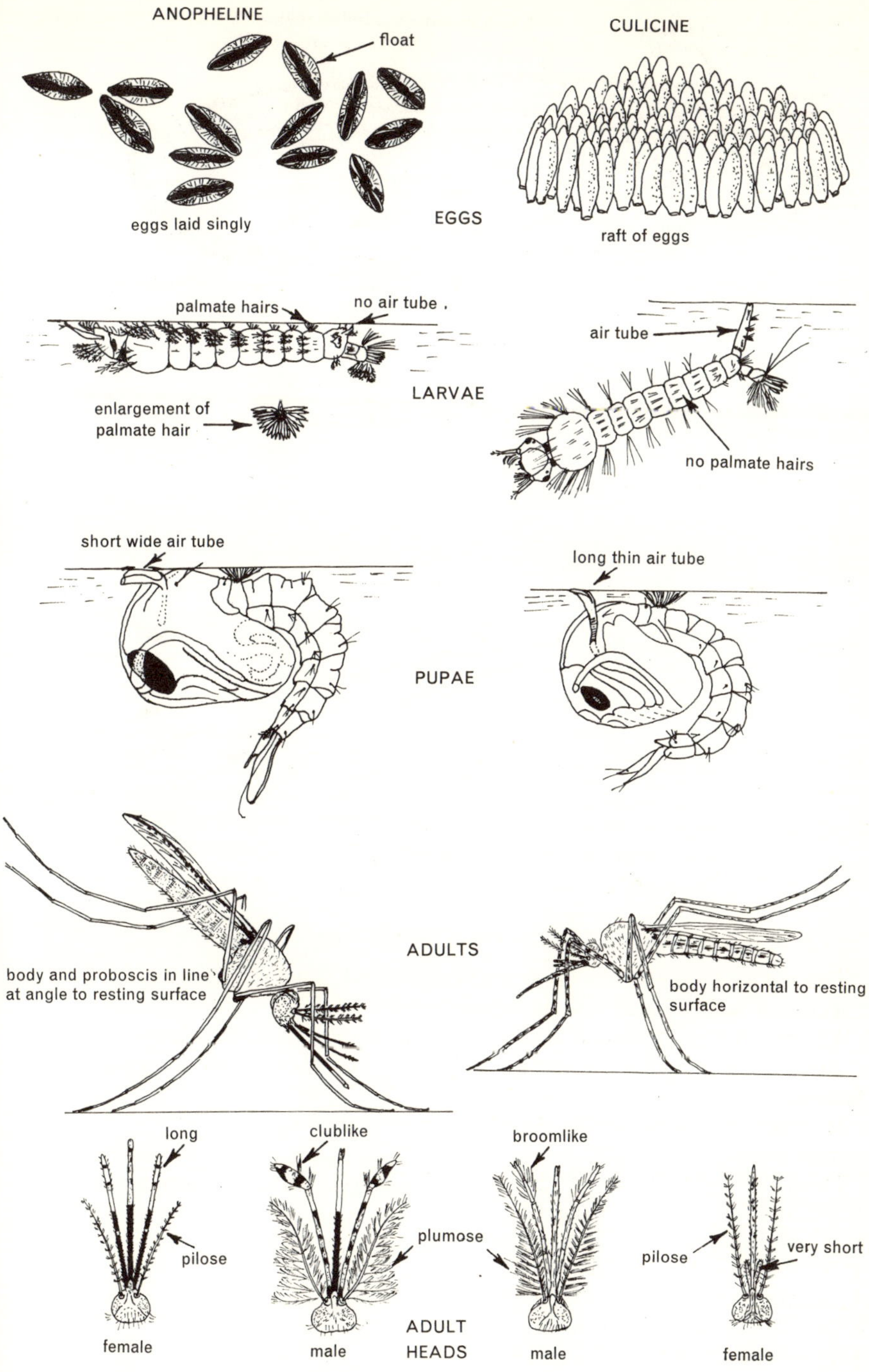

Fig. 3.6 The differences between anopheline and culicine mosquitoes

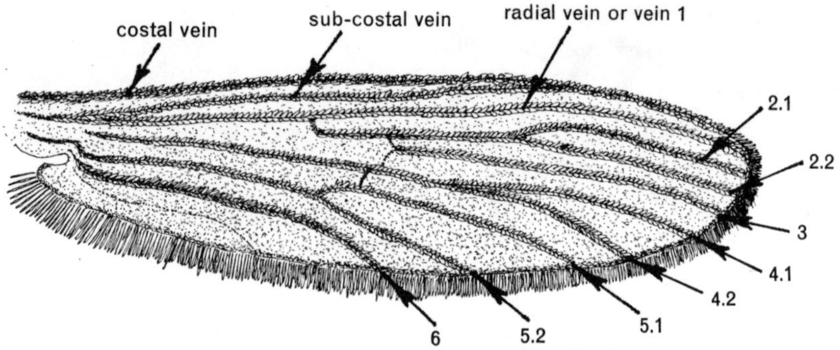

WING PATTERN OF *ANOPHELES* MOSQUITO

VEIN 3 STRAIGHT AND BETWEEN TWO FORKED VEINS

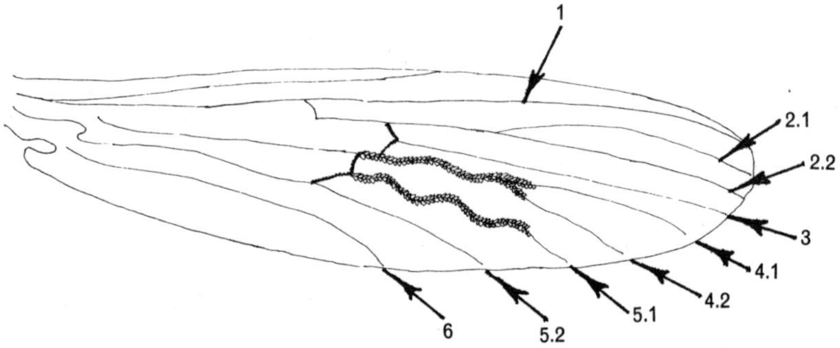

WING PATTERN OF *BIRONELLA*

SAME AS *ANOPHELES* BUT VEINS 4 AND 5.1 ARE WAVY

Fig. 3.7 Wing pattern of *Anopheles* and *Bironella*

The genus *Bironella* lives only in Niugini and some nearby islands. It is not a vector of malaria and seldom attacks man.

All the vectors of human malaria belong to the genus *Anopheles*. This genus is divided into two subgenera; Subgenus *Anopheles* and subgenus *Cellia* (Seal-i-a).

Members of the subgenus *Anopheles* have dull wings and are not vectors of human malaria in Niugini.

There are eight species of the subgenus *Cellia* in Niugini. They are:

A. (*Cellia*) *longirostris*	Brug 1928
A. (*Cellia*) *subpictus*	Grassi 1899
A. (*Cellia*) *annulipes*	Walker 1856
A. (*Cellia*) *punctulatus*	Donitz 1901
A. (*Cellia*) *farauti*	Laveran 1902
A. (*Cellia*) *koliensis*	Owen 1945
A. (*Cellia*) *clowi*	Rozebloom & Knight 1946
A. (*Cellia*) *karwari*	James 1903

Table 3.4 Differences between Anopheline and Culicine mosquitoes

Stage of Development	Anopheline	Culicine
egg	laid singly; has a float; generally dark in appearance	generally laid in rafts sometimes singly; never has a float; generally light in appearance
larva	lies parallel to and feeds at surface of water; does not have a breathing siphon; has palmate hairs.	lies at an angle to surface of water; feeds below the surface; has a breathing siphon; does not have palmate hairs.
pupa	has very short, wide breathing trumpet	has thin breathing trumpet.
	N.B. At this stage of development the pupae are difficult to differentiate.	
adult	stands at sharp angle to resting surface; scutellum oval; female has long palpi; male has clubbed palpi	stands parallel to resting surface; scutellum trilobed; females have very short palpi; males have broomlike palpi

Ref. also Fig. 3.6.

The species that are known to transmit human malaria (the proven vectors) in Niugini, the Solomon Islands and West Irian are the species *punctulatus*, *farauti* and *koliensis*. They are all closely related species and are often referred to as the *punctulatus* group of species, or the *punctulatus* complex, because the differences between them are very small. Only a skilled technician can tell the difference between these three species. (See Fig. 3.8.)

In addition, *A.* (*Anopheles*) *bancrofti* has been proven to be a vector in West Irian, but it has never been found infected in Niugini. And *A.* (*C*).*karwari* has also been proven to be a vector, but it has only been found in Kota Baru, West Irian.

A. (*C*).*clowi* has only been recorded from the type locality, Humboldt Bay, in West Irian. The other species of the sub genus *Cellia* are widespread, but have not been found infected with malaria so far.

The Life Cycle of Anophelines

The life cycle of all mosquitoes contains the following stages; egg, larva, pupa and adult.

The life cycle of Anophelines takes from 9 to 19 days. The length of the life cycle depends on two important environmental factors which are:

(*i*) *Temperature.*
The life cycle takes less time at warm temperatures than at cold temperatures. Extreme hot and cold temperature will kill mosquitoes, but within the range 15°C to 35°C they can live normally and, within this range, the higher the temperature the faster the life cycle is completed.

(*ii*) *The availability of food.*
Larval development is slowed down if the larvae do not have enough food. Of course, if there is no food, the larvae will starve to death. However, this probably seldom happens in nature.

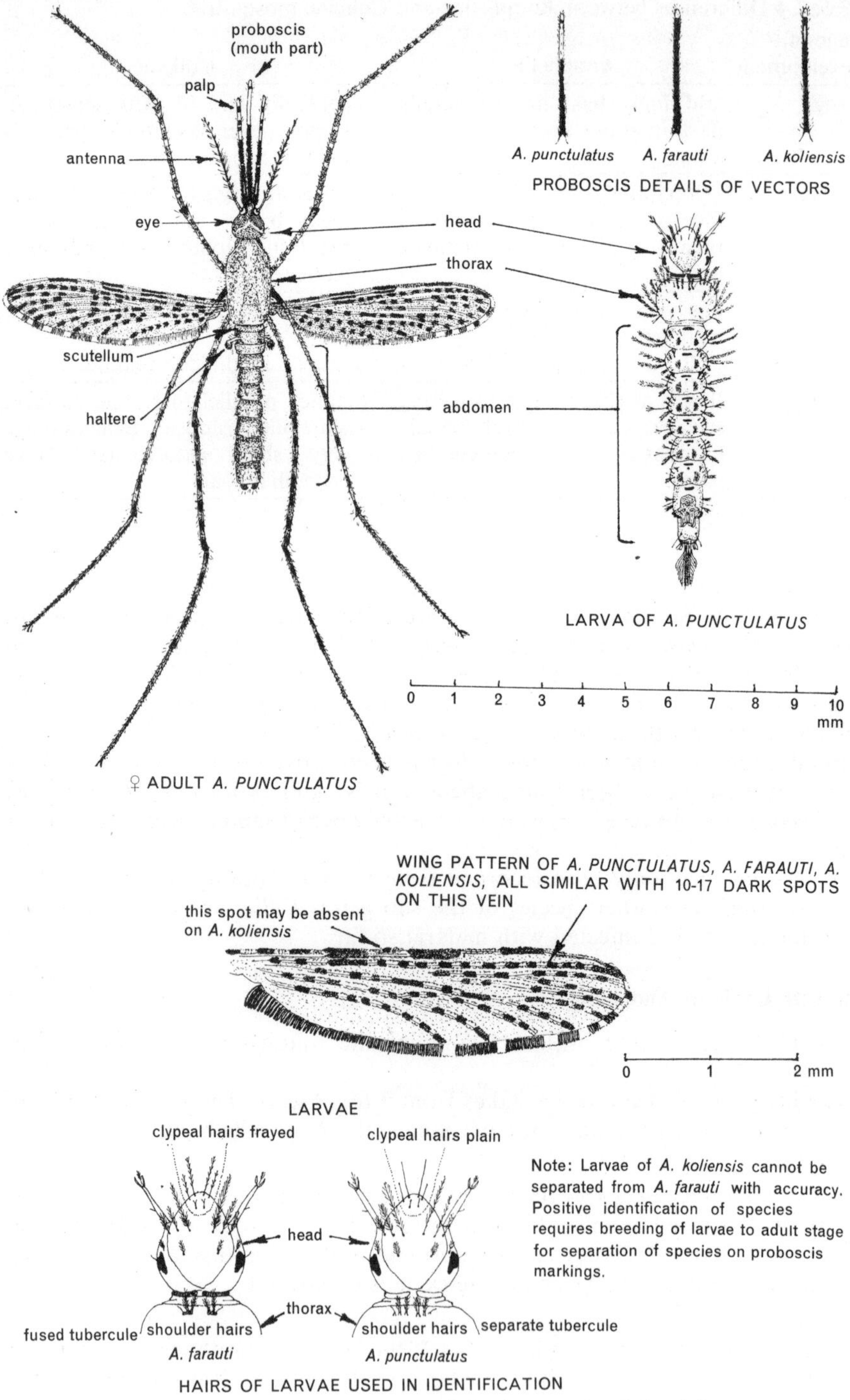

Fig. 3.8 *Anopheles punctulatus*, *Anopheles farauti*, *Anopheles koliensis*

The adult females cannot lay eggs until they have had a meal of blood. If the females cannot find a host from which to get blood, egg laying is delayed. The survival of the females depends a great deal on the humidity. They cannot survive for long, under dry conditions.

We will now consider the life cycle of the *punctulatus* group. Unfortunately very little work has been published on them.

The Eggs

Each female lays about 150 eggs in a single batch (range 100 to 350). The eggs hatch in 2 to 6 days. The incubation period is longer in the Highlands than on the coast, because the temperature is lower.

The Larvae

The speed of development of the larvae depends mainly on the temperature. When the larvae first hatch from the eggs they are called first instar larvae. After they have grown a bit, they moult, or shed their skins, and become second instar larvae. They moult three times in all, so there are four instar stages.

The time spent in each stage is as follows.

first instar larvae	1 to 4 days
second instar larvae	1 to 3 days
third instar larvae	1 to 4 days
fourth instar larvae	2 to 5 days
TOTAL	5 to 16 days

Anopheline larvae lie with their bodies at the surface of the water. They feed by brushing food into the mouth with mouth brushes that lie on each side of the head. They feed continuously, day and night, on tiny particles of food which float on the surface of the water.

The larvae of the common culicine, *Culex halifaxi* eat the larvae of anopheline mosquitoes in Niugini.

The Pupa

The fourth instar becomes a pupa. It cannot move very quickly. It bobs up and down in the water as it comes to the surface to take in air through its air trumpet. It does not feed. The pupal stage takes about 2 to 3 days on the coast, and 5 to 8 days in the Highlands.

The adult develops within the skin of the pupa. This process is known as meta-morphosis. (meta = change;morph = shape or form.)

The Adult

Emergence

When the adult has formed within the pupal skin, the pupa takes in extra air and floats to the surface of the water. The pupal skin splits open at the top of the head and the adult struggles free. This process takes 5 to 15 minutes. The adult then remains standing on the pupal skin, or the surface of the water for some time. When its wings have set hard it flies off. Emergence usually takes place in the early morning.

Mating

The newly emerged adults hover, like helicopters that hang in the air without moving, about 6 to 12 feet above their breeding site. When a large number has gathered, they start a special swarming dance. Perhaps two hundred mosquitoes fly around in circles in a mass about a yard wide. The females make a loud buzzing noise which can be heard some distance away. They dash into the swarm and grab a male and mate with him as they fall out of the swarm. Mating is completed before they reach the ground. This mating sing-sing goes on for 5 to 10 minutes.

The females only mate once. Large numbers of the *punctulatus* group have been examined, but an unfertilized female has never been found in nature.

Feeding

The males do not bite man. They feed on nectar from flowers. We do not know how long they live, but it seems that they may have a shorter life than the females.

The female needs one or more feeds of blood before her eggs mature. What happens is this. When the female has a full meal of blood, her abdomen becomes distended and this stimulates a pair of glands in the neck, which are called corpora-allata glands. These glands then produce a hormone, or chemical messenger, which makes the eggs in the ovary start to mature. The blood in the stomach is digested and provides the materials needed to make the eggs.

When a female Anopheline feeds, she pushes her sharp mouth parts through the skin and cuts one or more tiny blood capillaries. Before she starts to suck the blood, she injects saliva (or spit) into the wound. This contains an anti-coagulant, or substance which prevents the blood from clotting. If this substance was not present the blood would clot in her mouthparts and set hard, and she would not be able to feed again.

Females of the *punctulatus* group usually feed at night. However, they will feed during the day if they are very hungry.

Fig. 3.9 gives data on the feeding behaviour of *A. farauti* in the D'Entre-casteaux Islands before and after spraying with DDT.

This sort of data is collected in the following way. A technician catches and counts all the vector mosquitoes that bite one man, every hour, throughout the night from 6 pm to 6 am. He then expresses the catch for each hourly period, as a percentage of the total number of vectors caught that night. For example, if 24 vectors were caught between midnight and 1 am, and a total of 200 were caught for the whole night, $\frac{24}{100} \times \frac{100}{1} = 12\%$ of them, would have been caught during this hour. This data is then presented in a histogram graph so that we can see what is happening more easily.

The data in Fig. 3.9 is typical of the picture in other parts of Niugini. Before the area was sprayed with DDT, most of the vectors came to bite man between 9 pm and 3 am. The number gradually increases from early evening until about midnight and then gradually declines until early morning.

After the area had been sprayed with DDT the feeding pattern changed. Most of the vectors came to bite man early in the evening. At this time most of the people are outdoors cooking, eating and talking.

In some other parts of Niugini the same sort of thing has happened. Most

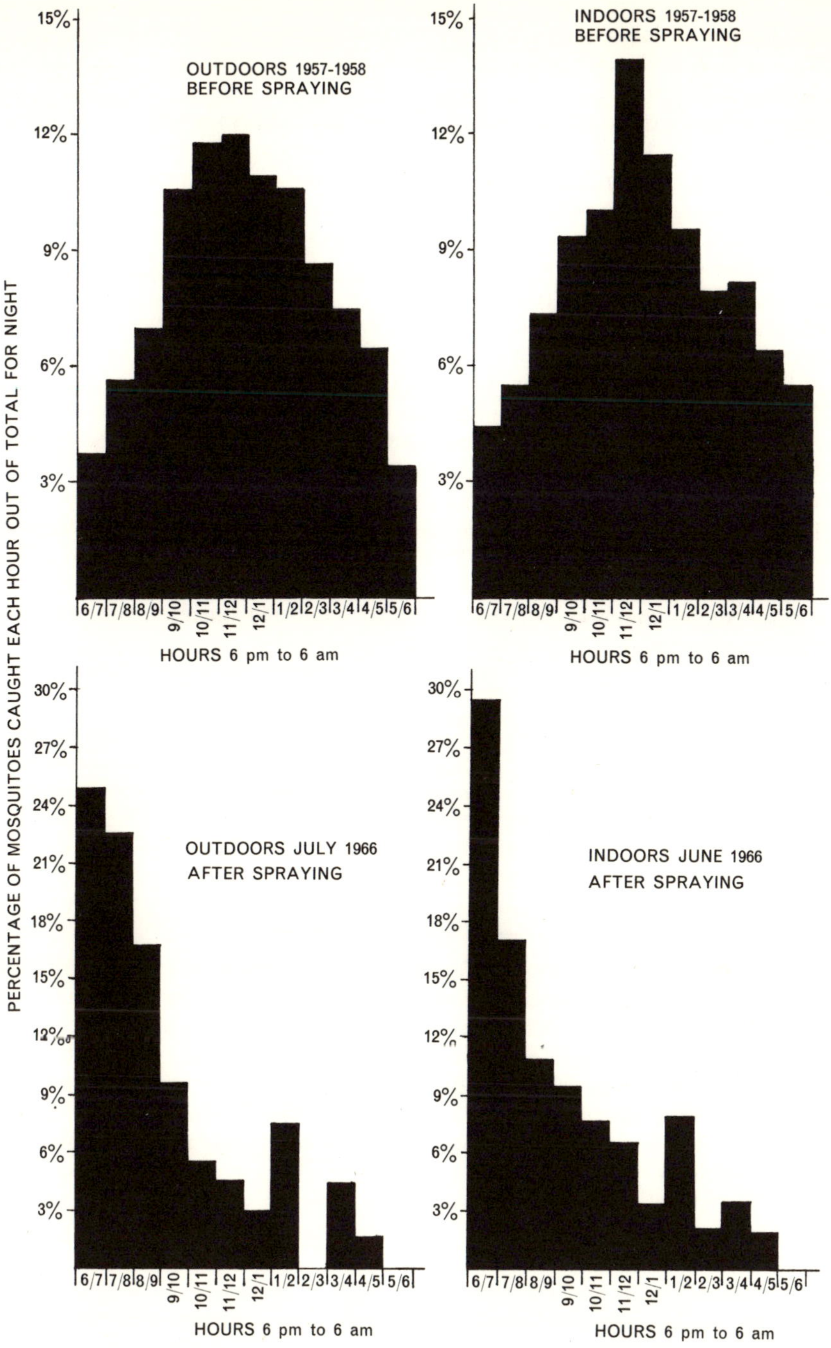

Fig. 3.9 Feeding behaviour of *A. farauti* before and after spraying with DDT

of the mosquitoes are coming to bite man early in the evening and in some cases also about dawn.

We will discuss the importance of this change in the behaviour of mosquitoes, in Section 3.8.

When a female mosquito has taken a blood feed, she is very heavy and cannot fly far. We do not know how much blood our vectors take in one meal. But other species take 1 to 3 milligrams. After feeding they weigh nearly three times their unfed weight. No wonder they cannot fly far.

When they have fed, the females fly to a nearby surface, which is usually within a few feet of their host, and rest for a period of several hours. During this time, their food is partly digested.

Most mosquito vectors feed inside houses. And because they cannot fly far after feeding, we can kill most of them by spraying the surfaces inside houses with DDT. This is discussed in more detail in Section 3.8.

Some mosquitoes feed on people while they are outside the houses. They will never be killed by DDT that has been sprayed inside the houses. It is important to know how much outdoor biting is going on, because if there is a lot, malaria cannot be eradicated by the spraying of DDT.

The vectors of malaria in Niugini feed mostly on human blood but they will sometimes feed on other animals. This is shown in Table 3.5. The blood from freshly fed mosquitoes can be tested to see what host they have been feeding on. The table shows that most of the mosquitoes had fed on man, but some had fed on dogs and pigs.

Table 3.5 Percentage of freshly fed *A. farauti* that had been feeding on various mammals.

Host	Minj 1958	Fergusson Is. 1958
man	71	75
pig	28	4
dog	1	16
?	—	5

One observation concerning the feeding habits of vectors is worth mentioning. Mrs M. Spencer found that babies on Fergusson Island, under five months old, were not bitten by vectors, while they were with their mothers. She recorded eight cases where the babies were not bitten at all, and their mothers were bitten from 1 to 106 times (average 29 times) during one evening. We will talk about the importance of this observation when we discuss Infant Parasite Rates in Section 3.8 (p. 129). The reasons for this observation need investigating. Perhaps there is some substance, which attracts mosquitoes, in the skin of adults, which is not present in babies. Or perhaps babies' skin gives off some substance which is actually a mosquito repellant.

Resting

During the daytime the female mosquitoes rest in shady sheltered places. They prefer dark, humid, warm places. They soon die if they are exposed to direct sunlight, because the heat of the sun causes them to lose water and they die of desiccation.

Females of the *punctulatus* group are most often found resting in the following kinds of places:

(i) On clean wall surfaces. They will not rest on walls that are black with smoke. They are most often found near the floor, but they sometimes go near the roof.
(ii) Under the floors of houses that are built up off the ground.
(iii) Under beds and on clothing, mosquito nets, and furniture.
(iv) In holes in the walls.
(v) In plants such as crotons and bananas that grow close to houses.

In the Highlands of Niugini, anophelines often stay inside houses for long periods. They crawl into cracks in the wall between bark, grass and leaves.

In most parts of the world, malaria transmission does not occur above an altitude of about 5000 feet, because vector mosquitoes seldom live above this altitude. However, in Niugini, vectors are found above 5000 feet. They have been found in Gembogl, on the side of Mt Wilhelm, at a height of 7500 feet. Mosquitoes infected with sporozoites have been found at Minji, Kundiawa and at Kainantu and other parts of the Highlands, at altitudes of more than 5000 feet.

In very cold areas, such as Northern Europe and Canada, mosquitoes hibernate, or go to sleep, during the long cold winter. They live on fat that has been stored within their bodies.

Vectors in the Highlands of Niugini appear to be able to stay alive for long periods also. They may live on fat stored within their bodies in the same way as hibernating mosquitoes do. During this time they do not lay eggs.

Flight

Hungry anophelines in other countries fly up to about half a mile from their breeding sites. As far as we have seen, the vectors in Niugini do not fly more than about twenty yards before having a rest. They are seldom found more than a few hundred yards from their breeding sites. Of course, they may be blown longer distances by the wind.

Our vectors make short flights in the evening and early morning to and from their daytime resting places, which are often only a few yards from houses. They fly close to the ground.

Provided the female can get blood as often as she needs it, her eggs will mature regularly. This happens every two days in warm coastal areas of Niugini and every four days in the Highlands.

The female is like an egg laying machine. She takes in blood and, two or four days later, produces another batch of eggs. Her whole life is spent in this way. She will fly to get blood, to lay eggs and to find a resting place.

Egg Laying

The life cycle of the mosquito is completed when the female lays eggs. Eggs are laid one at a time on the surface of water. The female first selects the right kind of place for egg laying. She will not lay her eggs in places where the larvae are unlikely to survive. When the female is ready to lay, she hovers above the surface of the breeding site and then flies low over the water and drops an egg so that it falls only a few millimeters. Her legs just touch the water as she rises. She then repeats this performance so that all the eggs are laid near one another. She lays about 150 eggs every 2 to 4 days.

Members of the *punctulatus* group lay their eggs in shallow pools of temporary water. *A.punctulatus* prefers muddy water, with a clay bottom and clay in suspension in the water. *A.farauti* prefers places with a sandy bottom, and *A.koliensis* will breed in either kind of place.

These species often breed in man-made breeding sites. For example in pools made by footprints, wheel ruts, roadside drains, canoes that contain water and borrow pits. A borrow pit is a hole that is left where soil has been taken (or borrowed) for gardening or road building. They also breed in hoof prints and pig and cattle wallows.

The *punctulatus* group seldom breed in swamps or moving water. They may breed in brackish water (that is, salty water which is not as strong as sea water).

These vectors lay their eggs in open water that is exposed to sunlight. They do not lay eggs in shady places.

Length of Life

We do not have a great deal of information about how long our vectors live. The length of life depends mainly on the temperature, humidity and availability of blood. It almost certainly varies from place to place and throughout the year.

We can tell how long a mosquito can live by keeping it alive in the laboratory. But this tells us nothing about how long most mosquitoes live in nature.

Under laboratory conditions *A.punctulatus* will live up to 56 days in the Highlands.

We can estimate how old a female mosquito is, by examining its reproductive organs. A mosquito has two ovaries, and each contains about a hundred ovarioles. Each time a batch of eggs are laid, one egg is produced by most ovarioles. When this egg moves out of the ovariole, a small lump is left in the ovariole. It is called a shrunken follicle. If a number of ovarioles are examined (at least six) and the greatest number of shrunken follicles is determined, by counting, it tells how many times the female has laid. (See Fig. 3.10.) We know that laying takes place every two days on the coast, and every four days in the Highlands, so we can estimate the age of the female. For example, in an examination of six ovarioles, the highest number of shrunken follicles was six. If the mosquito took blood on the first day after emerging and took blood regularly thereafter, it must be at least $1 + (6 \times 2) = 13$ days old, if it came from the coast. And $1 + (6 \times 4) = 25$ days old if it came from the Highlands.

Transmission of malaria can only occur if the females live long enough after taking infected blood, for sporozoites to develop. This period is known as the extrinsic cycle.

Experiments have been done on the coast, and in the Highlands, to determine the length of the extrinsic cycle in our vectors. They are as follows:

	Coastal Areas	*Highlands*
P.falciparum	11 days	21 days
P.vivax	8 days	16 days
P.malariae	19 days	28 days

If a vector on the coast becomes infected with *P.vivax* on the day it emerges it would have to live through four lots of egg laying, before it could transmit malaria. It must feed on blood at least four times. If it feeds on man inside a house and rests on a surface that has been sprayed with DDT, it will be killed

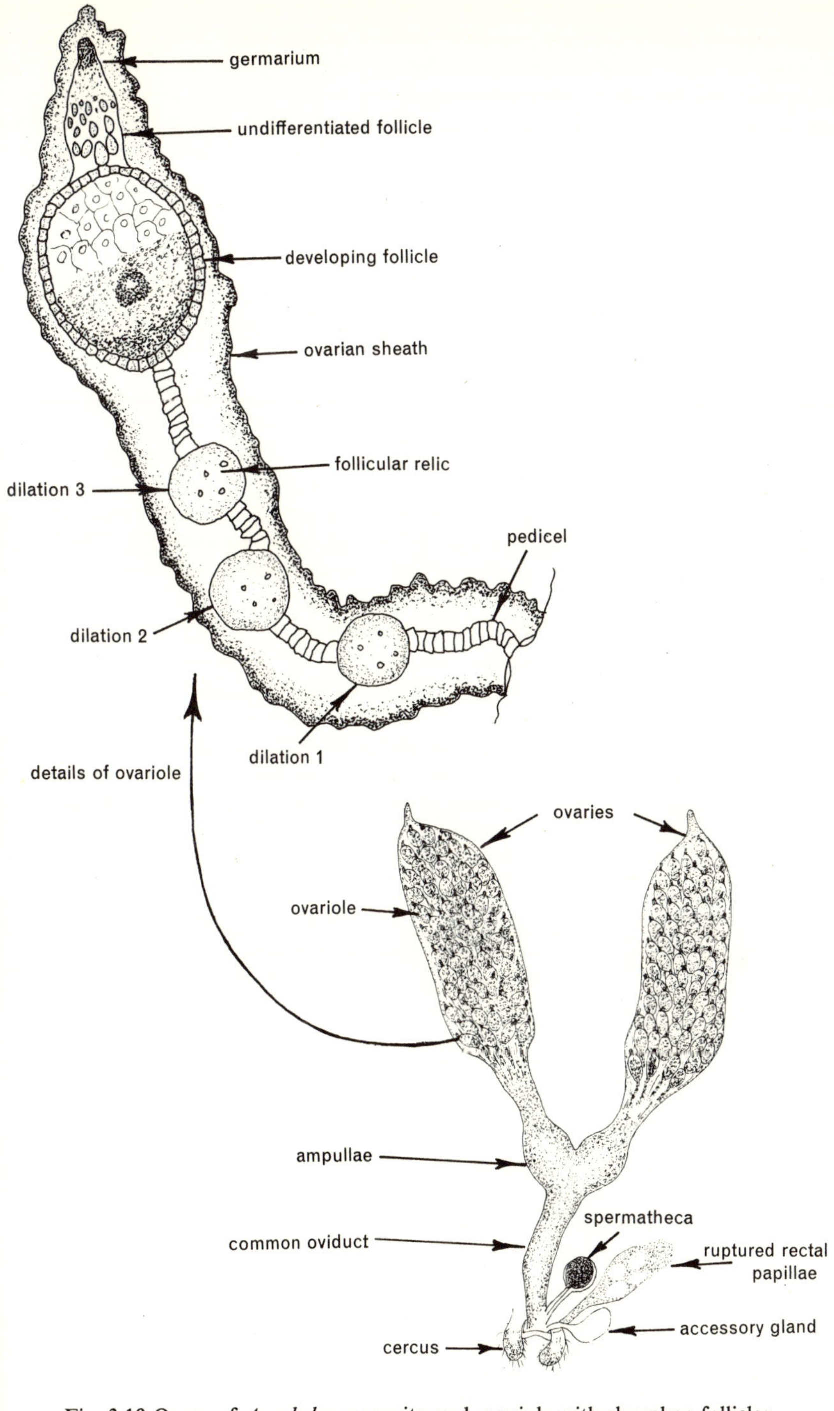

Fig. 3.10 Ovary of *Anopheles* mosquito and ovariole with shrunken follicles

before it is infective to man. In this case the mosquito has four chances of being killed by DDT.

If we determine the average age of the female vectors after malaria control has started, we can see if they are living long enough for transmission to take place. If the use of insecticides is going to interrupt transmission, the expectation of life must be reduced to a period less than the length of the extrinsic cycle.

3.7 The Epidemiology of Malaria

INTRODUCTION

Epidemiology (epi-deem-i-ology) is the study of the way disease affects communities, or groups of people that live together.

Epidemiologists try and find why the amount of disease varies in people of different ages, from time to time, and from one place to the next. This often provides information which is important in controlling disease. It is also important to have measurements of the amount of disease in various places, so that the authorities can decide where it is best to spend money to control disease.

In programs designed to control or eradicate malaria, or any other disease, the epidemiologist continually assesses progress. That is, he gathers figures which enable him to see how well the program is working.

In a certain village there might be an increase in the number of people who are sick with malaria during certain months of the year. An epidemiologist would try to find out why this happened. He might find that the mosquitoes which carry malaria could only breed at the end of the wet season and that an outbreak of malaria occurred each year, soon after the number of vectors increased. A comparison of the patterns of disease in different districts often provides important clues about the way disease spreads.

SURVEYS

Surveys are made to gather information or measurements for a particular purpose.

When dealing with parasitic diseases, three kinds of measurements may be important. Firstly, we want to know how many people are infected, or more particularly, what proportion or percentage of the total population are infected. Secondly we want to know how badly they are affected by the disease and thirdly, we want to know how quickly the disease is being spread.

The incidence of a disease is the proportion of a particular population or community that is infected with it, at a certain time. It is usually expressed as a percentage. (See p. 62 for statement on malaria terminology.) For example, we might say that the incidence of malaria in people from Bontong village was 35 per cent in July, 1970. This means that 35 people out of every hundred were infected at that time.

If we examined blood from 116 people and found that 35 of them were infected, the incidence of infection would be; $\frac{35}{116} \times \frac{100}{1} = 30\%$
Malariologists usually refer to the incidence of infection as the Overall Parasite Rate (OPR). This measurement is obtained by taking blood from lots of people, examining it for parasites, and calculating the percentage that are infected or positive.

In communities that are continually infected with malaria, the incidence of infection varies with the age of the people. It is usually much higher in children than in older people. This is because the older people have developed an immunity to the disease. (Immunity has been discussed in Section 3.5.) It is important therefore, to determine the incidence of infection in various age groups within each community.

In malaria surveys the incidence is usually measured in the following age groups:

0 to 11 months
12 to 23 months
2 to 4 years
5 to 9 years
10 to 14 years
15 years and over

Sometimes when the samples are small, two adjacent groups are put together or 'pooled'. For example, if we only had 20 people in the first age group and 17 in the second, we might pool them and give the incidence in people less than two years old.

The incidence in infants less than one year old is called the Infant Parasite Rate (or IPR). It is an important measurement in malaria surveys, because it gives some indication of the amount of new infection that is going on. Cases of malaria in older children and adults can be caused by new infections or relapses of old infections (see Section 3.4), but cases of malaria in infants less than one year old must be due to bites from infected mosquitoes that occurred within the last year. We shall see later that when malaria is being controlled so that transmission has stopped, the incidence in infants should fall to zero. However, the incidence in older persons could remain high for a year or two because of relapses.

Another measurement of the number of people who are, or have been infected with malaria, is the Spleen Rate. This is the percentage of people who have enlarged spleens (see p. 132, Section 3.8). It is usual to determine spleen rates only for children 2 to 9 years old.

So far we have considered various measurements of the amount of malaria that is present in the community.

It is often useful to have a measurement of how badly people are affected by malaria; this is more difficult to measure accurately because of the very nature of the disease. Malaria starts off with mild symptoms which become worse and then the patient gradually recovers.

However, there are two kinds of measurements which give some indication of the severity of the disease in the community. The first one depends on how much the spleen is enlarged. A doctor can measure how much the spleen is enlarged in each child and then calculate the average amount of enlargement for all the children examined. (See Section 3.8.) The second kind, provides a measurement of the number of parasites in the blood. There are two ways of determining this. Firstly, the percentage of red blood cells that are infected with parasites can be determined by counting them, or secondly, the number of parasites in a given volume of blood can be estimated. It is difficult to measure very small quantities of blood accurately and counting takes a long time. However a little trick helps to save time and provides a fairly useful measurement. The number of white blood cells in a given volume of blood is fairly constant, and is known to be about 7000 per cubic millimeter, provided no other diseases are present. Thus, if we count the number of parasites in an area under the microscope which contains say 1000 white blood cells, we can estimate the approximate number of parasites in one cubic millimeter of blood. If we have these kinds of measurements for many people, we can calculate an average measurement for the whole population.

The transmission of malaria is indicated in two ways: by the Infant Parasite Rate and by sudden increase in the Overall Parasite Rate.

Malaria surveys are discussed in more detail at the end of Section 3.8.

VARIATION IN THE INCIDENCE OF INFECTION

Appendix II, page 232, gives malaria survey figures for various places in Niugini. It gives all the figures, published and unpublished, that we have been able to find. It is a long list, but we believe it will be a useful one because these figures have never been gathered together before. All figures collected after malaria control started, have an asterisk (*) in front of them.

Niugini is still a very malarious country, although malaria is being controlled in many parts. The situation has been much worse in the coastal areas and the islands than in the highlands. In fact, most of the highlands were probably free of malaria until roads and airstrips were built and infected people were brought in from other areas. We shall tell the story of malaria in the Highlands presently.

In highly malarious areas, the incidence of infection and the spleen rate is highest in children. In these areas, people of all ages are constantly being bitten by infected mosquitoes. Many infants and small children die of malaria before they have a chance to develop much immunity to it. If they survive long enough to develop immunity, they are seldom seriously affected by attacks of malaria. But they are never really well while they are infected. In the Maprik area, before malaria control started, about 60 per cent of the children died before they reached their first birthday. Most of them died from malaria.

In 1967-8, 109 people died of malaria in administration hospitals in the Territory of New Guinea; 82 of these were children under 10 years old; of these 11 were less than one year; 58 were between one and five years and 13 were between 5 and 10 years old. Probably many more died in their villages because they did not go to hospital.

In areas where there is little malaria, or where it spreads among a small proportion of the population after every wet season, the incidence of infection and the spleen rate may not vary much with age. The people do not get much of a chance to develop a strong immunity to it and people of all ages get acute attacks of malaria.

The same thing happens when non-immune people move into a malarious area. Malaria made life very difficult for the first Europeans who came here, because they had no satisfactory malaria suppressives and no immunity. Gavin Souter, in his book *New Guinea—the last unknown* (published by Angus and Robertson) gives some details of what happened. For example, from 1828 to 1835, at the Dutch settlement at Triton Bay, 75 Europeans and Indonesian soldiers died, most of them from malaria, and the settlement had to be abandoned. In the 1880s the headquarters of the German Administration was at Finschhafen. Most Europeans there suffered attacks of malaria once a month. The fever was treated with quinine and it was customary to take far larger doses than were necessary. In large doses it causes ringing in the ears, dizziness, deafness and blurs the vision. Between 1886 and 1898, 224 officials worked for the New Guinea Company, 41 died on the job and 133 resigned or were given the sack. One of them wrote 'I am one of the few to get out of that malaria hole, Finschhafen, with a whole skin, because I treated the fever with alcohol instead of quinine and the orders of the Company similarly, with alcohol instead of respect'. It is a wonder they were able to work at all. The German administration moved its headquarters several times because of malaria.

The London Missionary Society brought more than 200 Polynesian teachers to Papua in the 1880s. Lawes reported that more than half of them died of malaria within three years, and all of them suffered from fever. They came from areas where there was no malaria and they had no immunity to it.

Europeans escaping as the Japanese entered Madang in World War II, employed 252 carriers from a malaria free village in the Bismark Mountains. They went into the Ramu and Markham Valleys which are highly malarious. The carriers were away from their villages for only six weeks, but 160 died of malaria.

The data in Appendix II, page 232, shows that the amount of malaria, as measured by the parasite rate and the spleen rate, varies considerably from district to district and from place to place within the same district. In most parts of the world, the proportion of people infected with malaria in a particular place varies with time. Usually there is more malaria at certain times of the year. The incidence of infection often reaches a peak a month or two after the biggest rains. In many places, there is more malaria in some years than in other years.

Many factors influence the amount of malaria in a complex way. Some of them interact with each other and it is often impossible to tell why the amount of malaria has changed without having accurate measurements of a large number of factors.

We will discuss some of the factors which influence the incidence of malaria. They can be divided into factors concerned with people, factors concerned with the vectors and factors concerned with the interaction of the two.

(1) Factors concerned with people

The number of people who have malaria at a particular time will depend on the number of people who have been infected in the previous months. It will depend on the number of them that had gametocytes in their blood and the density of gametocytes in their blood. Malaria can only be transmitted to an uninfected person by a vector which has become infected with gametocytes from the blood of man. If the density of gametocytes in the blood is very low, mosquitoes are unlikely to become infected.

The number of new cases of malaria will also depend on the level of immunity within the population. If a large proportion of the population have recently been infected with malaria they will be immune to sickness from that strain, for some weeks or months. As they lose their immunity, they can become sick again.

Relapses of malaria occur from time to time. It often is very difficult to tell if an attack of malaria is due to a new infection or a relapse. Obviously the number of relapses that occur, will play an important part in the incidence of malaria. In some cold countries a large proportion of the population sometimes goes down with relapses of malaria at about the same time. This seems to be associated with the start of warm weather. This rise in the incidence of infection is often followed by a second rise, a month or so later, which occurs as a result of new infections. The relapses provide a reservoir of infection and the vectors transmit malaria to uninfected people.

The incidence of malaria is likely to change if new people come into an area. If these people have no immunity they may be very susceptible to malaria. If they are already infected, or if they suffer a relapse, they may start a series of new infections in other people.

(*2*) *Factors concerned with the vectors*

New infections cannot occur without vectors. They must feed on blood containing gametocytes; they must then live long enough for the sporozoites to develop within them and they must then inject sporozoites into an uninfected person.

Thus a change in the incidence of infection may be associated with a change in the number of vectors or in their longevity (length of life).

The number of vectors is greatly influenced by rainfall. Rain provides a breeding place for mosquitoes. Too much rain may wash the larvae into streams and lakes where they cannot survive. Very heavy rain may kill many of the adult mosquitoes that have no shelter from it. If there is too little rain, the breeding places will dry up. The number and kind of breeding places will depend on other factors such as the slope of the ground and the kind of soil. Because the rainfall varies seasonally, malaria transmission is often seasonal.

The longevity of mosquitoes is also very important. They cannot transmit malaria unless they live long enough for sporozoites to develop within them. That is, their length of life must exceed the length of the extrinsic cycle. In Niugini the extrinsic cycle takes at least eight days on the coast and sixteen days in the Highlands. In many areas, adult mosquitoes are present throughout the year, but transmission only occurs during relatively short periods when the longevity of the mosquitoes exceeds the length of the extrinsic cycle. Longevity is influenced a great deal by the humidity. Mosquitoes cannot live long under dry conditions. The humidity is highest during the wet season because there is a lot of water lying on the ground. This water evaporates, so that the amount of water vapour in the air is kept high. Plants take water from the soil and give off water vapour from their leaves. This process is called transpiration. It works much faster when there is a lot of water in the soil and slows down during the dry periods.

The length of the extrinsic cycle depends on the temperature. It is shortest when the temperature is high and takes longest under cold conditions. Temperatures in the highlands are lower than those on the coast and because of this, the extrinsic cycle is longer there. In Niugini the temperature is very constant throughout the year. The average temperature during the warmest month is only a few degrees higher (about 5°F in Port Moresby) than the average temperature in the coldest month, so that seasonal changes in the temperature will not cause very much of a change in the length of the extrinsic cycle.

(*3*) *Factors concerned with the interaction of man and mosquito*

Lastly the distribution and numbers of vectors in relation to the distribution and numbers of people, will be important in determining the amount of malaria. The habits of each, in relation to the other, will also be important because transmission can only occur when the two hosts come together.

In Niugini, the vector mosquitoes mostly feed on man, although they may feed on certain domestic animals as well. In some other countries, the vectors prefer to feed on domestic animals, such as cows, and the amount of transmission is often low in places where there are lots of cows.

In Niugini, the vectors usually feed on man in houses. In places where most of the people work in the gardens during the wet season, it is possible that a lot of transmission may occur outside. This is very important when we attempt to control malaria by spraying houses with DDT (See section 3.8).

People from areas where transmission does not occur, frequently become infected when they travel to malarious places. In one area in the Highlands of West Irian, people became infected every time they travelled to collect salt. Other cases are known of people becoming infected when they moved from the edge of the Highlands down into the lowlands to collect betel nut or to hunt. In some cases, they have brought malaria back into areas where there were plenty of vectors but which had been free of malaria.

SEASONAL VARIATION IN THE INCIDENCE OF MALARIA IN NIUGINI

In most places in Niugini the amount of malaria increases towards the end of the wet season and reaches a peak about two months after the month with the highest rainfall. Figures 3.11, 3.12, 3.13 and 3.14 illustrate this for Rabaul, Minj and Port Moresby.

These seasonal changes may be connected with the numbers of vector mosquitoes but other factors may be much more important. Unfortunately, we do not have any data on seasonal changes in the number of mosquitoes in these areas and we do not have any other data from Niugini which we can analyse to see how the incidence of malaria is related to the number of vectors. Figs. 3.24, 3.25 and 3.26 show seasonal changes in the number of vectors and monthly rainfall for three other areas (Wankum, Kundiawa and Maprik).

In general, these figures show that the number of mosquitoes is high after wet periods and low after a dry period. But there are exceptions to this. A period of relatively little rain is sometimes followed by a marked increase in the number of vectors. In many places the amount of transmission probably depends more on the longevity of the vectors than on the number of them that are present.

The data in Figs. 3.11 to 3.14 are based on the number of cases of malaria occuring in people coming to hospital. Blood slides are taken from all people with fevers who come to hospital. This data probably gives a good idea of the seasonal variation in the incidence of malaria in areas where the people come to hospital when they are sick, but it would be better to have data based on monthly surveys of the whole population.

One feature of these graphs is that the incidence of malaria in the worst (highest) month is many times greater than it is in the best (lowest) month. This is a common situation in many places. However, in some places where the spleen

Fig. 3.11 The monthly rainfall and total number of malaria cases from all hospitals in the

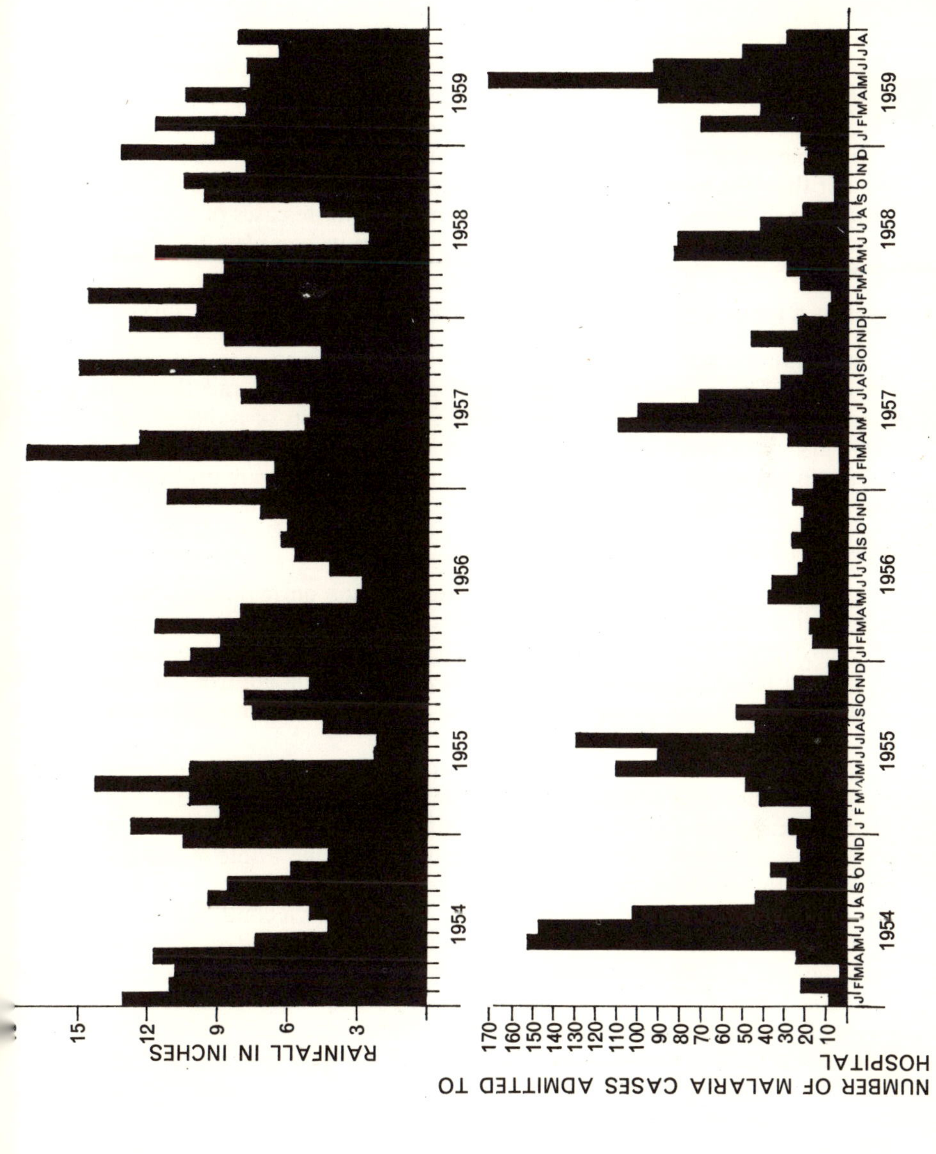
RAINFALL IN INCHES
NUMBER OF MALARIA CASES ADMITTED TO HOSPITAL
1954
1955
1956
1957
1958
1959

rate of children is constantly over 75 per cent, the people are constantly being bitten by infected mosquitoes and there is no great seasonal change in the incidence of infection.

Fig. 3.14 shows changes in the number of cases of each kind of malaria at the Port Moresby hospital. It can be seen that in 1967 the greatest number of cases of falciparum malaria occurred during May and June and that during the months before this, the number of cases of this kind gradually increased and then they gradually decreased. The situation for malariae and mixed infections is similar, with a peak in July. But the number of cases of vivax malaria was about the same throughout the year with small increases during February and May.

In some other places in Niugini, the number of cases of vivax malaria is far greater than the number of cases of falciparum malaria in certain months of the year. The situation then reverses so that there are more cases of falciparum than vivax. The reasons for these changes are likely to be complex. We do not know of any detailed study of these changes in Niugini.

LONG TERM CHANGES IN THE INCIDENCE OF MALARIA

In certain countries the incidence of malaria seems to increase greatly every few years. These long term fluctuations are called cycles and they have been found to occur regularly about every eight years in parts of India, every 11 or 12 years in certain parts of South America, and every 20 years in Malaya. Not much is known about the causes of these long term cycles. They are probably complex and vary from one place to the next. We do not know if they occur in Niugini.

MALARIA SICKNESS IN NIUGINI

Appendix II provides a great deal of information on malaria in Niugini, but it gives little indication of the amount of sickness caused by this disease. This is very hard to measure. It is obvious from the data in Appendix II that the amount of malaria has decreased greatly in areas where malaria control programs have started. We will now examine data on malaria sickness in hospitals to see how this has changed over the past few years.

Table 3.6 gives the numbers of people admitted to administration hospitals in Niugini with malaria and the number of deaths in these hospitals due to malaria, for each year from 1952-3 until 1967-8. Mission hospitals admit nearly as many people as administration hospitals, so the figures for the whole of Niugini would be about double those given. The number of people admitted to hospital with malaria has not decreased very much, despite the malaria control program which started in the 1960s. One reason for this is that there are now more hospitals and more people who are sick with malaria can get to a hospital. Before, they just stayed at home. For example, from 1960 to 1967, the number of administration hospitals increased from 98 to 143 and the number of mission hospitals increased from 204 to 277. Another reason is that there are now many more technicians who can accurately diagnose malaria, so that people who come to hospital with malaria, are more likely to have their sickness correctly diagnosed.

Table 3.6 also gives the percent of all hospital admissions that were due to (diagnosed as) malaria. In Niugini in the early 1950s malaria accounted for about 15 or 16 % of all hospital admissions. In the late 1960s it accounted for only about 9%. This means that, compared with other diseases, malaria is not as important as it was. There has been a similar decline in the relative importance of malaria in Papua, but it is not as great. Malaria still remains an important cause of sickness in Niugini.

TABLE 3.6 Data on malaria cases in Administration hospitals 1952-1968. (From Annual Reports of Territory of Papua and Territory of New Guinea)

	PAPUA				NEW GUINEA			
Year	Number of Admissions for Malaria	Total Admissions for all kinds of sickness	% of Admissions due to Malaria	Number of Deaths Malaria	Number of Admissions for Malaria	Total Admissions for all kinds of sickness	% of Admissions due to Malaria	Number of deaths Malaria
1952–3	2307	19,881	11.6	35	12,155	76,343	15.9	117
1953–4	1882	21,298	8.8	41	13,134	79,863	16.4	127
1954–5	No Data							
1955–6	2786	27,091	10.3	47	15,080	96,483	15.6	166
1956–7	3853	29,060	13.3	55	15,747	97,040	16.2	149
1957–8	2657	23,300	11.4	61	11,579	91,467	12.7	132
1958–9	2759	22,239	12.4	57	14,631	88,815	16.5	169
1959–60	2487	23,220	10.7	41	12,502	82,588	15.1	110
1960–1	2269	23,478	9.7	39	10,117	77,136	13.1	105
1961–2	1754	20,427	8.6	20	8,192	76,239	10.7	109
1962–3	2168	25,060	8.7	26	8,632	82,668	10.4	93
1963–4	2149	30,769	7.0	27	8,253	83,968	9.8	81
1964–5	2325	25,436	9.1	29	6,412	73,529	8.7	79
1965–6	2082	29,729	7.0	34	7,573	84,721	8.9	90
1966–7	1740	30,377	5.7	24	7,510	92,045	8.2	113
1967–8	2628	31,119	8.4	19	8,539	92,357	9.2	109

In certain areas where malaria control has been effective for a long time, malaria has become much less important than it was. For example, in the Niugini Islands, the number of deaths from malaria dropped considerably. Here, the spraying of DDT started in 1959 and was supplemented by the distribution of the drugs pyrimethamine and chloroquin at the time of spraying. (In Niugini, pills containing these two drugs have been called Erone, (see Section 3.8)).

In the period 1952-57, 133 people in every 100,000 people in Niugini Islands died of malaria. In the period 1962-7, the figure was 17 per 100,000 people.

MALARIA AND ANAEMIA

Malaria causes the destruction of the red blood cells so that the concentration of haemoglobin in the blood is lowered. This may lead to anaemia.

We have already discussed this in Section 3.4. Table 3.7 gives some data which suggests that malaria is an important cause of anaemia in Niugini, but we must go over some other information before we discuss this.

Now, we have already mentioned that enlargement of the spleen follows malaria infections (see Section 3.4). This is so characteristic of malaria that surveys to determine the percentage of children with an enlarged spleen and the average amount of enlargement, are frequently used to measure the amount of malaria (see Section 3.8).

Spleens that cannot be felt are called size 0. Those that can be felt, are measured on a scale ranging from one to five. The largest are size five. In areas where the people are not constantly being infected with malaria, the size of the spleen depends on the severity of the attack and the time since the last attack. In areas where the people are constantly being infected, the size of the spleen depends a

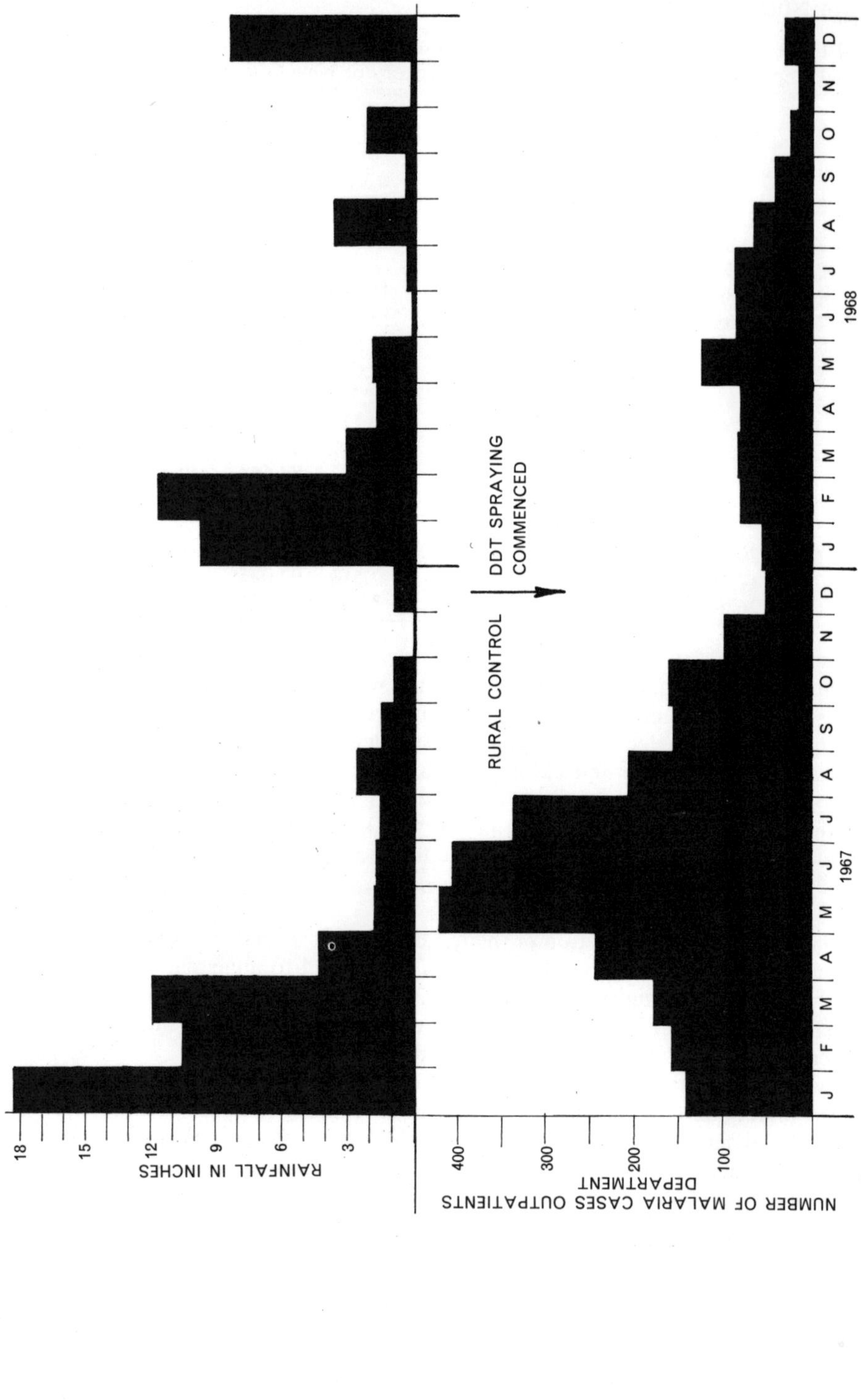

Fig. 3.13 Port Moresby General Hospital number of malaria cases and monthly rainfall

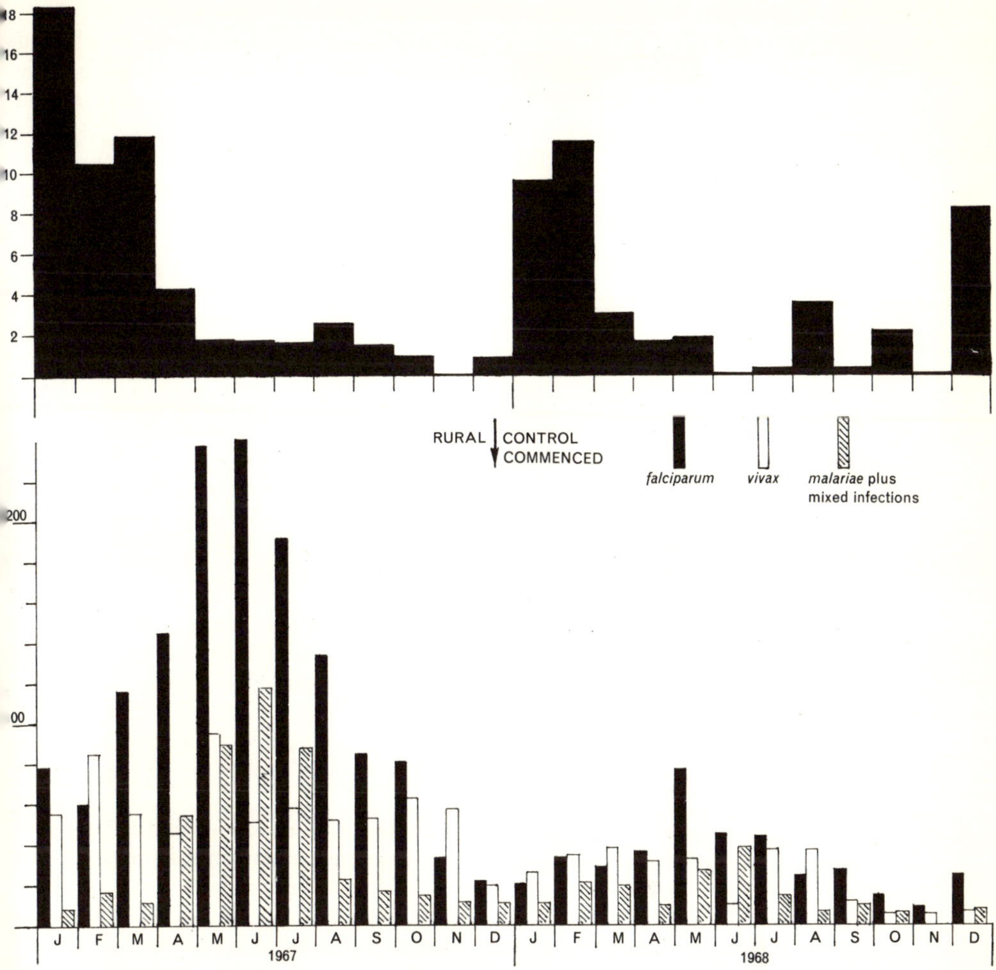

Fig. 3.14 Port Moresby General Hospital, malaria species distribution and rainfall, before and after spraying

good deal on age. The spleens of children are greatly enlarged, but they slowly become smaller after the age of nine or thereabouts as immunity develops.

Table 3.7 gives data on the haemoglobin values in people of different ages with different spleen sizes from the North Coast-Mainland of Niugini. Haemoglobin values are measured in grams of haemoglobin per 100 milliliters of blood. The table shows that people with enlarged spleens have lower haemoglobin values than people with normal (size 0) spleens. As a rule, the people with the largest spleens have the lowest haemoglobin values. Furthermore, the haemoglobin values of women who have started to menstruate (aged about 15 to 44) are usually lower than those of men of the same age (menstruation is the loss of blood from the female organs, which occurs every month).

The haemoglobin values of both males and females with size 3, or larger, spleens are about two grams per 100 milliliters lower than those with unenlarged spleens. This amounts to a difference of about 18 per cent. Because malaria is the only common cause of enlarged spleens in this country, this means that it is an important cause of anaemia.

TABLE 3.7 Haemoglobin values for people of different ages and with different spleen sizes from the North Coast-Mainland of Niugini (after Vines 1966)

SPLEEN SIZE	AGE							
	< 1	1–4	5–9	10–14	15–29	30–44	> 45	Ages
MALES								
0	9.3	10.2	10.6	11.4	13.5	12.2	11.3	11.7
1–2	9.2	9.8	10.6	9.6	11.3	11.3	10.0	10.6
> 2	6.7	8.5	9.2	9.4	11.3	10.6	9.7	9.6
FEMALES								
0	8.8	10.6	11.4	11.0	11.3	10.9	11.8	11.1
1–2	9.1	9.1	9.7	10.4	10.2	10.3	10.2	10.0
> 2	7.4	8.2	9.2	9.6	9.4	9.3	9.7	9.2

Anaemias resulting from malaria are often most severe in young children. This may seriously affect their growth and development. We will illustrate this with data from work done by Dr T. E. T. Spencer in the D'Entrecasteaux Islands. One survey done in 1957 showed that the average haemoglobin values of infants (under one year) suffering from malaria was 8.3 grams, compared with a value of 10.3 grams in infants who did not have malaria. Values for infected children, 2 to 4 years old, averaged 8.9 grams, compared with a value of 10.1 grams for uninfected children of the same age.

Data from another survey is given in Table 3.8. This shows that a high proportion of young children are anaemic. The incidence of anaemia decreases as the children get older and this coincides with a decrease in the incidence of malaria parasites and the development of immunity. Spencer suggests that the differences between males and females is probably due to the fact that the people value male children more than females and give them better food. If this is true, it is most unfortunate. It is just as important for the future of our society that girls and boys get the best food that is available.

TABLE 3.8 Incidence (%) of Anaemia in children in the D'Entrecasteaux Islands (after Spencer, 1967)

	Incidence (per cent)	
Age	Males	Females
< 1 yr	19	12
1–2 yrs	13	31
2–4 yrs	11	5
5–9 yrs	1	3
> 9 yrs	0	0

We discuss the importance of haemoglobin values in Niugini again in Section 4.3.

MALARIA IN THE HIGHLANDS

Through the centre of the island of Niugini there is a great range of mountains which divides the island into three regions—the North Coast, the Highlands and the South Coast. This range is from 50 to 150 miles wide and rises from a height of 5000 feet to more than 15,000 feet. The highest mountain in Niugini is Mt Wilhelm (14,762 ft). The Highlands are drained by a number of isolated valleys which run eventually to the North and South coasts.

Nearly half of the population of Niugini lives in the Highlands. Until the area came under the control of the Administration, the peoples of the area were divided into small language groups, which were continually in conflict with their neighbours. This state of affairs restricted the movement of people so that most of them never moved more than a few miles from their birthplaces.

The first European to get into the Highlands was a German called Detzner, who went there in 1918. The first Australians were Dwyer and Leahy who crossed Niugini from Lae to the Gulf of Papua in 1930. In 1933 Leahy and Taylor went through the Western Highlands, Chimbu and Central Highlands Districts. In 1935 Hides and O'Malley, and in 1938, Taylor and Black led expeditions into various parts of the Highlands.

As far as we know, none of them definitely reported malaria in the Highlands peoples. Perhaps in the future, an historian will re-examine their records in detail to see if they made any observations about this.

All of these explorers were accompanied by carriers and often policemen as well, who had come from malarious areas on the coast. They may have played an important part in introducing malaria into parts of the Highlands of Niugini. Some of their accounts tell of cases of malaria among members of the expeditions. Many of them suffered from relapses brought on by the change in climate and the hard conditions during the expedition. It is possible that they may have become infected while on their journeys but this seems less likely.

We do not really know whether or not malaria existed in the Highlands before the white man came. We shall see that the evidence suggests that it was probably present in certain parts of the Highlands, but was probably absent or very rare in other parts.

One of the first reports of malaria in the Highlands concerns Korn Farm. This farm was established during the war by the Army near Mt Hagen, in the Western Highlands. Vegetables were grown there and then flown to the coast. A number of native foremen were brought from the coast to work on the farm, but most of the workers were Highlanders who lived nearby.

In 1945, there was an outbreak of malaria among workers at the farm. Examination of blood slides in October 1945, showed that 13 per cent of them were infected with malaria. They had a spleen rate of 21 per cent.

This survey was made by Mr S. H. Christian, MBE. He spent much of his time during the next five years doing malaria surveys in the Highlands. (Most of the data are included in Fig. 3.15 to 3.19 and details of them are given in Appendix II. We are grateful to him for letting us gather these data from old reports.) He must have worked very hard. For example, during one month he and one assistant, examined 15,060 people for enlarged spleens and made 7809 blood slides and examined some of them for parasites. He must have walked many miles over bush tracks as well.

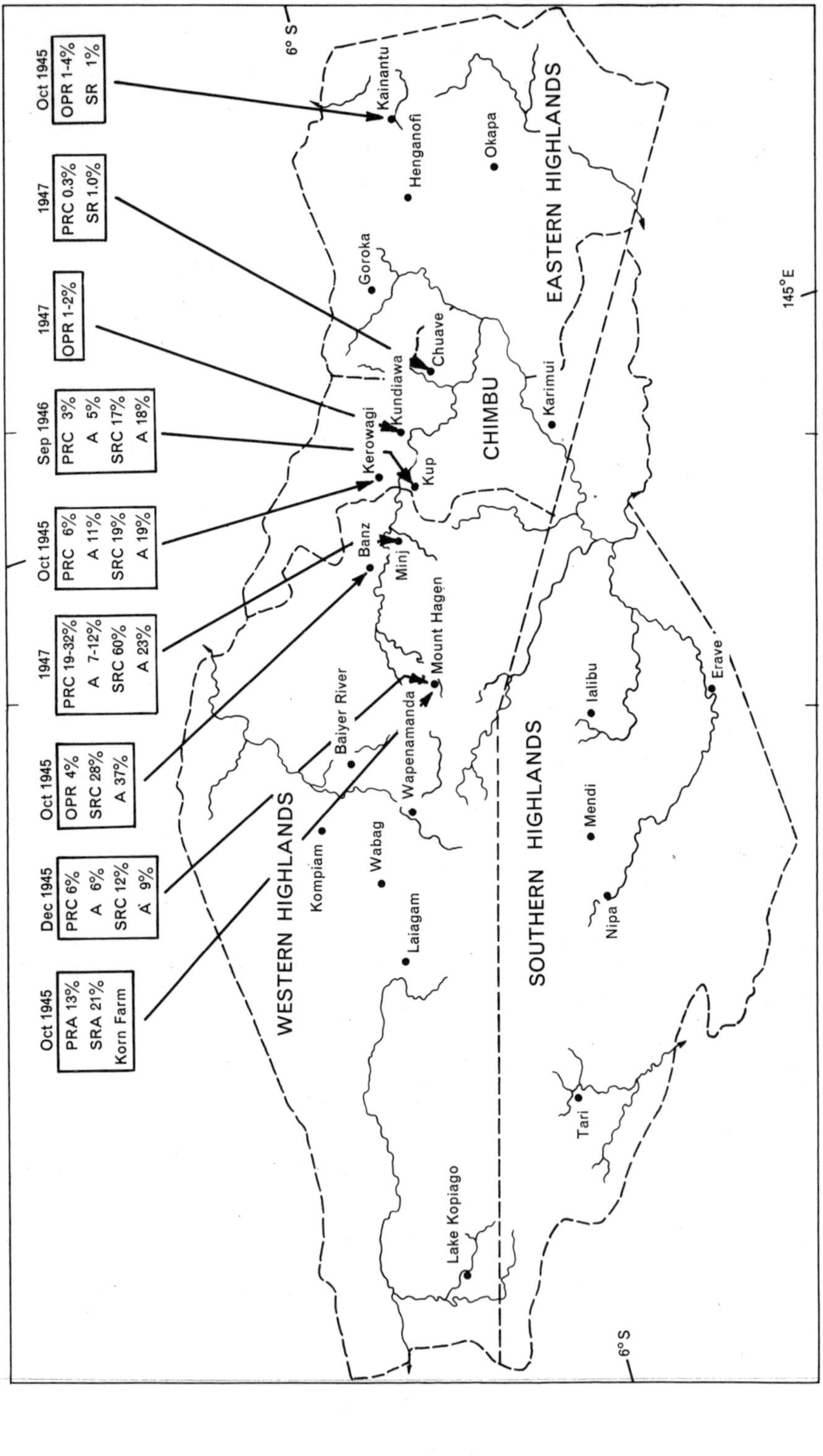

Fig. 3.15 Malaria in the Highlands 1945-1948

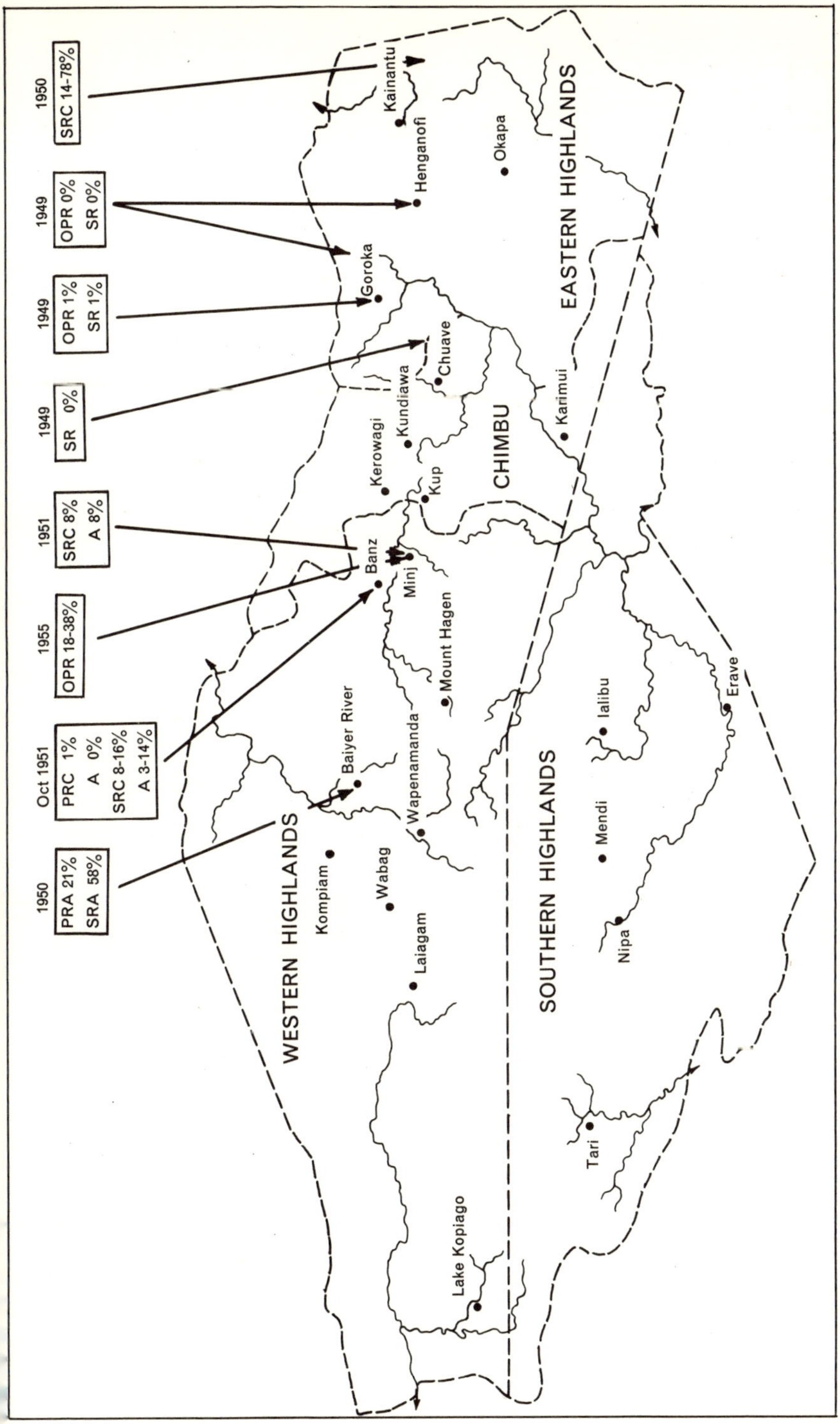

Fig. 3.16 Malaria in the Highlands 1949-1955

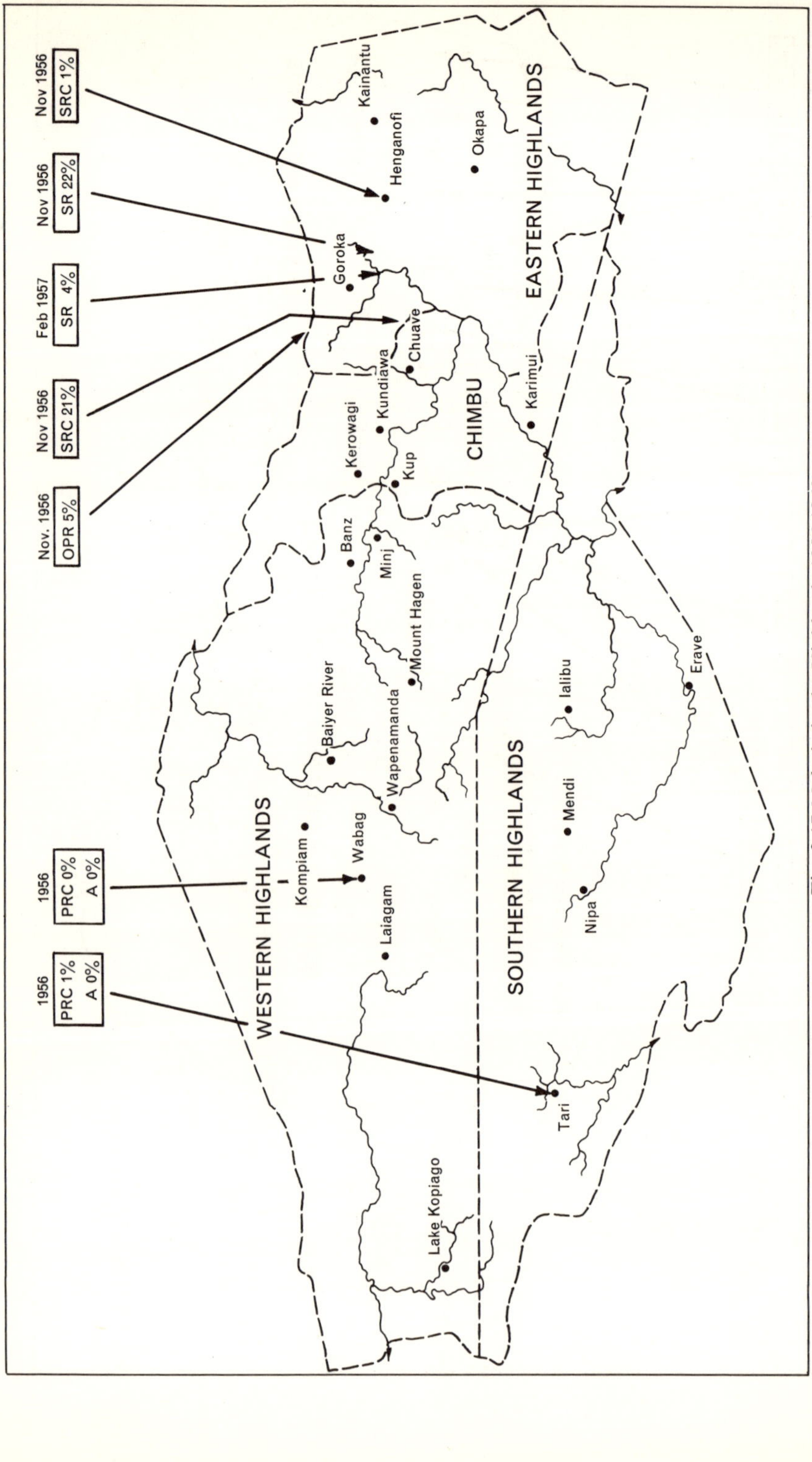

Fig. 3.17 Malaria in the Highlands 1956-1960

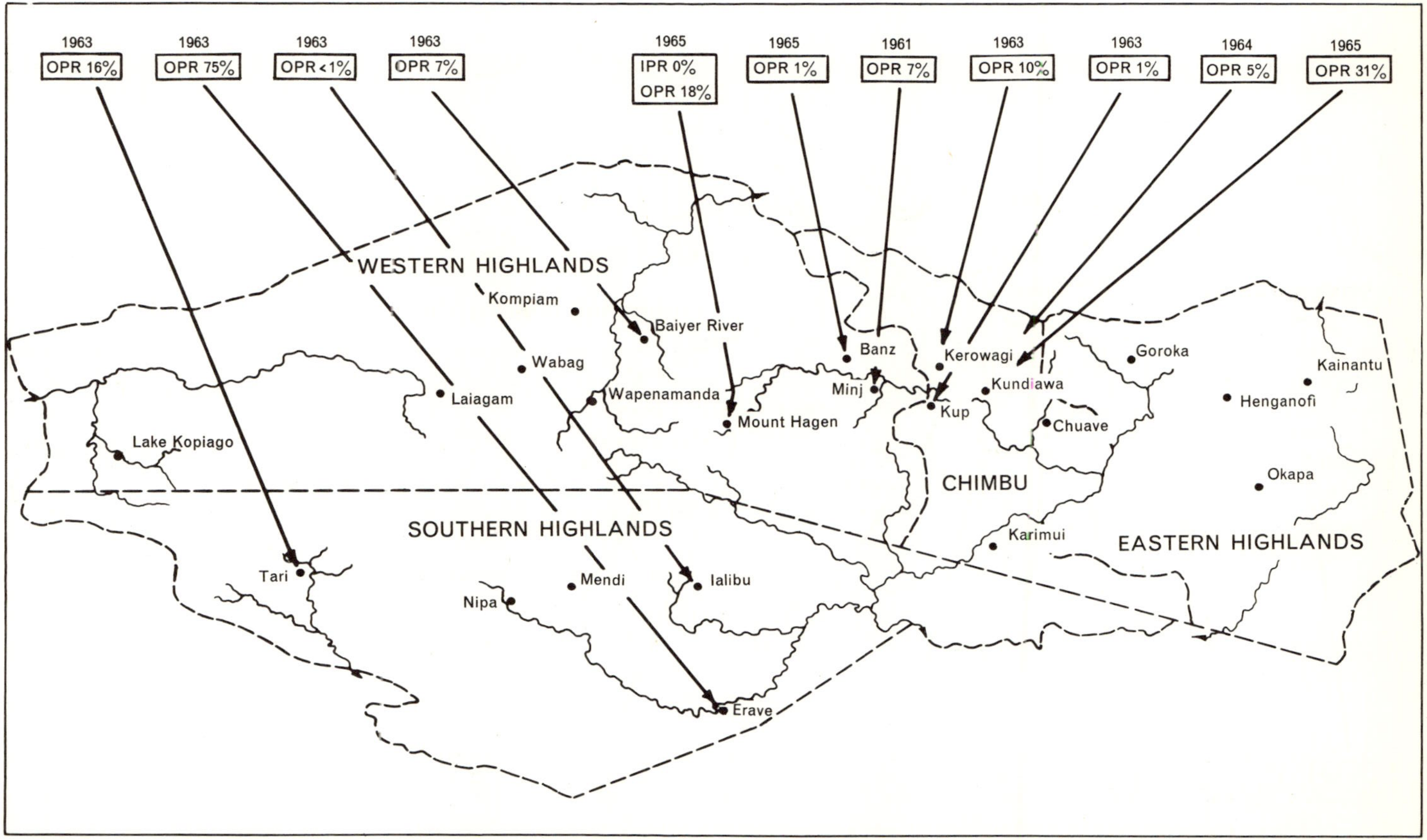

Fig. 3.18 Malaria in the Highlands 1961-1965

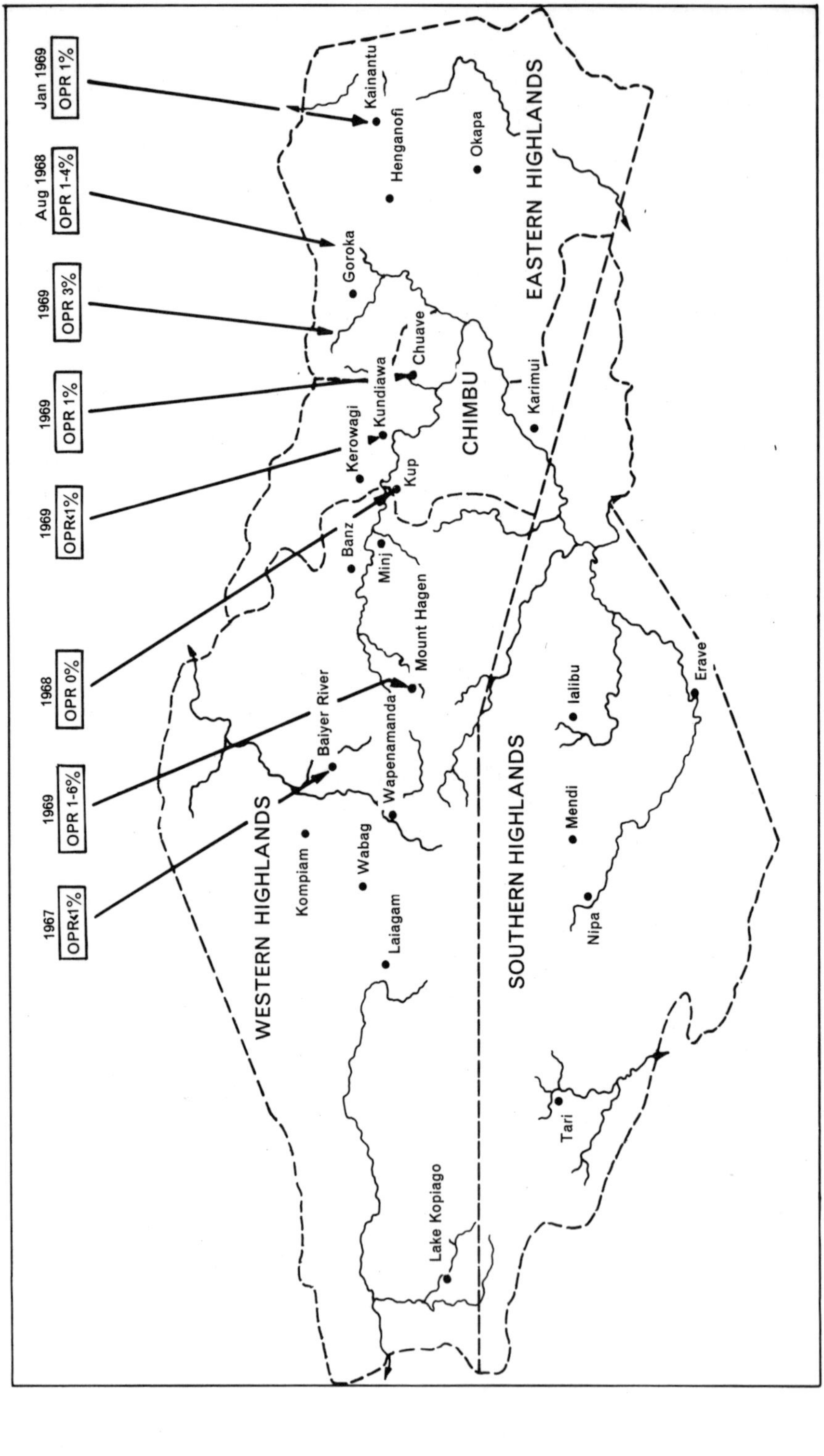
Jan 1969
OPR 1%
Aug 1968
OPR 1-4%
1969
OPR 3%
1969
OPR 1%
1969
OPR‹1%
1968
OPR 0%
1969
OPR 1-6%
1967
OPR‹1%
Kainantu
Henganofi
Okapa
EASTERN HIGHLANDS
Goroka
Chuave
Kundiawa
Kerowagi
CHIMBU
Karimui
Kup
Banz
Minj
Mount Hagen
Baiyer River
Wapenamanda
WESTERN HIGHLANDS
Kompiam
Wabag
Laiagam
Lake Kopiago
Erave
Ialibu
SOUTHERN HIGHLANDS
Mendi
Nipa
Tari

Malaria seems to have spread from Korn Farm to the villages in the Mt Hagen area. No doubt the local labourers had relapses when they went home and their infections were passed on to others who lived near them by vectors that were already present. Surveys done in 1945 showed a parasite rate of 6 per cent in both children and adults in the Mt Hagen area. The spleen rate was slightly higher, indicating that a higher proportion of the population had recently been infected.

The Wahgi Valley is a wide, flat valley to the East of Mt Hagen. Along its sides are steep mountain slopes. Professor R. H. Black, of the School of Public Health and Tropical Medicine, Sydney University, conducted a malaria survey in this area in 1954. He says that the Wahgi people have a traditional belief that bad spirits lived in the valleys near the rivers. A long time ago these spirits are said to have frightened the people from the floor of the valley to the mountain slopes along its sides. The spirits were said to be able to cause a sickness that is now recognised as malaria. Most of the people in the Wahgi valley lived on the mountain slopes until recent times.

The people of Minj (in the Wahgi) and nearby at Banz and further away at Kundiawa, have special words for malaria in their languages. It seems that people in the Baiyer River area and in the Northern Jimi Valley also had special words for malaria. These places lie to the north of the Wahgi Valley. In these places there is a folk history of epidemics of malaria and of the extinction of people who went to live in certain places at an altitude of about 2-3000 feet. In recent times, people from high altitudes have become infected with malaria when they have gone hunting Bird of Paradise in the lower Jimi River. In parts of the Wabag area, west of Baiyer River, the Enga and Ipili people also have a folk history of malaria in the lower valleys.

The existence of local words for malaria and a folk history of the disease suggest that it was probably present in parts of the Western Highalnds, prior to the coming of the white man.

In October 1945, Christian found an overall parasite rate of four per cent at Banz, in the Wahgi Valley. The spleen rate was much higher (28 per cent in children and 37 per cent in adults). This means that there had been an epidemic of malaria a few months previously.

East from Banz, at Kerowagi, Christian found a similar picture. In October 1945 the spleen rate was 19 per cent in both children and adults and the parasite rate was lower than this in both groups.

A little further east again, in the Chimbu area, it looks as if malaria may have been present for a long time. Dr Heydon found a spleen rate of 3.3 per cent in children at Mogay (near Chimbu) in 1934-5. He also found anopheline mosquitoes there.

In 1946, Christian found spleen rates of 17 per cent at Kup, a little south of Kerowagi and west of Kundiawa. In 1947, he found a low parasite rate (1 to 2 per cent) in the Chimbu district. He found little or no malaria east of Kundiawa to Kainantu (see Fig. 3.15; 3.16).

Thus, it seems that malaria was well established in many places from Kundiawa westwards to Wabag during the mid-1940s. But it was absent or very rare in the Eastern Highlands from Kundiawa eastwards to Kainantu. Spleen rates and parasite rates in these parts were seldom higher than 1 to 2 per cent, and often there were no signs of malaria at all. Many thousands of people were examined for enlarged spleens and many thousands of blood slides were examined for parasites. At Kainantu, (October 1945) Christian found a parasite rate of 1 to 4 per cent in certain villages. In others there was no malaria. In most villages none of the children had enlarged spleens. Christian suggested that malaria had spread from infected policemen (who came from the coast) to the nearby villages. Adults who came to work in the town of Kainantu probably took malaria back to their villages.

In 1950, Christian surveyed the people south east of Kainantu on the edge of the Highlands. He found a high spleen rate in both adults and children from the villages of Wompur, Koranga and Arau (see Fig. 3.16 and Appendix II). People from these villages regularly went to trade in the Markham Valley, which is highly malarious. He also reported that most of the people of Dampit and Mirir were said to have died of malaria and the few survivors had joined the people of Wompur.

Christian reported that vectors of malaria were present throughout the Highlands.

Thus by 1950, the situation was ripe for the introduction of malaria into the Eastern Highlands. It was already present in some of the villages around Kainantu and it was present in the Chimbu area, but the area in between was free of malaria. The people in this large area were highly susceptible because they had no immunity.

Great development took place in the Highlands during the 1950s. New airstrips and new roads were built. This kind of activity always creates breeding places for anophelines. Many coastal people came to the Highlands to work in the Administration and no doubt many of them had relapses of malaria. Coastal strains of mosquitoes may even have been brought into the Highlands by aeroplane or in trucks.

In 1950 the Highland Labour Scheme was started. Labourers went from the Highlands to work for eighteen months or two years, on plantations in other parts of Niugini. During the early 1950s most of these labourers came from the Chimbu area and the Eastern Highlands. In 1958 the scheme was extended into the Western Highlands. They were going from areas that were non-malarious to areas where there was a lot of malaria. The scheme required employers to give anti-malaria tablets to all workers while they were away from the Highlands. Despite this, many of these labourers returned to the Highlands with latent infections in their livers. A survey done in 1956 on labourers who had returned to the Highlands, showed that 19 per cent had enlarged spleens. On returning to the Highlands all labourers were given a course of drugs to kill any schizonts that might be in their livers. In the early years this treatment was probably not as effective as it was thought to be. The treatment was later changed so that it would be more effective. Relapses probably occurred in some of the labourers after they returned to their villages. These provided a source of infection in areas that had previously been free of malaria.

We have not been able to find out how many labourers went from the Highlands under the Labour Scheme during the period 1950 to 1960. About 10,000 people went in the first year, and in 1957, about 4000. So that perhaps 40-50,000 people were involved in the whole period. If only a few per cent of them had relapses after returning to their villages, their effect on the spread of malaria in the Highlands would have been very considerable.

During the early 1950s, epidemics of malaria swept through parts of the Western Highlands. It seems that malaria was worse then, than ever before in the memory of living people.

There was an epidemic in the Minj area in 1952, but no figures are available. Another epidemic occurred there in 1954, and over half of the patients admitted to the hospital had parasites in their blood. In 1955, malaria was bad again and the incidence in the area between Banz and Minj was 18 to 38 per cent. People who lived in the foothills of the Wahgi Valley were being infected as well as those who lived in the valleys.

By the mid and late 1950s, malaria had become established in the eastern Highlands in places that had previously been malaria-free. We do not know how many people died, but many thousands were probably sick, because the people had no immunity. The changing situation is shown in Figs. 3.15 to 3.19.

Malaria control started in the Wahgi Valley in the 1950s, but did not extend very far. In the early 1960s, projects were started to control malaria in other parts of the Highlands and now, happily, the parasite rate is less than 1 per cent in most of the Highlands. This great reduction in the amount of malaria has not come about without a great deal of effort. Almost every house in the Highlands, between Mt Hagen and Kainantu, is now sprayed with DDT at least once a year.

Malaria can only be kept at a low level in the Highlands if malaria control continues. There is now a good road from Lae (on the coast) which goes right through the Highlands. Highlanders are moving to and from malarious areas in increasing numbers and must be continually bringing infections back into their villages. If malaria control measures are relaxed we can expect epidemics of malaria to return to the Highlands.

Control has been more successful in the Highlands than in many coastal areas of Niugini. One of the reasons for this is that most of the transmission of malaria is restricted to a period of two or three months following the end of the wet season. This contrasts with the situation in many coastal areas where transmission probably occurs for six or more months of the year.

We will describe the seasonal changes in the amount of malaria and the number of vectors at Minj during the 1950s. There is little data available for other places in the Highlands, but the situation at Minj was probably typical of most Highland areas.

Most of the cases of malaria occurred in the dry season. This is illustrated in Fig. 3.12. which shows the number of cases of malaria at the Minj hospital each month from 1954 to 1959. In some years there were many more cases of malaria than in others. For example, in 1956, very few cases were recorded. In most years there were 80 or more cases per month, during May, June and July, or during two of these three months. This suggests that conditions are right for transmission during April, May and June.. Over the period 1955 to 1962, a few infected mosquitoes have been found during most months of the year, although in some years no infected mosquitoes were found for several months at a time. This means that transmission can occur throughout the year, but at a very low level. There are always a few cases of malaria in every month and we cannot tell if they are due to new infections or to relapses. However, because the number of cases is very high in May, June and July, it would appear that most transmission is restricted to the months of April, May and June. (Development in the mosquito takes at least 16 days and the development of gametocytes usually takes about two weeks after man has become infected, so that transmission must reach a peak about a month before the peak in new cases of malaria.

During the wet season (December to April) the number of vectors builds up to a peak. May, June, July and August are relatively dry months (with only 4 to 7 inches of rain per month). Vector mosquitoes are usually hard to find in July and August. Their numbers slowly decline as the dry season proceeds and their breeding sites dry up. As their numbers decline the amount of transmission decreases. This probably accounts for the decline in the number of cases of malaria that occurs as the dry season proceeds.

3.8 The Control and Eradication of Malaria

INTRODUCTION

The spread of malaria is like a chain in which one link connects the parasites that are living in man with the mosquito vector. And another link joins the parasites that have developed within the mosquito with other people. If we can break either of these links we stop the transmission of malaria. This can be done by:

(i) killing the parasites in man with drugs,

or

(ii) preventing mosquitoes from transferring parasites from infected to uninfected people.

If we can stop transmission completely we can eradicate malaria. If we reduce the amount of transmission we only control it.

The first large scale demonstrations of malaria control were carried out by Gorgas in Cuba and in the Panama Canal Zone between 1899 and 1914; and by Watson in Malaya from 1901 to 1927. They controlled malaria by draining the land so that the mosquitoes had nowhere to breed.

Malaria was a serious problem in tropical areas during World War II and special units were trained to control this disease. They used various methods; all servicemen had to take drugs and chemicals were used to kill the adult and larval mosquitoes.

In this chapter we will deal with the various methods, or techniques, that have been used in the fight against malaria. In the last thirty years some new techniques have been developed which have made it possible to completely eradicate malaria from large areas. This has led to a change in our way of thinking about this disease. In the past, malaria could be controlled; it still existed, but it could be controlled so that few people became sick from it. No one ever thought that it would be practical to completely eradicate it. With the new techniques that are now available, we know that this is possible. Malaria can be completely eradicated—wiped out forever.

THE DIFFERENCE BETWEEN CONTROL AND ERADICATION

People have become confused about the differences between malaria control, mosquito control, and malaria eradication.

Malaria control is a continual process which reduces the amount of malaria in the community. Many methods can be used to achieve this. One method is mosquito control. This aims at reducing the number of mosquitoes so that there are not so many available to transmit malaria.

Malaria eradication is the complete removal of malaria from a country. This can only be done if there is a well planned Malaria Eradication Program, which is properly carried out by trained people who are enthusiastic about their work. It must use methods that have been proved to be successful. It must have enough money from the government and it must aim at completing the job of eradication within five years.

We will now discuss the differences between malaria control and malaria eradication in some detail.

(i) The objective or aim of a control program is to reduce the amount of malaria sickness. Such programs are usually considered to be successful if malaria ceases to be an important public health problem. We could even set a more definite objective by saying that our aim is to reduce the number of cases of malaria so that they account for less than 1% or even 0.1%, of all hospital admissions.

The objective of an eradication program is to completely stop the transmission of malaria.

(ii) We measure the success of control and eradication programs in different ways. Control is successful if there is a reduction in the amount of malaria. We can measure this by parasite surveys, or by measuring changes in the incidence of enlarged spleens, or by the number of malaria cases being admitted to hospitals.

Malaria eradication is successful when no more malaria parasites can be found in the blood of people. This means that we must examine the blood of all people with fevers. One undetected case of malaria can soon lead to many new cases, a lot of sickness and some deaths.

(iii) A control program is usually employed in areas where there is a lot of malaria. Places where there is only a little malaria are often neglected. However, people from these neglected places may introduce malaria into other areas where there is no malaria.

An eradication program must completely cover the whole country.

(iv) A control program goes on forever. It continually costs money. Some people are always sick and they will always give rise to new infections.

An eradication program has been successful when all malaria transmission has been stopped for a period of three years. The Public Health Department now takes over the duty of watching and treating any relapses, or cases of malaria that are introduced from other countries. An eradication program costs a lot of money, but it is a great investment. Most of the money is required in the first few years. Once malaria has been eradicated a little money must always be spent to make sure that malaria does not become established again.

(v) An eradication program requires a stable government that is determined to provide the money and other things that are necessary to get the job finished.

(vi) An eradication program requires the co-operation of every one within the country. They must all want to see malaria eradicated. They must be told why it is important, and they must be kept informed of the progress of the campaign.
(vii) An eradication program requires perfection. There is no place for lazy people. The supervisors must be able to sack anyone who does not work properly. All workers must be well paid for the job that they do.

The World Health Organisation (WHO) of the United Nations Organisation, has laid down detailed procedures for Malaria Eradication Programs. If these can be followed it should be possible to eradicate malaria throughout the world.

Any anti-malaria program which does not follow WHO procedures is probably only a control program, or it soon becomes one. Some countries think they cannot afford the money for a malaria eradication program, so they have a control program.

With control programs, continual reassessment of the program is necessary. We should continually ask ourselves:

What are we doing?
Why are we doing it?
Is it worth doing?
Can it be done better?
Can we afford *not* to have an eradication program?

THE STAGES IN A MALARIA ERADICATION PROGRAM

The World Health Organisation has laid down five stages for a Malaria Eradication Program. They are:

(i) *The Pre-Eradication Stage*

During this stage surveys are carried out throughout the country to determine the incidence of malaria parasites in the blood and the incidence of enlarged spleens. Both measurements are necessary. Together they show how much malaria is present, and the amount of immunity that has developed to it.

Pilot projects are set up to find out which techniques are effective under local conditions. These projects also provide experience and training for malaria workers.

It is important that this stage is completed before the next stage is started.

(ii) *The Preparatory Stage*

Detailed maps must be made showing the position of every house, in every village or clan line. Every house must be numbered and records made of the number of people living there. Examples of these maps are given in Fig. 3.20 to 3.22.

The government must provide the necessary money for the whole program. Men must be trained for the work ahead and all the necessary materials must be bought and stored in readiness. Offices and laboratories must be established and equipped. Vehicles, spraying equipment, insecticides, spare parts and a thousand other items must be bought. A detailed plan must be made for every district showing what is to be done and when.

All this must be done within one year.

(iii) *Attack Phase*

This lasts from three to five years.

The pilot projects will have shown which techniques are successful and these are now put to work. Most programs, so far, have depended on the regular spraying of all buildings with DDT or other residual insecticide, to stop the transmission of malaria within three years.

Maps that were made during stage (ii) are continually updated and the success of the campaign is continually measured by surveys in selected villages.

When the incidence of malaria in all age groups falls below 2%, these surveys cannot accurately measure the incidence of malaria. Special techniques must now be used. They are called surveillance (sir-vey-lance) mechanisms. When somebody is under surveillance they are being watched closely, because they are not trusted. This is exactly the situation with malaria. It has been successfully eradicated and a close watch must be kept to make sure that it does not get away again. It cannot be trusted.

Surveillance consists of two kinds of watching which are referred to as Active Case Detection and Passive Case Detection. (It is unfortunate that Malariology is infected with a heavy load of cranky terms. Some of them are too long, but we use them because they have been established by WHO.)

Active Case Detection means visiting EVERY house, EVERY fortnight and making a blood slide from EVERY person with a fever. These slides are then quickly examined by technicians to see if they are positive, that is, contain malaria parasites. All positive cases are treated with drugs. A list is kept of the names and addresses of all people that are positive and they are checked again on the next visit. The malaria visitor must ask four questions:

(i) Has anyone in this house had fever since my last visit?
(ii) Has anyone been sick?
(iii) Have you been away to another place since my last visit?
(iv) Has anyone from another place been to stay with you since my last visit?

The answers help trace every case of malaria to its origin and identify the remaining areas of malaria transmission.

If any malaria is found in a place, malaria workers are sent to respray and/or issue drugs in that place.

Passive Case Detection is a similar kind of watching which is done in all schools, hospitals, government departments, mission stations and village aid posts. Every case of fever must be reported and checked to see if it is due to malaria. Passive Case Detection requires the support of the whole population. Everyone in the country must co-operate in surveillance, to eradicate malaria.

Surveillance workers must be supervised continually. It is most important that they do their job properly.

At the end of the attack stage, people will have lost their immunity to malaria and infection will cause fever. Thus these surveillance mechanisms should detect all cases of malaria in the community.

When the incidence of malaria, as measured by surveillance, has dropped to 1 person in 10,000 per year, spraying is stopped and the attack stage is over. The links in the malaria chain have now been torn apart.

(iv) *The Consolidation Stage*

Surveillance mechanisms are continued to detect any new cases of malaria. If, after three years of consolidation, no new cases of locally transmitted malaria are found, the program enters the next stage.

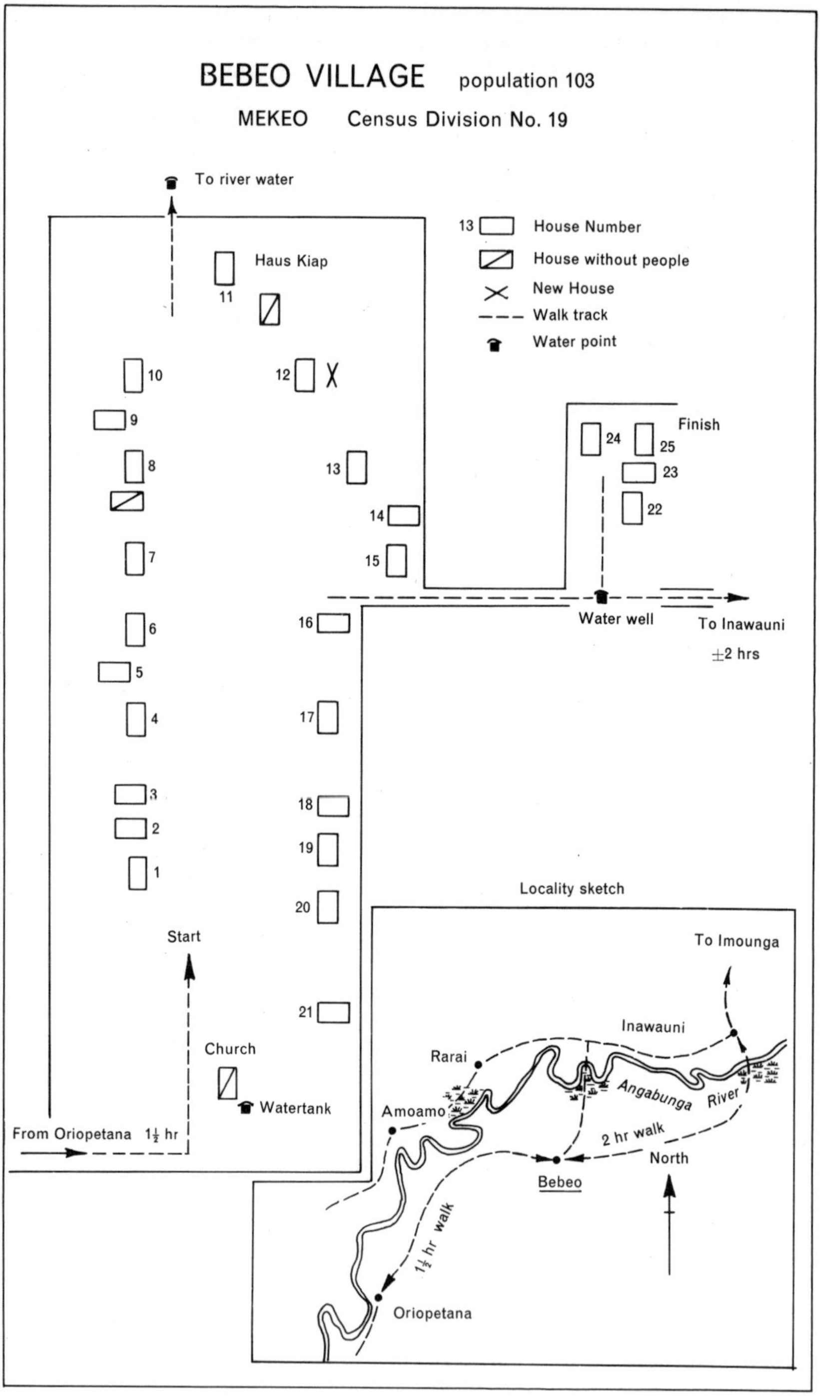

Fig. 3.20 Village map

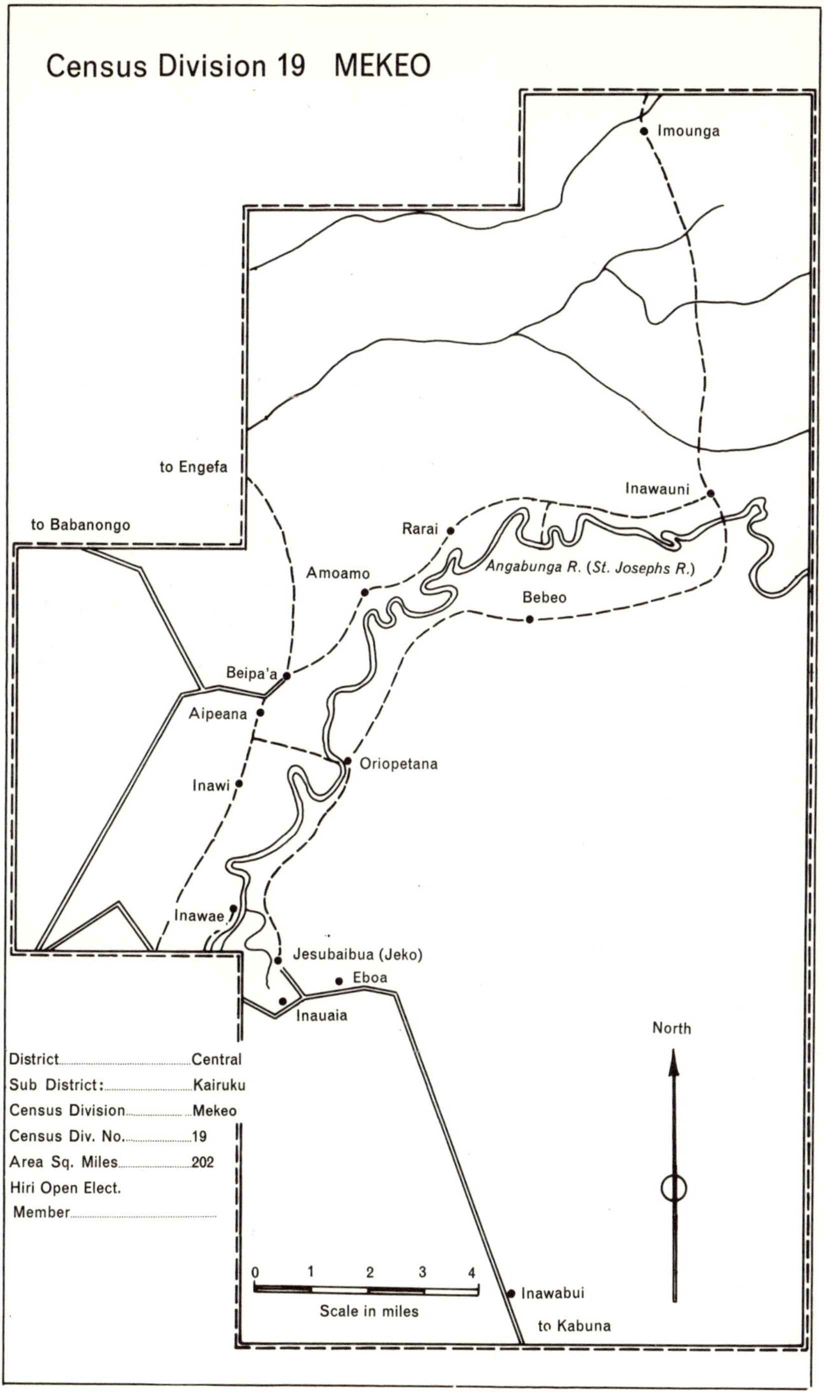

Fig. 3.21 Census Division sketch

During consolidation there is no widespread use of insecticides or drugs. However, in most countries there will be a few outbreaks of malaria which start from relapses or infections that have been carried in from a malarious area. These cases will be found by surveillance mechanisms.

As soon as a new case of malaria is found, an epidemiologist is sent to find out where the malaria came from. He will then organize spraying or give drugs to prevent further fevers and relapses.

These small pockets of local transmission are treated like something very evil. They must be stopped before they spread and a careful watch must be kept, because these areas are not to be trusted.

(v) *The Maintenance Stage*

Malaria has now been eradicated. Most malaria workers are trained for new work. A small force is kept to deal with any emergency. The public health service must now keep a careful watch forever, to make sure that malaria never becomes re-established. If malaria is allowed to sneak back unnoticed it can quickly start an epidemic.

THE ERADICATION OF MALARIA FROM THE WORLD

The World Health Organisation aims to eventually eradicate malaria from the world. They have developed detailed procedures which, if followed, will achieve this aim.

However, there are many difficulties. Two of the most important are a lack of money and political instability.

Many countries think that they cannot afford an eradication program. In the long term view, they cannot afford not to have one. The rich countries must help the poorer countries to eradicate malaria.

Political stability is most important. Once an eradication program is started it must continue until malaria is eradicated. There is no hope of eradicating malaria in Vietnam until the war ends and the country becomes politically stable. Indonesia provides an example of a program that failed. Their program was financed by the World Health Organisation and aid from the United States of America. The Government of Indonesia withdrew from membership of the United Nations and decided to work on its own. Overseas aid was then withdrawn. Their malaria eradication program collapsed and will have to start all over again.

However, some countries have eradicated malaria completely or are very close to it. For example, some small tropical American countries, the United States of America, parts of Western Europe, Italy, Taiwan, Australia, Israel and Guyana. Fig. 3.22 shows the picture of malaria in the world. You will see the great task that lies ahead before malaria is eradicated. Malaria remains the most important disease in tropical countries.

The situation in countries near us is as follows: The Solomon Islands has started an eradication program. They have a population of about 150,000 people and it should not be a difficult job. They have WHO advisors helping them.

The Philippines and Malaysia have both started malaria eradication programs, with help from WHO. The Philippines gets extra help from the United States of America. Both countries have previously had control programs.

Indonesia has a population of about 120 million. Although just over half the population lives on the island of Java, the rest of the population is spread over more than 6000 inhabited islands. A malaria control program was started in 1952 and changed to an eradication program in 1959 with aid from WHO and the United States of America. In 1958, it was estimated that there were 30 million attacks of malaria a year causing a loss of 250 million working hours per year and about 120,000 deaths per year. Difficulties started in 1965 when aid was withdrawn due to politics. In many areas the vectors have developed resistance to DDT and other insecticides. This has happened partly because DDT was used extensively to control pests in rice fields. The larvae of mosquito vectors breeding there were exposed to small amounts of DDT which did not kill them and they became resistant to it.

Malaria has been eradicated from Australia. Public Health authorities keep a careful watch to see that transmission does not start again. Malaria was never a serious problem in most of Australia although epidemics used to occur in the northern coastal areas.

We have been talking about the control and eradication of malaria and the great task that lies ahead. We will now discuss the techniques that can be used to control and eradicate malaria. The time has come when we must re-assess the techniques that can be used and try every method that might be useful.

In this country, many village people are objecting to having their houses sprayed with insecticides. As the control program in Niugini is based on DDT, it is time we had a good look at other methods of control to see which of them might be used here.

We will discuss these methods under the following headings:

I. ACTION AGAINST THE LARVAE
(a) methods which destroy mosquito larvae
(b) methods which make water unsuitable for breeding.

II. ACTION AGAINST THE ADULT MOSQUITO
(a) insecticides
(b) methods which prevent mosquitoes feeding on man.

III. ACTION AGAINST THE PARASITES IN MAN

I. Action against the larvae

The vectors of malaria in Niugini breed mostly in temporary shallow waters that are exposed to sunlight. Many of their breeding places are man-made, for example, tyre ruts, footprints, hoofprints, roadside drains, borrow pits (these are holes left where soil has been removed or borrowed, for road making or gardens), canoes filled with water and pig wallows.

Every member of the community should be aware of the danger of man-made breeding sites. If we took care not to make new sites and to fill in small temporary waters, the numbers of vectors might be considerably reduced. However, it would be an almost impossible task to eliminate all breeding sites because every shower of rain makes more.

(a) Methods which destroy mosquito larvae

Larvacides are chemicals which are added to the water in which mosquitoes breed in order to kill them.

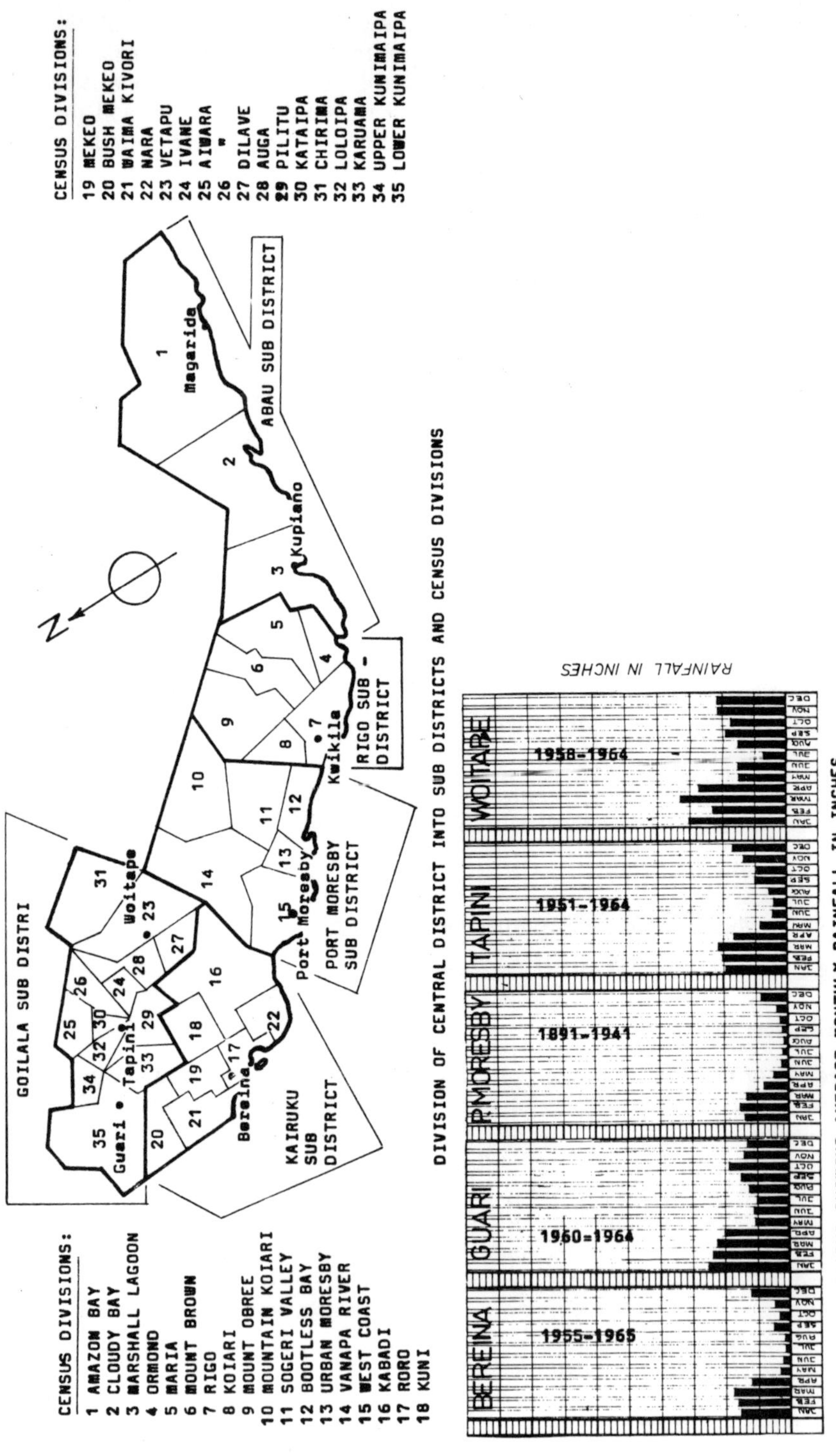

Fig. 3.22 District location map and rainfall

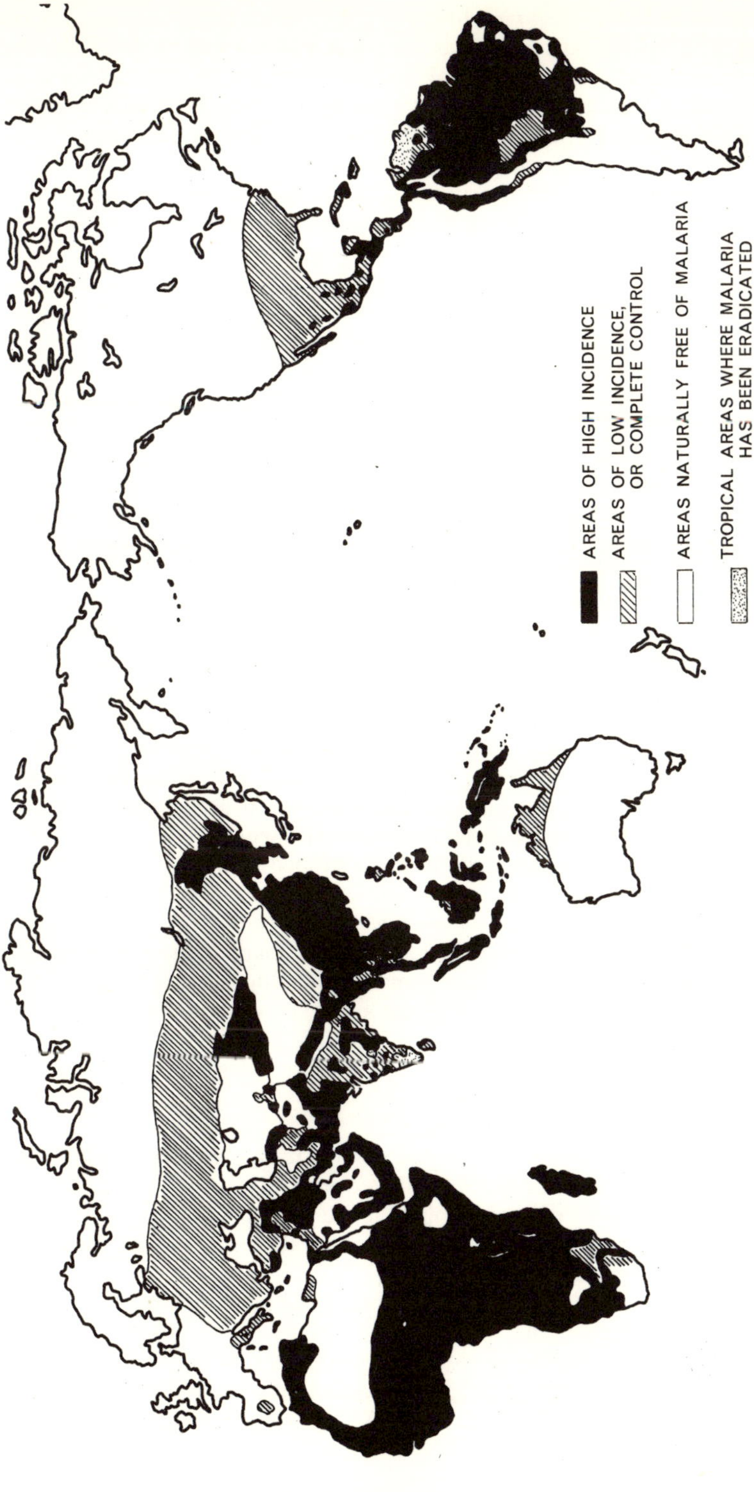

Fig. 3.23 Geographical distribution of malaria

Before World War II, larviciding was the main method of mosquito control. This method has the advantage that it acts at a time when many larvae are concentrated in a small body of water. Large numbers can be killed at the same time with one lot of larvacide. As the adults emerge, they fly in all directions and are thus harder to control, because they are scattered over a wide area.

The two most common larvicides are oil and Paris Green. When oil is added to water it spreads in a thin film over the surface. When the larvae and pupae come to the surface to breathe, the oil blocks up their breathing organs and they die. Petroleum oils are also toxic to the larvae. Distillate (or diesoline) and old sump oil that has been discarded from motor engines, can be used as larvicides. Distillate is better than sump oil. The effectiveness of both is greatly increased if 1% of vegetable oil (for example, peanut or castor oil) is added.

Oils can be sprayed onto water with various kinds of pumps. However, oils can be used without expensive equipment. For example, a bottle full of sump oil with a piece of cane or pitpit stuck in the neck can be hung upside down over a breeding site. The oil will slowly drip out. Permanent drips can be made by punching a small nail hole in the bottom of a four gallon tin and inserting a piece of rag as a wick, so that one drop falls about every minute. The tin is then hung over the water. This method is good for slow moving streams and boggy ground.

Another good method is to fill an old bag with grass, saw dust or wood shavings. It is then soaked in sump oil and anchored on the bottom of a pond, with stones. The oil will keep coming up for months and kill mosquito larvae.

DDT and other such insecticides should not be used as larvicides because the mosquitoes are likely to become resistant to them.

Paris Green (copper-aceto-arsenite) is also used as a larvicide. It is applied as a fine powder to the surface of breeding sites and is eaten by *Anopheles* larvae, which then die. It does not kill fish, cattle or pigs if used in the right amount. (0.35 ounces per 100 square meters of water surface).

Other chemical larvicides have been developed, but they have not yet been thoroughly tested in Niugini.

(*b*) *Methods which make water unsuitable for breeding.*

The larval stages of mosquitoes that breed in small streams may be killed by flushing. The stream is dammed and once a week the water is released suddenly. This flushes mosquito larvae and pupae away. This method only works if there is enough water in the dam to make a big wave.

There are many simple ways of making a place unsuitable for larvae. We will list a few examples:

(i) Continual changes in water level

(ii) Removal of shade if the vector breeds only in shady places, and the planting of shade plants if the vector breeds only in open places. All the vectors in Niugini breed in small shallow ground pools that are exposed to the sun.

(iii) The introduction of fish which eat the larval stages, for example *Gambusia affinis* and *Tilapia mosambique*.

The best and most permanent way of removing breeding sites is by filling low-lying ground or draining it.

In 1950-53 Mr S. H. Christian, MBE, started the Wahgi Valley Pilot Drainage Scheme. Several miles of deep drains were dug in the area around Banz, in the Western Highlands. The water table was lowered about eight feet. This not only reduced mosquito numbers considerably, but also provided a lot of new land for gardens. Since then tea and coffee plantations have been developed in this area.

There are many other areas in Niugini where drainage could be used to control mosquitoes. If the drains are well made they need little maintenance. No drainage project should be started without careful planning, and complicated schemes need the advice of an engineer. The National Malaria School, Kundiawa, has prepared an illustrated guide on drainage which is available to anyone who asks for it.

Nowadays, much less emphasis is placed on the control of mosquito larvae. The use of long acting (residual) insecticides has made it easier to control adult mosquitoes, than their larvae. However, we feel there is still an important place for drainage and the elimination of man-made breeding sites in malaria control. The people of a village could get together and drain mosquito breeding sites near their village. Oiling has never been widely used in Niugini, and it should be tried by local government councils. Used oil from motors costs nothing. There are probably many places in Niugini where such action would play an important part in malaria control.

II. Action against the Adult Mosquito

(*a*) *INSECTICIDES*

Residual insecticides are insecticides which continue to kill insects that come in contact with them several months after they have been sprayed onto surfaces.

When we talked about the ecology of the vectors of malaria, we said that after a blood meal the vectors usually rest on a nearby wall or surface inside houses. This means that they can be killed if the resting surface has been sprayed with a residual insecticide.

At present the best insecticide is DDT. It is best because it is the cheapest available, it kills mosquitoes at concentrations that are harmless to man and has been proved to retain its killing power for 4 to 7 months under most Niugini conditions.

DDT is a chemical called Dichloro-di-phenyl-trichloro-ethane. It is one of a group of insecticides called the chlorinated hydro-carbons. It was first found to be an insecticide in 1939 by Dr. Paul Muller, a chemist from Switzerland. He was later given a Nobel Prize because his discovery has been so important in the control of insect pests.

DDT is a fine white powder. Before it is sprayed on surfaces, it is thoroughly mixed with water. After spraying, the water evaporates and leaves tiny particles of DDT evenly spread over the surface. When a mosquito rests on the surface, tiny particles of DDT stick to its feet and kill it. The particles of DDT must be no bigger than 30 to 40 microns in diameter or they are too big to fit into the spaces between the hairs at the end of the mosquitoes feet. DDT passes through the chitinous skin of the mosquito and poisons the nerve cells. The following things then happen. The mosquito becomes excited and makes a number of short, dashing flights without settling down for very long; the muscles in the legs and wings twitch; the whole body shakes vigorously and the mosquito falls down and continues to shake. The legs and wings become paralysed and stop moving. Gradually the shaking stops. The insect is dead.

No mosquito here in Niugini has ever been seen to recover after the complete pattern of actions has taken place. Vectors in Niugini die after resting on DDT for only 2 to 4 minutes.

Entomologists have done experiments on vector mosquitoes which show that two grams of DDT must be sprayed on each square meter of surface in order to kill the vectors that rest on it.

Residual insecticides do not last for ever. Surfaces must be sprayed about every 4 to 7 months. The following factors influence how often spraying needs to be done.

(i) The time of transmission. In areas where transmission goes on throughout the year (for example, the wet coastal areas of Niugini), surfaces must always be covered with active DDT. In areas where transmission only goes on for a few months at the end of the wet season (for example, in the Highlands and dry parts of the Papuan coast), surfaces only need to be sprayed once a year, before the transmission season begins.

(ii) The amount of smoke, salt from sea air, or dust, that covers the DDT and makes it inactive.

(iii) How quickly the DDT is removed by people in the houses. Some DDT is accidentally rubbed off on people's clothes and skin. A few misguided people actually wash the walls of their houses soon after the spraymen have left. They do not realise that this action may lead to serious sickness, or even death, within their own family.

(iv) DDT gradually decomposes or changes, to an inactive chemical. This process is probably speeded up by warm temperatures and the action of sunlight.

The importance of the above factors will vary from area to area and the effectiveness of DDT must be determined by special tests in each area.

The concentration of DDT needed to kill anopheline vectors, probably also varies from place to place and tests must be done to determine the best concentration to kill the mosquitoes in different parts of the country.

We will now give an example of the effectiveness of DDT against the vectors of malaria in Niugini.

Mr H. van Leeuwen of the Malaria Service, Kainantu, has kindly provided us with details of the following experiment which shows the effect of spraying houses with DDT on the numbers of anophelines.

Wankum is a village in the Markham Valley, about 900 feet above sea level. It has a population of about 4000 people.

In July, 1968, before any malaria control had started, the incidence (OPR) of malaria at Wankum was 48% ,and 70% of all infants were infected.

In 1967, two houses, A and B, were built in the village especially for this work. They were similar to other village houses. From July 1967, onwards, the number of mosquitoes caught in these houses per unit time, was recorded. Separate data was kept for anophelines and culicines.

In July and November, 1968, and again in March, 1969, all village houses were sprayed with DDT. House A has never been sprayed, house B was sprayed in July, 1968 and again in March, 1969.

Fig. 3.24 shows the number of anopheline mosquitoes caught per unit time, each month. It shows that very few anopheline mosquitoes can be caught in areas that are regularly sprayed with DDT.

Other examples from Maprik and Kundiawa are given in Figs 3.25 and 3.26.

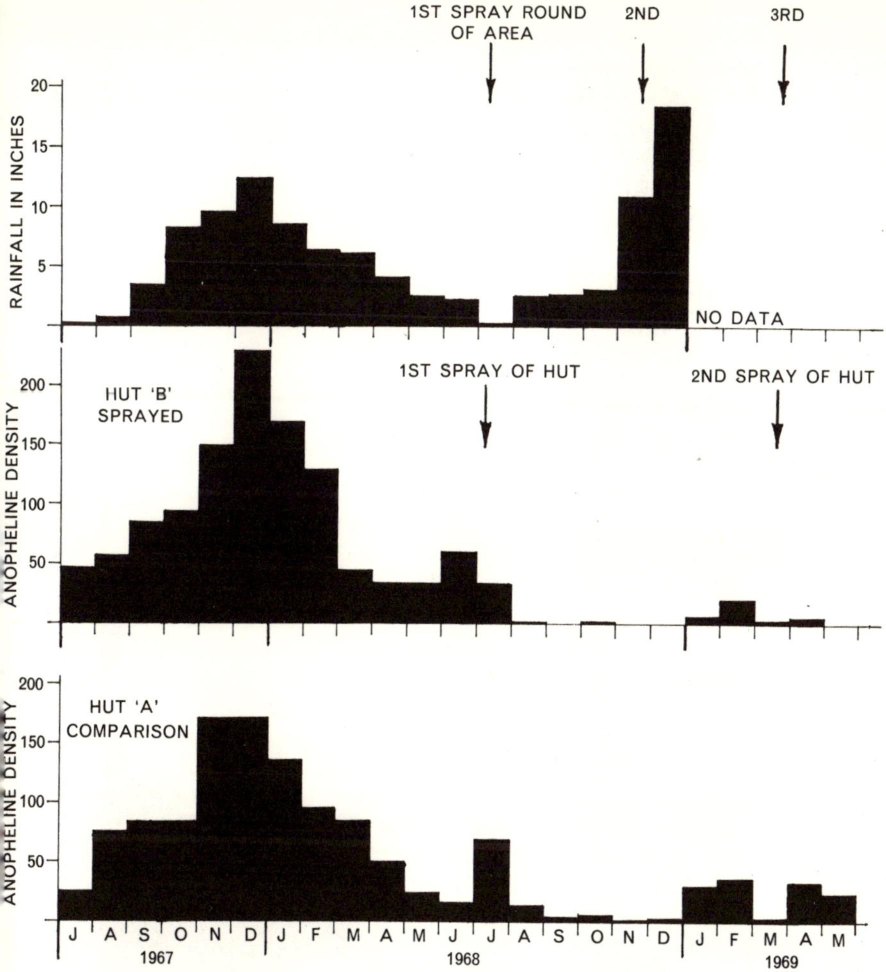

Fig. 3.24 Anopheline density at Wankum before and after spraying DDT

Other residual Insecticides

There are a number of other residual insecticides that can be used to control malaria. Most of them are more expensive than DDT and some of them are toxic or poisonous to man. Some of them are being used in anti-malaria programs in other countries because the mosquitoes have developed resistance to DDT, that is, the mosquitoes are no longer killed by the usual concentrations of DDT.

Other Chlorinated Hydrocarbons

Dieldrin is unsuitable because it is about five times as toxic to man as DDT. In addition it soaks into sago thatch so that very little remains on the surface. It is also five times more expensive than DDT.

Benzene-Hexachloride (also called Gammexane, Lindane, BHC or HCH) lasts less than three months. It costs about the same as DDT and is less toxic.

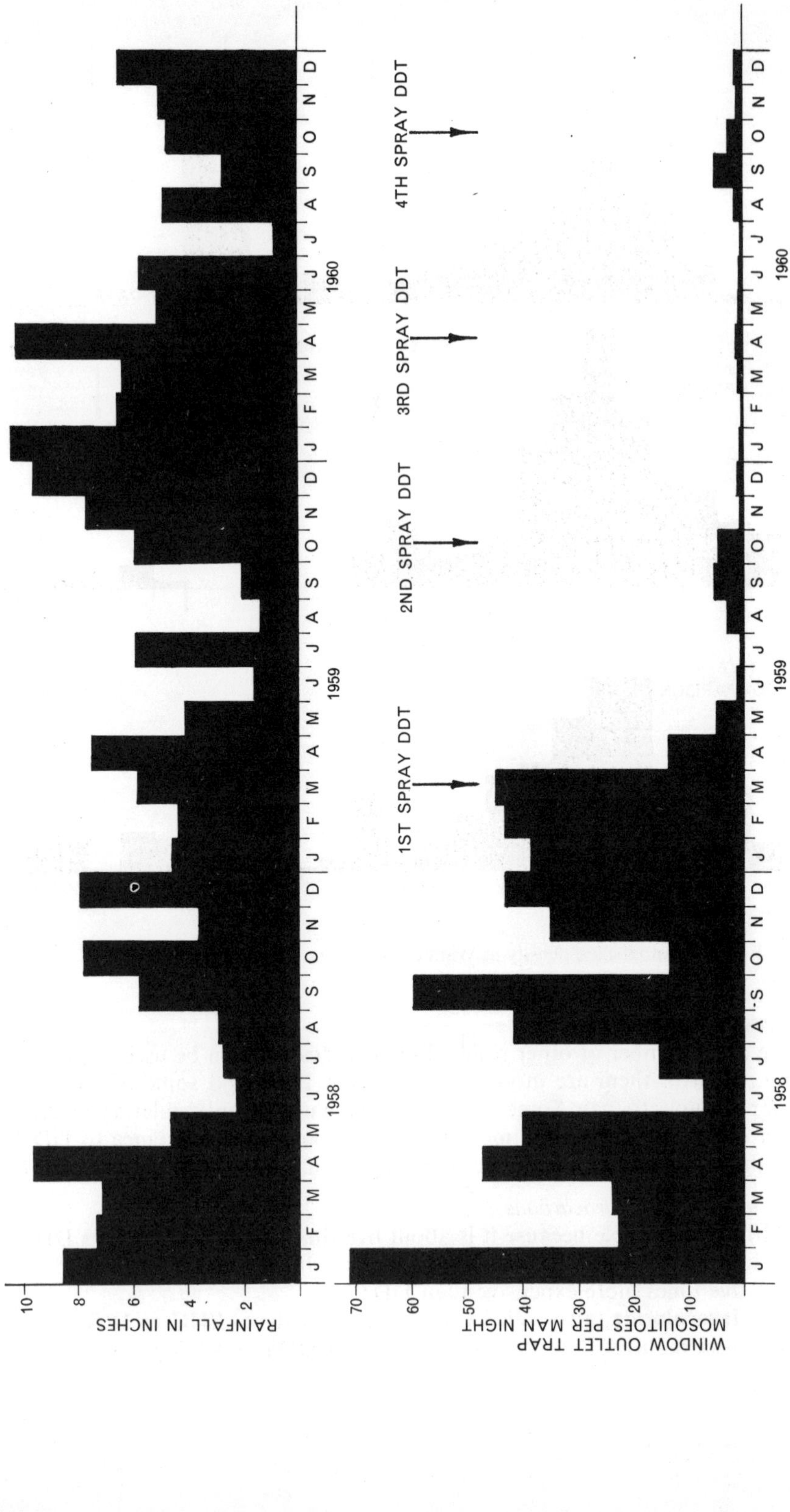

Fig. 3.25 Monthly rainfall and mosquito vector density at Maprik before and after spraying

Chlorinated Naphthalene Hydrocarbons

Aldrin and Endrin are two insecticides of this group, but they are very toxic.

Organo-Phosphates and Carbamates

These groups of insecticides are expensive and often toxic to man.

Malathion is one of the organo-phosphates, but it is not very toxic. It is now being used in malaria programs where the mosquitoes have become resistant to DDT.

Pyrethrins

Some insecticides have little residual effect, but when suspended in kerosene, diesoline or water, and sprayed out in tiny droplets, they knock down insects and kill them. The insecticides are absorbed through the skin of the insects.

Pyrethrins are often used in this way. These chemicals are extracted from chrysanthemum flowers which are grown in the Highlands of Niugini. The sprays contain less than 0.2% pyrethrins. They are not toxic to man or other vertebrate animals.

Pyrethrins cannot be used as residual insecticides in houses that are well lit because they decompose in the presence of sunlight. However, they last for 4 to 6 weeks in houses without windows. It might be worthwhile conducting field trials to see how effective they are against mosquitoes in dark houses in Niugini. Pyrethrins also kill bed bugs.

Knockdown insecticides can be dispersed by special machines that pump out fog or smoke. They are used to fumigate buildings.

The Dangers Associated with the use of DDT

DDT is poisonous to many animals if enough of it enters the body. It is a good insecticide because the concentrations that are used to kill insects have no effect on man. However, when man and other amimals are continually exposed to DDT, it gradually accumulates in certain tissues of their bodies, especially fatty tissues.

In countries where a lot of DDT has been used to kill agricultural insect pests, everybody has small amounts of DDT in their fatty tissues. This has entered their bodies on fruit and vegetables and in meat that they have eaten. People who work continually with DDT have much larger amounts of it in their bodies. No one knows what long term effects this will have on them.

There have been several cases in Niugini where people have stolen DDT powder and used it to make cakes or added it to their food. They became very sick.

Some malaria workers in Niugini have been constantly handling and spraying DDT since 1957. They have probably been exposed to as much DDT as any group of workers in the world. They probably have accumulated large amounts of DDT in their bodies. Research should be done to find out how much DDT is present in their tissues and if it is having any bad effects on them.

Fatty tissues from workers in DDT factories have been found to contain up to 648 parts per million of DDT. These workers appeared to be healthy.

People who live in houses that are sprayed with DDT probably take only very small amounts of DDT into their bodies. They would take in far less than spraymen or factory workers and they are unlikely to be affected at all.

We know that in America, where DDT has been used extensively to control agricultural pests, the whole environment has become contaminated and large numbers of fish, birds, mammals and useful insects have been killed.

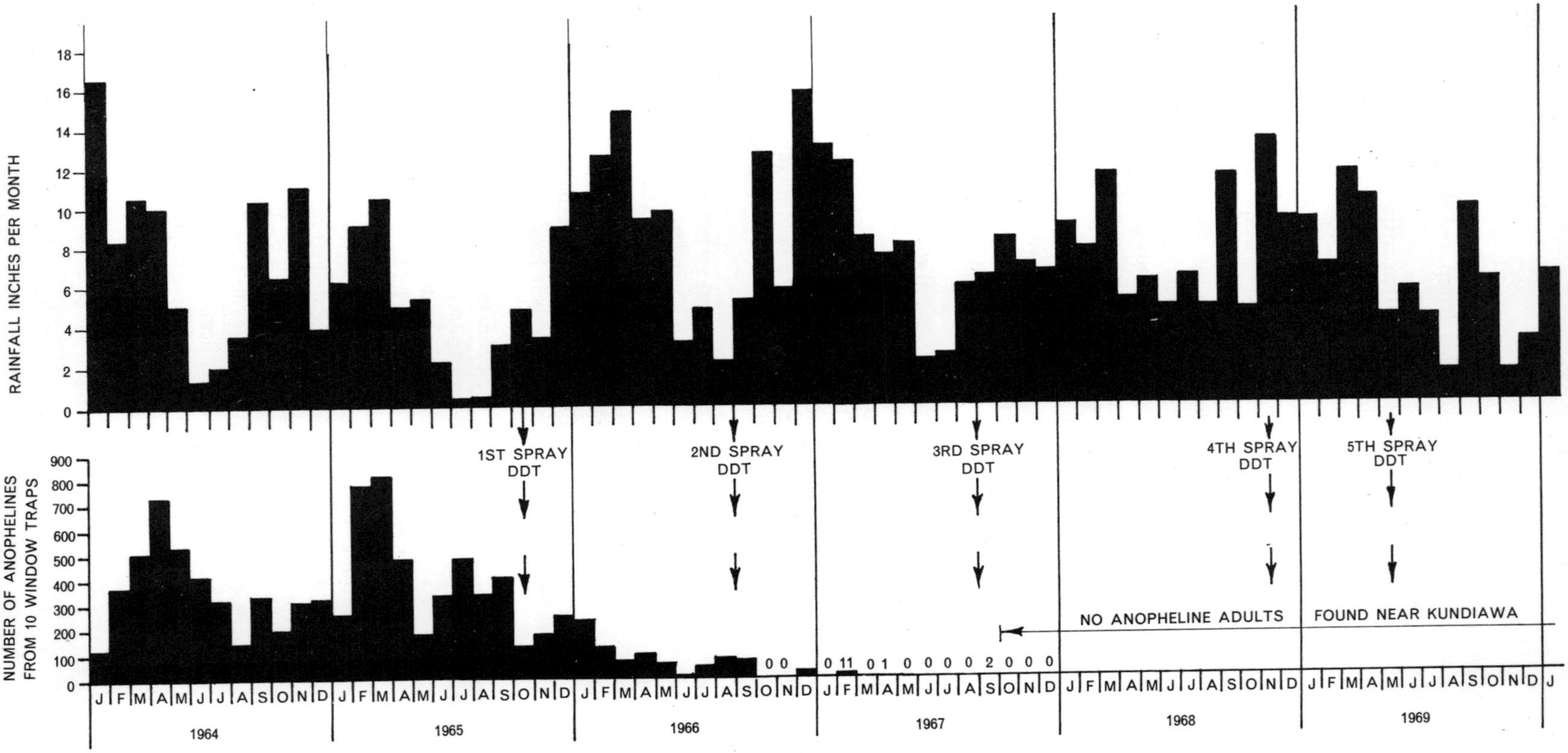

Fig. 3.26 Showing monthly rainfall and number of anophelines from window outlet traps
Malaria School Kundiawa

In Niugini, DDT has been used extensively only to spray houses to control malaria. The environment has not been polluted because houses made from palm fronds and kunai do not last long and the old house is burnt when a new one is built. When DDT is burnt, it is changed into other chemicals which disappear in the smoke.

However, there are some problems associated with the use of DDT for malaria control in Niugini, and we will discuss them now.

Problems Associated with the Use of DDT in Niugini

There is a good deal of circumstantial evidence that domestic animals have been poisoned with DDT. It often happens that most of the cats in a village die a week or two after DDT has been sprayed. It is likely that the cats rub against walls that have been sprayed, then lick their fur and thus eat DDT. Fowls and ducks are sometimes affected in the same way. It appears that they get DDT from eating dead insects that are plentiful around houses soon after spraying. As far as we know, no one had ever examined tissues from dead animals in Niugini to see if they contain DDT, and no one here has conducted any experiments to see how much DDT is needed to kill fowls and cats.

There is further circumstantial evidence that spraying with DDT causes an increase in the numbers of bed bugs, and the larvae of a moth, called *Herculia negrivitta*. This eats sago thatch that is used in most Niugini houses. We think what happens is this: when houses are sprayed with DDT, many small insects are killed, but bed bugs are not killed by the concentrations of DDT that are used in malaria control. The larvae of *Herculia* are not killed because they do not come in contact with the DDT. They live inside the walls and roof and they seem to avoid surfaces that have been sprayed. Experiments have shown that they are more susceptible to DDT than mosquitoes. The numbers of these two pests probably increase because DDT has killed the lizards, wasps and spiders that normally eat them and control their numbers.

We will be dealing with bed bugs in a later section of this book (6.3).

Herculia is a problem in all tropical countries with malaria control programs. In Niugini the life cycle is as follows. (See Fig. 3.27.) The females lay 10 to 20 eggs in a cluster or group, at the one time. We do not know how many clusters are laid by each female. Laying takes place in the late afternoon and the eggs are laid on the unsprayed exterior surfaces of houses. Caterpillars hatch from the eggs in about three days and feed for about twelve days. When they are about an inch long, the caterpillars spin a thin, egg-shaped, silk cocoon which lies between two pieces of sago leaf. Pupation takes place within this cocoon. The pupae are about half an inch long. The adult moths emerge in about six days. They are small and have a wingspan of about half an inch. The wings are pale purple-grey in colour.

In Niugini, houses that are sprayed with DDT last about three years, but they last about five years if they are not sprayed. The difference is thought to be due to damage due to *Herculia* caterpillars. In one survey, done in the Sepik District in March, 1961, one square foot of roof was taken from 20 sprayed houses in the village of Kalabu, and an average of eight caterpillars was found per square foot. Similar counts were also made on roof samples from 20 unsprayed houses in the village of Bonohoi, and an average of four caterpillars were found. Because the two lots of samples came from different villages we cannot say definitely that the difference between them was due to spraying, but the results are suggestive.

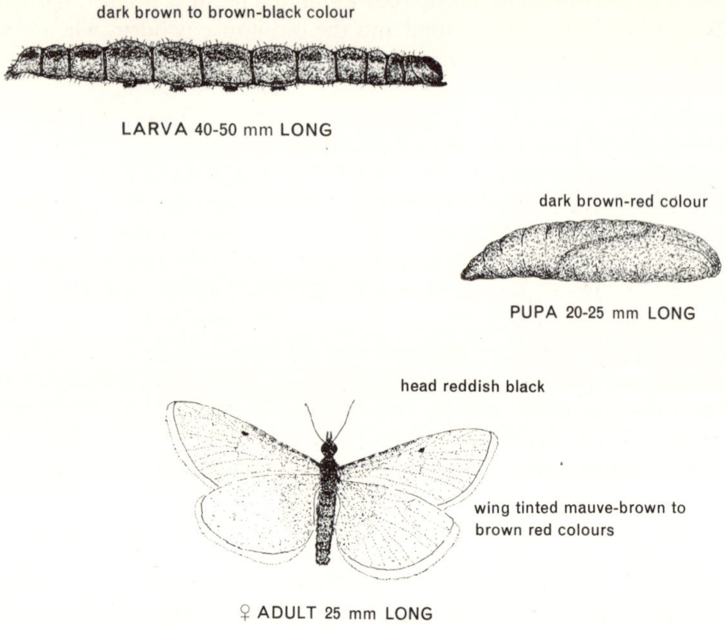

Fig. 3.27 *Herculea negrivetta* (Walker)

If DDT spraying continues for a long time, some people refuse to have their houses sprayed. Each time their house is sprayed, they must take out all their cooking pots, sleeping mats, clothing and food, so that they will not become contaminated with DDT. These people may believe that spraying has killed their cats and fowls, or they may believe that more bed bugs bite them at night, or more caterpillars are eating the walls and roofs of their houses because of spraying. In some places, people who have refused to have their houses sprayed have been taken to court and fined. There is a law which says that if a sprayman comes to spray your house, you must let him.

We can understand people who do not want to have their houses sprayed, year after year, forever. If malaria was being eradicated instead of controlled, the spraying would be over in a few years. It would be well worth while putting up with these other things.

If enough people refuse to have their houses sprayed, we may need to use other methods to control malaria in Niugini. We will talk about these later. Meanwhile there is a need for greater co-operation between the people and the malaria workers. Everybody should understand why spraying is so important. Many people have forgotten how bad malaria used to be. In many areas malaria was never recognised as such; most cases of sickness and death were thought to be caused by sorcery.

The malaria service must do much more to publicise its activities. Everybody in the country must be convinced that the eradication of malaria is of the greatest importance and it needs their help. As the incidence of malaria falls, the people should be told about it, so that they can be proud of the progress that has been made in their place.

The malaria service should investigate the stories about bed bugs, *Herculia*, cats and fowls. If they are true, they should see what can be done about them. For example, there may be an inexpensive way of killing bed bugs. If the malaria service killed the bed bugs when they sprayed the houses, it would make people happy.

The use of DDT for spraying houses may lead to changes in the behaviour of the vectors. Two kinds of change have been observed in Niugini and other places. They might threaten the success of a malaria eradication program that is based entirely on the use of residual insecticides.

The first is that mosquitoes tend to come to bite man in the early evening when many people are outside the houses. We have already discussed this in Section 3.6. The mosquitoes can thus feed on man without settling on a surface that has been sprayed with DDT. This is known as outdoor transmission and if there is outdoor transmission, malaria cannot be eradicated by the use of residual insecticides alone. There is no easy way of measuring the amount of outdoor transmission. If infected mosquitoes are caught outside, by using a man as a bait, outdoor transmission is probably occurring.

The second kind of change that occurs in the behaviour of mosquitoes, is that they avoid settling on surfaces that have been sprayed with DDT.

The following experiment illustrates this phenomenon. It was conducted by Mr A. W. Sweeney of the Malaria Service, at Rapola, near Rabaul.

Two identical test houses were made of native materials. A trap was put on some of the windows so that a proportion of the mosquitoes leaving the house would be caught. A man slept inside each house. Catches made over a period, showed that each house was equally attractive to vector mosquitoes. That is, about the same number of mosquitoes entered each house. One house was then sprayed with the usual concentration of DDT. The other was left unsprayed as a control or comparison.

It was found that vectors caught leaving the unsprayed house were almost always alive next morning (one or two died). About 20% of the mosquitoes caught leaving the sprayed house were dead by 6 am each morning, or died within the next day. They were being killed by the DDT. But 80% of the mosquitoes that were entering the sprayed house were leaving unharmed.

Counts were made of the number of vectors found resting on the walls of both houses at 6 am every morning and during one night each week. Over a 6 month period the results were:

	Sprayed House	Unsprayed House
Number of vectors on walls at 6 am	25	626
Number of vectors on walls at night	50	319

This showed that the vectors tended to avoid contact with sprayed surfaces. This was confirmed by experiments conducted in the laboratory. Vectors that were put in a chamber lined with paper that had been sprayed with DDT, took off again quickly after settling on the DDT. They had 6 to 9 times as many flights over a period of 15 minutes as mosquitoes in another container (where conditions were the same, except that the paper had not been sprayed with DDT). Other experiments showed that the mosquitoes were killed by the usual concentrations of DDT if they settled on it.

We think that these two kinds of changes in behaviour are the result of natural selection (see Section 1.1, on evolution). In the first case, many of the vectors that instinctively feed early in the evening will get blood without going indoors. Those that tend to feed at a later time must go inside houses because people sleep in their houses at night. Once inside, they are likely to settle on DDT and be killed. The early (outdoor) feeders will leave more offspring than the later (indoor) feeders. If the tendency to feed early in the evening is inherited, the number of vectors with this characteristic will increase.

In the second case, vectors that possess an inherited tendency, or instinct, to avoid settling on surfaces that smell (of DDT) will survive, while others will die.

(b) METHODS THAT PREVENT MOSQUITOES FROM FEEDING ON MAN

There are certain chemicals called repellents that mosquitoes do not like. They can be rubbed onto the skin to keep mosquitoes away. Examples are oil of citronella (Lemon grass oil), and di-methyl-phthalate. Citronella is effective for less than half an hour, and di-methyl-phthalate for about 3 hours. Both burn the eyes and sometimes the skin. Di-methyl-phthalate dissolves plastics. They should be used with care.

Smoke also repels mosquitoes. Smoke from dry cow dung and mosquito coils keeps mosquitoes away. Mosquitoes cannot bite through heavy clothing, but heavy clothing is unsuitable in this country.

Mosquito nets that hang over the bed are a good idea. Provided they are well tucked in at the bottom they keep the mosquitoes out. Most mosquitoes nets bought from trade stores are no good because the mesh is too large. Female anophelines can get through a hole 0.045 inches square (or 0.064 inches or 1.63 millimeters across the diagonal). Nets must be kept repaired.

Windows and doors can be screened with aluminium or nylon mesh, but this is useless in houses made of bush materials, because there are many holes in the walls.

There is evidence that certain mosquito vectors prefer to feed on cattle and pigs rather than on man. If cattle and pigs are kept close to houses these mosquitoes may be diverted from man. The Niugini vector *A. farauti*, seems to prefer pig blood, at least in some places.

III. Action against the Parasites in Man

From ancient times quinine was used to treat malaria fevers. It is extracted from the bark of the cinchona tree which originally grew in parts of South America. It was later cultivated in many areas of the world to provide quinine.

During the 1920s, chemists made the first artificial drugs which were effective against malaria. Today there are several different kinds of these drugs.

Sometime in the past somebody must have made the important discovery that the drugs that were used to cure fever, would, if taken regularly, prevent the fevers from occurring. We have been unable to find out who made this discovery, but it was a most significant one. These drugs would suppress the disease or prevent it. The prevention of disease is called prophylaxis, and a prophylactic is a drug or agent that prevents the development of a disease. Hence these drugs are called prophylactics or suppressives.

Quinine was used as a suppressive as long ago as 1844, to prevent malaria in workers who were making a new railway line in America. Quinine is no longer used as a suppressive because it is toxic.

We will now deal with the most commonly used suppressive drugs in Niugini. Every drug company has its own name for the anti-malarials that it produces. We will list only the brand names that we have seen here. They are:
Chloroquine salts (Chloroquin, Aralen, Nivaquine, Resochin) and Amadiaquine (Camoquin).

These drugs act by killing trophozoites and schizonts in the blood. They have little or no effect on gametocytes or liver stages. They are not toxic to man if taken in the correct dose. The drugs should be taken on the same day every week.

Because chloroquine salts taste very bitter, small children should be given 'Infant Formula Camoquin' tablets, which taste nice. The recommended doses for prophylaxis are as follows:

INFANT FORMULA CAMOQUIN

Age less than 1 year	½ tablet per week
Age 1 to 3 years	1 tablet per week
Age 4 to 5 years	1½ tablets per week

(each tablet contains 100 milligrams of drug)

CHLOROQUIN OR CAMOQUIN:

Age 1 to 5 years	½ tablet per week
Age 6 to 9 years	1 tablet per week
Age 10 to 16 years	1½ tablets per week
Over 16 years	2 tablets per week

(each tablet of Chloroquin contains 150 milligrams of drug and each tablet of Camoquin contains 200 milligrams of drug).

There are two other groups of drugs that are useful in malaria programs. The first group is called the 8-amino-quinolines. They have the following brand names: Pamaquine, Pentaquine, Isopentaquine and Primaquine. They all act by killing liver stages and gametocytes in the blood. Primaquine is the only one at present used in this group. It is used in malaria eradication programs to kill liver stages and gametocytes, so that relapses and transmission cannot occur. It is given during surveillance. This drug is moderately toxic.

The drug pyrimethamine belongs to the second group which are called the Pyrimidines. It is known by the brand names Daraprim and Malocide. When taken by man it affects the gametocytes so that they do not develop in the mosquito. It is very toxic if the recommended dose is exceeded. This drug has been used in Niugini for malaria control, in combination with chloroquin, in tablets known as Erone. These tablets should not be taken as weekly suppressives. They were designed to be taken not more than once a month (adult dose—4 tablets). They are toxic if taken more often.

Chloroquin and Camoquin are used to treat malaria fevers. All cases of fever should be treated at an aid post or hospital. In an eridacation program blood slides are taken from all fever cases.

A new method of malaria eradication is at present being tested. It consists of suspending an anti-malarial drug in oil and injecting this into the muscles of the buttocks. The drug is slowly absorbed from the oil into the blood, so that the person is protected from malaria for many months. These drugs are called repositories because a repository is a storage place, and the body contains a store of anti-malarial drug. In recent field trials, this method has given protection from malaria for only a few months. But it is hoped that a drug will be found that will last for a year or more.

Until recently, yaws was a very common disease in Niugini. During the 1950s there was a big anti-yaws program. Yaws was almost completely eradicated by the injection of penicillin. The eradication of malaria would be just as easy if we had a suitable long acting drug with which people could be injected.

In certain parts of the world malaria parasites are not killed by certain anti-malarial drugs. This is known as drug resistance. In Vietnam, the malaria parasites are resistant to all commonly used anti-malarials except quinine. In West Irian, some species of malaria have become resistant to pyrimethamine. As far as we know, no drug resistance has developed here yet, although drug resistance is suspected in certain areas from time to time.

SURVEYS IN MALARIA PROGRAMS

Surveys are necessary to measure the amount of malaria before a program starts, so that progress can be assessed after control or eradication begins. If a program is being properly assessed we can tell at any stage if it is successful or not. And if it is not successful, we can usually tell what is going wrong.

The WHO has laid down standard procedures for assessing malaria programs. These involve making measurements of the amount of malaria and the numbers of vectors, etc. Most of the kinds of measurements that are made are estimates. For example, we might estimate the incidence of malaria in a particular place by examining 10% of the population to see if they are infected. The accuracy of our estimate depends a good deal on the size of the sample that we examine. An estimate based on 50% of the population is more accurate than one based on only 1% of the population. The smaller the sample, the less accurate our estimate is likely to be. The most accurate measurement involves the examination of the whole population. However, this is often impractical because of the amount of work involved and we can seldom catch the whole population. The WHO procedures specify the size of the sample that must be taken and how often observations must be made, so that the measurements are accurate enough to be useful, without involving more work than is necessary.

Surveys can be divided into those that provide information about the vector and those that provide information about man and malaria.

VECTOR SURVEYS

Before starting any program that is based on insecticides, it is important to know:

(i) the species that are vectors
(ii) how long they live
(iii) the amount of DDT that is necessary to kill them
(iv) the density of adult vectors.

We will discuss each in turn.

(*i*) *The species that are vectors*

This is determined by dissecting a large number of anophelines that bite man and examining their salivary glands for sporozoites. The importance of the vector is indicated by the sporozoite rate. This is the incidence of infection in a particular vector species, at one time, in one area.

The sporozoite rate is always very low (usually less than 5%) even in highly malarious areas. This is because most mosquitoes do not live long enough for sporozoites to develop within them and at any one time most mosquitoes are young.

Table 3.9 gives sporozoite rates from the Solomon Islands, West Irian and Niugini. These figures are taken from the reports of the third Inter-territorial Malaria Conference in Honiara, 1963.

In many places the sporozoite rate will vary throughout the year. And if pre-spraying data is to be compared with data collected after spraying, a series of measurements will be necessary.

TABLE 3.9 Sporozoite Rates (per cent)

	Solomon Is.	*West Irian*	*Niugini*
A. farauti	0.3	0.4–3.1	0.6–1.4
A. punctulatus	0.3	1.1–5.9	1.7–3.8
A. koliensis	2.5	0.5–5.0	2.6
A. bancrofti		0.2	
A. kawari		3.1	

(ii) The length of life of the vectors

The transmission of malaria can only go on if the females live long enough for sporozoites to develop within them. The aim of a spraying program is to reduce the average length of life of the female vectors so that transmission does not occur.

We have already discussed one method of estimating the age of females in Section 3.6. A quicker method, is to dissect a large number of females and determine the proportion (per cent) of them that have already laid eggs, (it is necessary to dissect about 200 a month to accurately assess this figure). The length of time from one egg laying to the next must also be known. A complicated mathematical formula is then used, which estimates the average length of life of the females in the population. If the average length of life exceeds the period required for the development of sporozoites within the vector, then most of the mosquitoes would be able to transmit malaria.

If the average length of life of the vectors is only a couple of days only a very small proportion of the mosquitoes will live long enough to transmit malaria.

The average length of life of the vectors in one place varies from time to time. It will be short during the dry season and longer during the wet when the humidity is high. It will also vary from time to time, with variations in the number of new adults that are emerging to join the vector population. If there is a sudden emergence of many adults the average age will fall for a while and will then rise again.

In order to be sure that changes in the average length of life of the vectors are due to control measures we must have measurements of the average length of life for several consecutive months, or preferably a whole year, before control starts. These measurements will provide a proper basis for comparison with measurements collected after control starts.

In many cases control measures are so successful that it is hard to collect enough mosquitoes to make accurate measurements of the average length of life.

We have been unable to find examples from Niugini of changes in the average length of life, month by month, before and after control measures have started. What little data we have been able to find is often unreliable because it has been based on samples that are too small (that is less than 200 mosquitoes per month).

TABLE 3.10 Average length of life for vector mosquitoes in Niugini.

Place			*Average length of life* (*days*)
Several Papuan villages		1960	4 -9
Fergusson Island		1960	6
	March	1961	14
Nissan Island	April	1960	6
		1961	6
East New Britain		1960	2–16
East Sepik		1960	14
		1969	3
West Irian		1957	4
		1959	3
		1960	5
		1961	2–11

(*iii*) *The amount of DDT that is necessary to kill the vectors*

The WHO has developed a standard test kit for determining this. It consists of a number of tubes which are lined with paper impregnated (or soaked) with various concentrations of DDT. At least ten females are put in each tube for half an hour. They are then removed to a clean tube and stored for 24 hours. The number surviving is then counted. From this test the concentration of DDT necessary to kill the vectors can be determined.

The paper in one tube is not impregnated with DDT. This is the control tube. Allowances are made for accidental deaths in the control tube. If all the mosquitoes in the control tube die, the test cannot be evaluated and it must be repeated.

It is important to do this test regularly to see if the vector mosquitoes are becoming resistant to DDT. Vectors have developed resistance to DDT in many countries, but so far there are no signs of resistance in Niugini.

(*iv*) *The densities of adult vectors*

Several measurements are used. They all tell whether spraying is effective or not. Spraying should reduce the number of vectors and their length of life. The measurements that are used are:

(a) The number of vectors which bite man each hour between 6 pm and 6 am, for four nights every month. One set of data is collected indoors and another outdoors on the same nights. This enables a comparison to be made of the density of mosquitoes biting man inside and outside the house. (See Fig. 3.9.)

(b) Window traps are put in test houses to catch mosquitoes that have entered the house and then tried to leave. The number surviving after 24 hours is determined and this tells the effectiveness of spraying. Spraying with insecticides is successful if mosquitoes are killed after entering a house.

The numbers of mosquitoes in any place vary seasonally and, as with all other measurements which vary in this way, data must be collected at regular intervals throughout the year.

HUMAN SURVEYS

Three different kinds of survey are commonly used. They provide the best assessment of the effectiveness of spraying. They are:

(*i*) *The Infant Parasite Rate* (*IPR*)

This is the incidence of malaria in infants less than one year old. It gives a guide to the amount of transmission going on in an area. Many of the cases of malaria in older children and adults are due to relapses. But malaria in infants must be due to recent new infections. If a malaria program is successful in stopping transmission, the Infant Parasite Rate should quickly fall to zero.

However, an Infant Parasite Rate of zero does not necessarily mean that the transmission of malaria has stopped. Babies are less likely to become infected than older people because they are often wrapped up in bedclothes which protect them from mosquito bites. Infants less than five months old are seldom bitten by vectors when they are with their mothers (see p. 78, Section 3.6) and furthermore, babies are small and have smaller surface area than other people and because of this, are not likely to be bitten as often as adults. Furthermore, babies in malarious areas get antibodies against malaria from their mothers and malaria parasites may not develop in their blood because of this. (See p. 67).

The Infant Parasite Rate should be determined about four times a year during malaria programs. Because there are few infants in the population, blood slides should be taken from every infant. When the Infant Parasite Rate becomes less than about five per cent it cannot be measured accurately unless it is based on a sample of about 7000 infants.

(*ii*) *The Overall Parasite Rate* (*OPR*)

This is the incidence of malaria in the whole population. It is best measured by taking blood from everybody in one or more villages, that are selected because they are representative of a large area. It should be determined twice a year at times when transmission is most likely to be occurring, for example, at the start of the dry season.

When the Overall Parasite Rate has fallen to less than 2% it cannot be accurately measured unless it is based on very large samples. It is usual, at this stage, to measure the amount of malaria in the population by surveillance techniques. These have already been discussed (see p. 107).

When a blood slide is found to contain malaria parasites, the species must be identified and recorded. Three species commonly occur in man in Niugini—*P. falciparum*, *P. vivax*, and *P. malariae*. *P. ovale* also occurs in Niugini but is very rare.

It is important to have records of the incidence of each species, because *P. falciparum* generally causes more sickness and many more deaths than the other species. *P. falciparum* is also the only species that does not have schizonts which stay dormant in the liver, and later cause relapses. However, recrudesences occur with *P. falciparum*. In these cases a few parasites may lie dormant in the blood and suddenly give rise to another lot of attacks of fever. But recrudescences of *P. falciparum* rarely occur more than six months after the initial infection. This has an important use in measuring the success of malaria control. If control is successful, the number of *P. falciparum* cases should quickly drop to zero. The only cases of malaria which occur, should be due to relapses of *P. vivax* and *P. malariae*. If the number of *P. falciparum* cases stay high, it is a sure sign that transmission is still going on. An example of the change in the ratio of the three species is given in Table 3.11.

TABLE 3.11 Changes in the ratio of the three species of malaria in one area near Kainantu where spraying started in 1965. (Data supplied by Mr H. van Leeuwen.)

Date	OPR	RATIO *P. falcip.*	:	*P. vivax*	: *P. malariae*	:	Mixed
1964	10%	1	:	1.7	: 0.07	:	0
1965	7%	1	:	4.8	: 0	:	0
1966	0.5%	1	:	1.5	: 0	:	0
1967	0.1%	0	:	0	: 0	:	1
1968	0.5%	0	:	9	: 0	:	0
1969	0.6%	0	:	6	: 0	:	0

The size of the spleen of each person is determined according to a scale devised by Dr Hackett. The largest spleens, size 5, extend to the bottom of the abdominal cavity; size three spleens extend to the navel and size zero spleens are not enlarged. Fig. 3.28 shows the Hackett Scale for determining spleen size. Fig. 3.29 is a photograph showing the outline of an enlarged spleen in a Niuginian child.

The average enlargement of the spleen (AES), is calculated for the age group 2 to 9 years, for each community, using a mathematical technique.

In a successful malaria program the Spleen Rate, Average Enlarged Spleen and the Overall Parasite Rate should all fall.

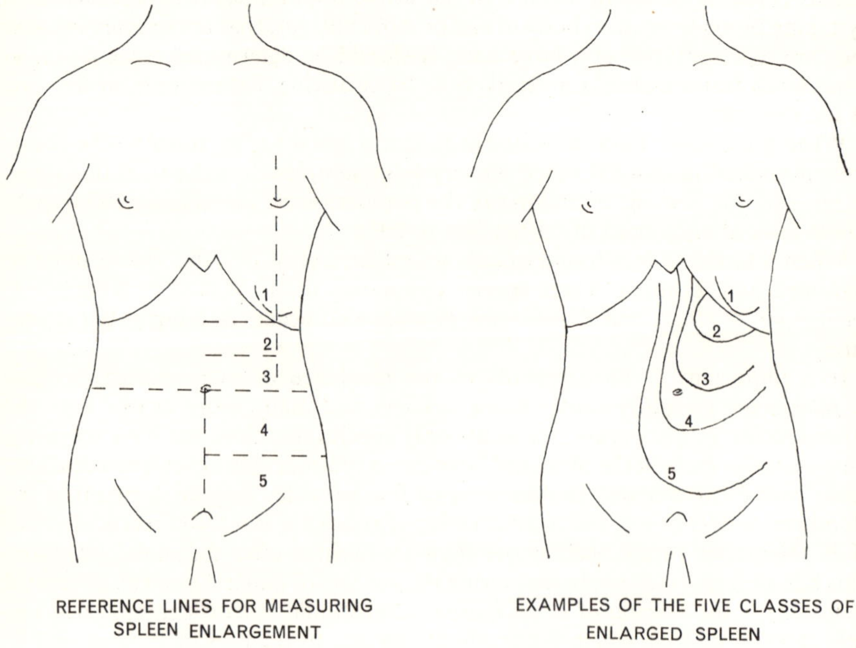

Fig. 3.28 The Hackett index for measuring enlarged spleens

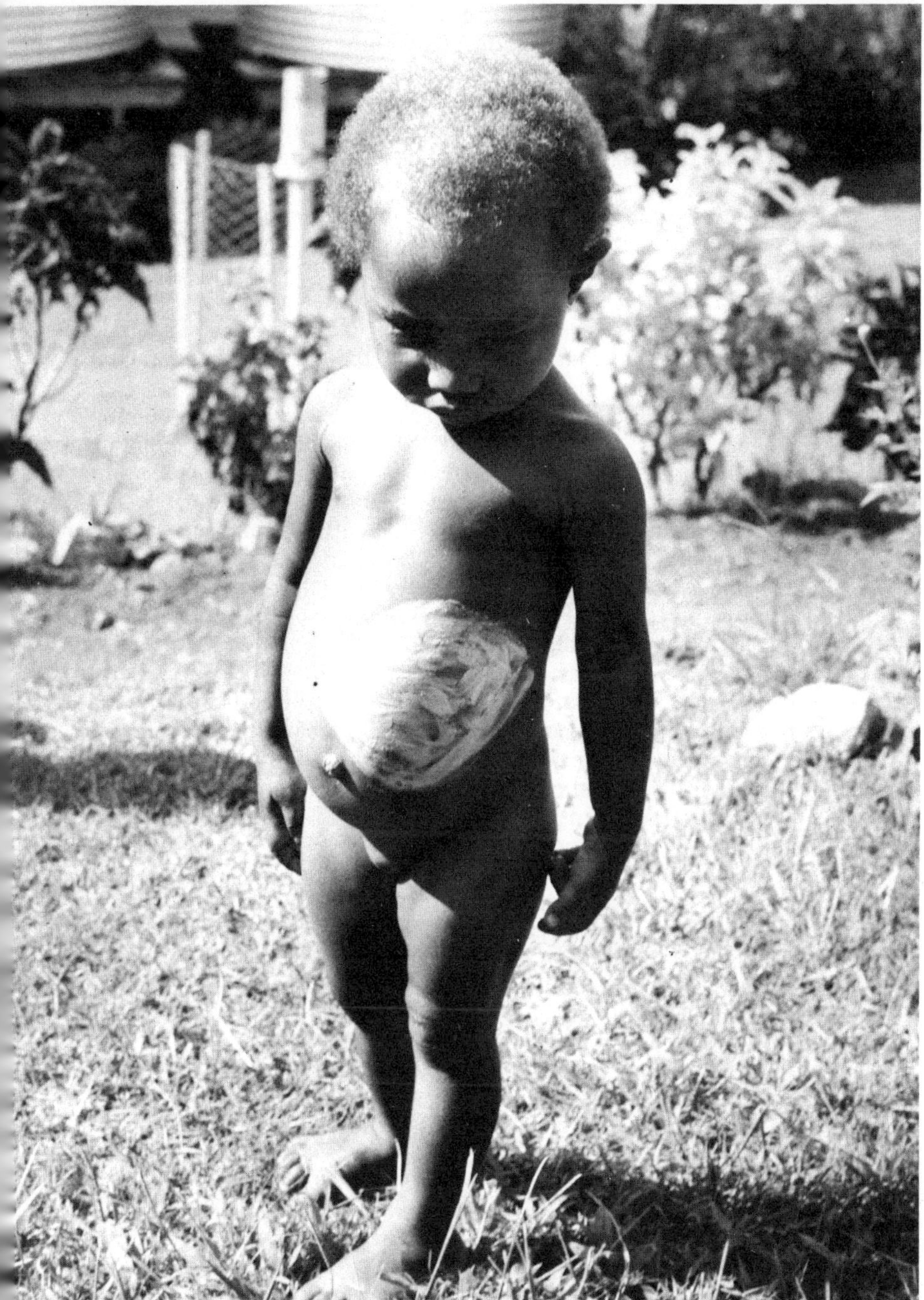

Fig. 3.29 Chimbu toddler recently returned from coastal area (size of spleen painted in)

In Section 3.7 we gave an example of seasonal changes in the incidence of *P. falciparum*, *P. vivax* and *P. malariae*. (See Fig. 3.14.) It is not unusual for the incidence of *P. falciparum* to be twice that of *P. vivax* and then for the situation to reverse within a few months. These changes can occur without any attempt at malaria control. This means that a reduction in the incidence of *P. falciparum* as compared with the other species, is not always a sign of successful malaria control.

In one control program in Dutch New Guinea the incidence of malaria fell from 46% to 6% over a period of six years. However, the proportion of all malaria infections due to *P. falciparum* nearly doubled over this period (see Table 3.3).

(iii) Spleen surveys

We have already said that enlargement of the spleen is a characteristic of people who have recently had an attack of malaria.

Spleen surveys are used to measure two things: the proportion of people who have enlarged spleens or the spleen rate (SR), and the average amount of enlargement within the population. For reasons that have already been discussed in Section 3.4, spleen surveys are usually done on all children 2 to 9 years old and about 10% of adults. Spleen surveys are done at the same time as the blood survey to determine the overall parasite rate.

This concludes the kinds of surveys used in malaria programs, however, we think that it is also essential to have some measure of the way that people are accepting the malaria program.

In Niugini, there are areas where house spraying has been going on twice a year for more than ten years, without stopping the transmission of malaria. In many of these areas the people are refusing to have their houses sprayed. Accurate data on the proportion of unsprayed houses may provide one reason for the failure of the program.

The time may come when it is obvious that spraying alone will not eradicate malaria in Niugini. The introduction of any new method, such as the distribution of drugs, will need the complete co-operation of all the people, if it is to succeed. It is important that properly conducted surveys be done to measure the acceptance of the malaria program, so that attempts can be made to alter public opinion if this is necessary. There is no point in introducing a new method to eradicate malaria if it is going to be rejected by the people. Perhaps every malaria program needs to employ specially qualified staff whose job is to promote or advertise the malaria program and to understand the people's reaction to it.

3.9 Malaria control in Niugini

Before we discuss the present situation, we will review the history of malaria control in this country.

In 1897 malaria was eradicated from the Island of Samarai by filling in a swamp to get rid of mosquito breeding grounds. In 1901, Professor Robert Koch conducted an experiment at Bogadjim on the use of quinine as a malarial suppressive.

Dr Heydon, in 1923, reported a reduction in the number of vectors at Rabaul following larviciding with oil.

The fish, *Gambusia affinis*, was brought to Kavieng, New Ireland in 1933, to control mosquito larvae. It was said to be very effective.

In 1944-45 a group of army workers did an experiment at Motu Motu No. 1 and No. 2 villages at the mouth of the Lakekamu River, about 150 miles west of Port Moresby. They tried to find out if DDT would control malaria there. The houses of one village were sprayed with DDT in kerosene applied at the rate of 100 milligrams of DDT per square foot. The other village was not sprayed and data collected from it were compared with similar data from the sprayed village. Spraying caused a slight decrease in the incidence of malaria and a large decrease in the number of vectors.

The concentration of DDT used in this experiment was very much lower than is used today. No further experiments using DDT were tried in Niugini for a long time, because this experiment did not produce very good results and because it was believed that the vectors did not rest on surfaces inside houses.

In 1950-3, Mr S. H. Christian, MBE, started the Wahgi Valley Pilot Drainage Scheme. This has already been discussed in Section 3.8.

Following reports of the use of DDT by the Dutch in West Irian, a Pilot Project was started at Maprik in 1957 to determine the effectiveness of DDT in malaria control in Niugini. Dr W. Peters was in charge of this project, and he was assisted by Mr H. Standfast and later by Mr W. T. Jeffrey. This project showed that when houses were sprayed with DDT there was a great reduction in the number of vectors and the incidence of malaria. Before the project started, the overall incidence was 75% and the incidence in infants was 95%. By 1961, the overall incidence had fallen to 27% and the incidence in infants to 26%. This project continued until 1962, but the transmission of malaria was never completely stopped.

Malaria control activities were gradually extended to other parts of the country. In 1959, units were established in New Ireland, Bougainville and the D'Entrecasteaux Islands. In 1961 the program was extended to New Britain, Manus and the Sepik Coast and, since 1962, there has been further expansion within the Highlands and coastal areas.

The program at present depends entirely on the spraying of DDT. Until recently however, anti-malarial drugs (ERONE, see Section 3.8) were distributed whenever a village was sprayed.

In 1953, the Malaria Control School was established in Minj by Mr S. H. Christian MBE. It moved to Kundiawa in 1963. Mr W. T. Jeffrey has been senior lecturer at the school since 1964. It is now called the National Malaria School. From 1962-5, the school had a field training program at Rabaul. This school has played an important part in training malaria workers. Up to the end of 1969, the following numbers of people have been trained for various tasks in courses lasting from three months to a year.

	Number trained
Squad leaders	343
Team leaders	129
Technicians	161
Area supervisors	62
Non-malaria personnel	101

THE PRESENT ORGANISATION

The malaria service is part of the Department of Public Health. All staff are permanent public servants and cannot easily be sacked. Spraymen are employed on a casual basis.

At the end of 1969, about 1000 workers were employed by the malaria service. Of these, 47 were expatriate officers mostly from Australia. The rest were Niuginians. Three of the expatriate officers have basic University degrees in medicine and one of them has an additional diploma. One other expatriate has a pass degree in Science. None of the Niuginians have University degrees, but it is hoped that some of the students from the recently established University of Papua and New Guinea will join the malaria service after they graduate. The service does not at present have a qualified entomologist.

In the five years 1962-8, about $6 million has been spent by the malaria service. The budget for 1969 was $1.37 million.

Progress So Far

At the end of 1969 malaria control operations are established in all Administrative Districts, except the Southern Highlands District, Gulf District and the Western District.

Fig. 3.30 shows the towns where the operations are established. At present very few administrative districts are completely covered by malaria operations. We look forward to the time when they will cover the whole country, but a great deal more money will be necessary before this can be done. At present the operations are very scattered and there is not enough senior staff to supervise them. In many cases the operations are not assessed properly. We know of only three areas which have adequate consecutive records from the initial surveys, which should have been done before spraying started, through until the present time.

In Appendix II, page 232, we have summarised all the data that we can find, both published and unpublished, that gives figures on the amount of malaria in Niugini. Items marked with an asterisk (*) are for surveys that have been done since malaria control started.

In many areas of Niugini there has been a great reduction in the amount of malaria since control programs started.

We will give a few examples:

At Maprik, the OPR was reduced from 68 per cent to 6 per cent in seven years of spraying. It has since increased to 16 per cent (1968). At Kainantu and Kundiawa, the OPRs before control started were 10 per cent and 31 per cent. The OPR is now less than 1 per cent in both of these areas. At Talasea, the OPR in 1961 was 45 per cent, in 1966 it was only 3 per cent.

We have already talked about the problems that have arisen in Niugini because our program is a control program based entirely on the spraying of houses with DDT.

It seems that the number of people who refuse to have their houses sprayed, is increasing. In the East New Britain District 12 per cent of the houses were not sprayed during 1969. In 1970 this figure jumped to 25 per cent. In 1962 there was no opposition to spraying.

The malaria control program in Niugini is not a great success. Nowhere has malaria been eradicated and there is no prospect of eradication in the near future using present methods. The use of DDT must be supplemented by other methods.

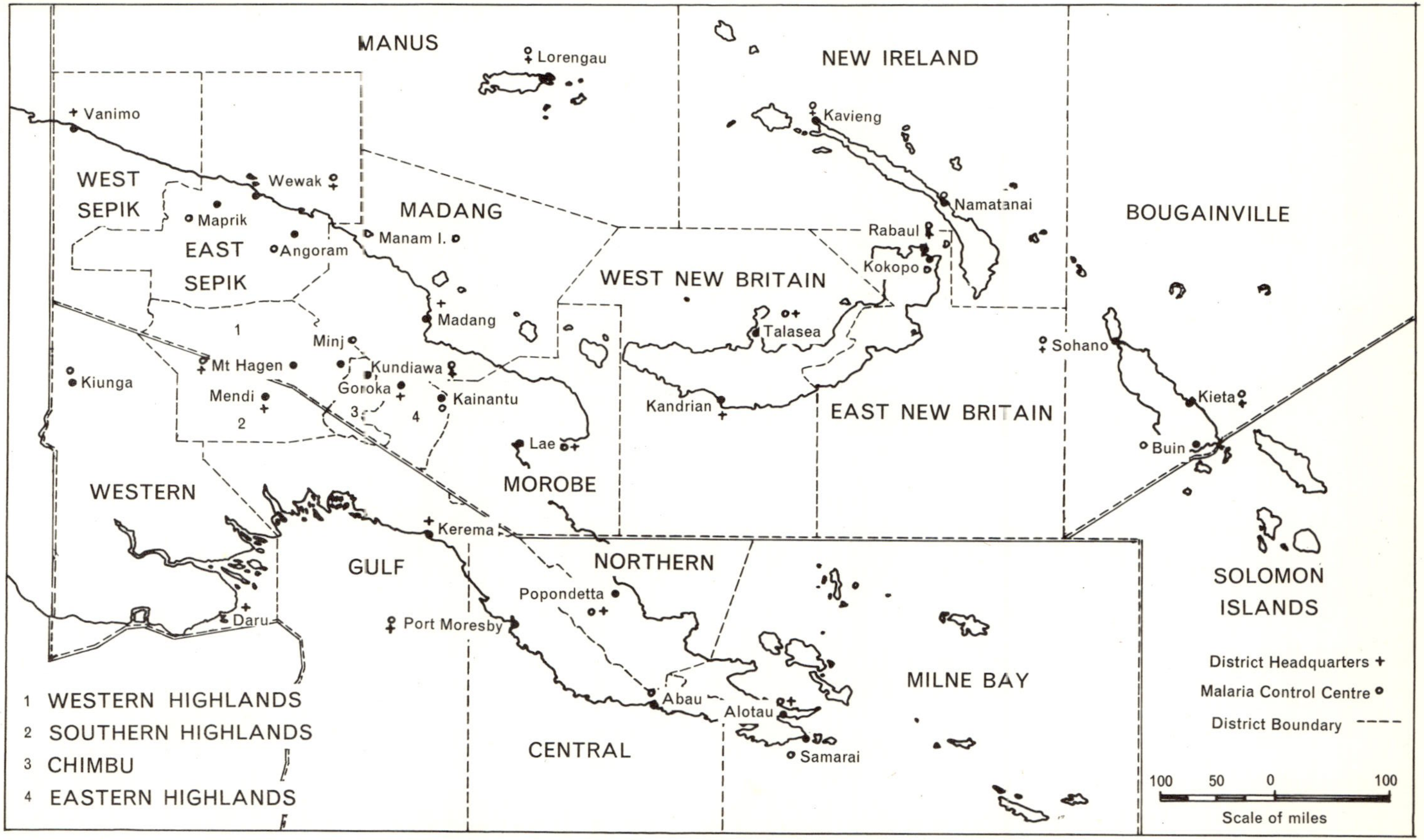

Fig. 3.30 Map of Niugini showing district boundaries, district headquarters and malaria control centres 1970.

Every possible method must be tried in properly designed trials to see if it will help.

The World Health Organisation and malariologists in general, think that the widespread use of suppressive drugs is of little use in malaria programs because it has been found that many people will not take them. However we feel that if only half the people take drugs the chances of malaria eradication in Niugini would be very much greater. We think that many people will take suppressives if they are convinced of the danger of malaria and importance of its eradication.

A national publicity campaign is necessary to make everyone aware of the dangers of malaria and to make everyone in the nation a worker against malaria. The publicity campaign should be run by a big advertising company with experienced and well qualified experts in charge. They should be assisted by local people who understand the customs of the country. It should not be run by The Malaria Service or the Health Education Section of the Public Health Department. The campaign would need the support of the press and radio.

A year's supply of chloroquin costs about $1 in the shops. If these drugs are bought in bulk, they are much cheaper. At present prices they cost 72 cents per adult per year. We think that if this drug was bought by the million, the cost of supplying the whole population of Niugini (2.5 million) would be about $660,000 per year. The present malaria program spends about $1.6 million a year and only covers about half the country. We would like to see suppressive drugs distributed by every possible means: through schools, local government councils, missions, aid posts, churches and trade stores. It would probably be worthwhile if trade stores sold the drug, because many people appreciate the things that they pay for, more than things that are free.

Every house should be provided with an attractive sign, printed in colour, saying; in pidgin: 'Mondei, dei bilong rausim sik malaria', or in Motu 'Monday be malaria gorere kokia dinana' or in English 'Monday is anti-malaria day'. In some areas it would be better to use signs in the local language.

A few years ago, the change-over to decimal currency was introduced by a most successful publicity campaign. It cost a lot of money. The eradication of malaria is far more important than the kind of money we use, and if malaria is to be eradicated, a lot of money must be spent telling people about it. A happy song should be written and sung in all schools urging everyone to get rid of malaria by taking drugs and spraying houses.

Niugini will soon be an independent country. It would be great if independence was accompanied by the eradication of malaria.

Chapter 4

GUT PARASITES

4.1 Introduction

Niugini is a lucky country because there are many gut parasites of man which do not occur here. In this chapter we will consider only the gut parasites that are found in man in Niugini. The important gut parasites in Niugini are mostly nematodes and Protozoa. The general characteristics of these have been discussed in Chapter 1. Among the nematodes, *Ascaris* and Hookworms are the most important. But *Trichuris*, *Enterobius* and *Strongyloides* also infect man here.

Among the Protozoa that live in the gut, *Entamoeba histolytica* is the most important. The related species *Entamoeba coli*, *Entamoeba hartmanni* and *Entamoeba polecki* are harmless, but because they also infect man, we must know something about them.

Giardia is another protozoan parasite which sometimes causes trouble in man. Several other species of Protozoa also infect man in Niugini. There is also a tapeworm, called *Hymenolepsis*, that infects man in Niugini, but it is rare.

A lung fluke called *Paragonimus westernmani* was recorded many years ago from a man in Rabaul, by Dr Backhouse. It has not been recorded since. The infection was found in a native sailor, who probably became infected during a trip to Asia. Dr Lambert, in 1920, recorded eggs from a Platyhelminth in the faeces of a man from Akauka in the Gulf District. They were probably from a tapeworm like *Spirometra*, which is common in cats and dogs in Niugini. These tapeworms have been recorded from man in other parts of the world. In addition, eggs of the flukes *Echinostoma* and *Opisthorchis* and an unidentified trematode, have been recorded in man from West Irian.

Man is infected by gut parasites because he often comes in contact with human faeces. We should always use properly constructed toilets. All faeces from small children should also be put in the toilets. People must be educated so that they understand why it is so important to have good sanitary habits. We should always wash our hands after going to the toilet and again before eating. Any campaign to control these diseases, gets off to a good start, if all the infected people are given drugs or medicines to kill the parasites in their bodies. If the life cycle of the parasite is then broken, so that re-infection does not occur, the people will soon be free of parasites. We will discuss the control of all the gut parasites in the last section of this chapter.

The Diagnosis of Gut Parasites

There is no way of telling from the symptoms that a person has a particular gut parasite. The most common symptoms are those of vague pains in the belly with occasional diarrohoea (runs) or constipation, and sometimes more severe symptoms. In order to find out if a person has gut parasites, it is necessary to find eggs or cysts of these parasites in the faeces. This is done by examining a small amount of faeces under a microscope, which magnifies about 400 times (400X). It helps if the faeces are mixed with a little iodine solution which stains the cysts and eggs, so that they can be seen more easily. Sometimes it is necessary to use other stains to see details in some of the smaller protozoa and a higher magnification (about 1000X) may also be necessary.

When faeces are mixed with a saturated solution of sodium chloride (common salt) the eggs of certain nematodes float to the surface. The eggs float because they are less dense than the salt solution. The faecal material sinks because it has a higher density than the salt solution. The salt solution must be saturated; that is, more salt must be put in the water than will dissolve after lots of shaking. This method concentrates the nematode eggs at the surface of the salt solution. It is called the 'Willis Technique' after the man who invented it. It is usually done in a small tin or glass bottle which holds about a half a gram of faeces and 5 to 10 milliliters of salt solution, although these quantities are not critical. The eggs are collected by placing a glass slide in contact with the surface of the salt solution and leaving it for about 20 minutes. The slide is then turned over and examined under a microscope (see Fig. 4.1). It is a good method for the diagnosis of *Ascaris*, hookworm and *Trichuris*.

There are other methods that can be used to separate eggs and cysts from faeces. Some of them use a centrifuge, which is a machine that rotates at high speed and separates substances with different densities.

The eggs and cysts are thus separated from other material. Centrifuges are often turned by an electric motor, but some models can be turned by hand. When

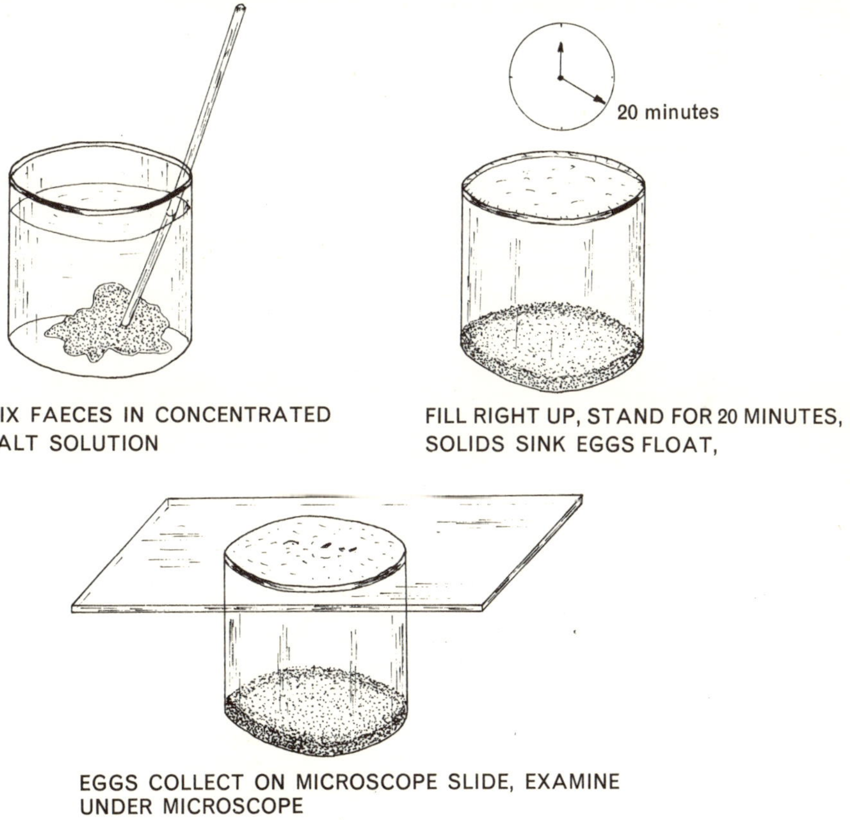

Fig. 4.1 Willis technique for collecting worm eggs

only a few worms are present in a person, there will not be many eggs in the faeces. These techniques increase the chances that an egg will be found. It is necessary to examine several lots of faeces from the one person before we can be sure that he is not infected with gut parasites. Special methods are used for the diagnosis of *Enterobius* infections. They are described in section 4.6.

Incidence and Load

When we discussed the epidemiology of malaria, we said that two kinds of measurements were important. They were; the proportion of people who were infected (the incidence %) and a measure of how badly they were infected. When we consider nematode infections in the gut of man, this second measurement is very important. A few worms do not usually produce severe symptoms, but a large number of worms can. The degree of sickness depends a lot on the number of worms present. We refer to the number of worms as the 'load' or burden that a person has to carry.

Measurements of load

Each female worm lays about the same number of eggs each day. so that if we could count all the eggs in the faeces from one person, for one day, we would have a measurement of the worm load. It would be very difficult to do this, but we can estimate the load fairly easily. We do this by estimating the number of eggs in one gram of faeces and multiplying this figure by the weight of the faeces in grams, produced by one person in a day.

The eggs produced by each worm are usually thoroughly mixed with the faeces, so that the number in any one gram of faeces is about the same as in any other gram. We can weigh out one gram of faeces and mix this with a known volume of concentrated salt solution and quickly put a small amount of this mixture in a special egg counting chamber of known volume. This container is called the McMaster egg counting chamber, because it was invented at the McMaster Laboratories of CSIRO in Sydney (see Fig. 4.2). The eggs will float to the surface

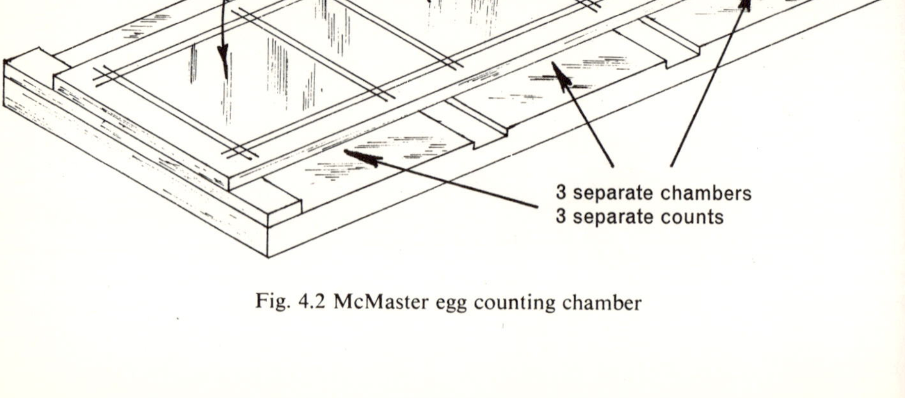

Fig. 4.2 McMaster egg counting chamber

of the container and can be counted under a microscope. This count gives a measure of the number of eggs in one gram of faeces. Because the average weight of faeces is about the same each day, we have a measurement, or index, of the number of worms present.

It is necessary to make allowance for the amount of water in the faeces. This will vary from time to time, depending on how fast the faeces pass through the gut and how efficient the large intestine is in reabsorbing water from the faeces. A gram of faeces containing a lot of water will contain fewer egsg than a gram of hard formed faeces. It is usual to multiply egg count estimates by the following factors to allow for this. For soft faeces which still have a cylindrical shape, estimates are multiplied by 1.5; for faeces which are mushy, that is, have no shape, multiply by two; and for faeces from people with diarrohea (where the solid matter sinks to the bottom of a watery solution) multiply by four. For hard faeces, we take the count as it is.

If we take a number of people who are infected with gut nematodes and determine the following measurements for each one, we can go one stage further, and calculate the average daily output of eggs for that species of nematode. The measurements are:

The average number of eggs per gram of faeces (A)
The average weight of faeces passed per day (B)
The number of female worms in each person (C)

The first measurement is made by using a McMaster egg counting chamber. The second varies with diet. Niuginians who eat lots of kaukau, taro, sago, bananas and other vegetables pass more faeces than Europeans who eat a lot of meat. (Adult Europeans on a mixed diet pass about 200 grams of faeces per day. The figure for Niuginians on a vegetable diet is about 300 grams.) The third measurement is determined by treating each person with a drug to kill all the worms and then carefully counting the number of dead female worms in the faeces (males do not lay eggs). It is necessary to examine the faeces after treatment, to make sure that no more eggs are being passed. If eggs are still being passed, some worms would still be alive and further treatment would be necessary. The number of female worms obtained from further treatments would be added to the number from the first treatment.

The average number of eggs per female per day would be calculated as follows:

$$\frac{A \times B}{C}$$

Usually there are about the same number of male and female worms in the host. Thus, once we have determined the egg count (A) for a person, we can estimate the number of worms present (2 × C), because we already have an estimate of (B).

Investigations on Gut Parasites in Niugini

Not very much work has been done on the gut parasites of man in Niugini. Most of the work has been concerned with the collection of data on the incidence of the various parasites. Our knowledge in this field is mostly due to the work of the following:

Dr S. M. Lambert who was in charge of the anti-hookworm campaign in 1920-1. His results are published in two survey reports which form part of the reports for the Australian Hookworm Campaign.

Dr G. A. M. Heydon who did a survey in the Highlands in 1934-5. The report of this work was published in 1940. Heydon was accompanied by Mr A. J. Bearup, who together with Dr J. J. Lawrence, did the parasitological work for the New Guinea Nutrition Survey Expedition in 1947.

Dr P. S. Venkatachalam, in 1957, reported the incidence and load of helminths in the Chimbu people.

Dr A. P. Vines and Mr A. Kelly conducted parasitological surveys in 1966 in three areas of Niugini; the Highlands, the New Guinea Islands, and the lowlands of the North Coast (listed here as North-Coast Mainland).

H. A. Bern and M. F. Hansen conducted a survey in the North Markham area 1944, while they were serving with the US Airforce.

In addition, Mr. S. H. Christian MBE, of Minj has supplied us with some unpublished data for the Niugini Highlands.

4.2 Ascaris

Ascaris lumbricoides is a nematode or round-worm, which normally lives in the small intestine of man. It is a common parasite in Niugini and in other tropical countries. It also occurs in cold countries. About 800 million people in the world are infected with it, that is, about one quarter of the people in the world. A closely related species called *Ascaris suis* is common in pigs. It also occurs in Niugini.

Morphology

The adults of *Ascaris* are between six and twelve inches long and nearly as thick as a pencil. They are the largest nematodes that live in the small intestine of man. The females are usually a little larger than the males. The posterior (tail) end of the male is curved and has two spicules or spines, about 3 millimetres long, which usually protrude from the inside of the curved part of the tail. These spicules can be withdrawn inside the body. They are probably used to hold the female during copulation or mating. (See Fig 4.3.)

The Life Cycle

The female worms lay eggs at a great rate. It has been estimated that they lay about 200,000 eggs each day. The eggs pass out of the gut mixed with the faeces. They have a thick wall which protects the organism inside the egg. Eggs can survive for up to six years in moist soil.

A tiny worm, or larva, develops inside each egg. It takes about 14 days to become fully developed and then the eggs are infective to man. The larva can be seen wriggling inside the egg if it is examined under a microscope.

If eggs containing developed larvae are eaten, the larvae hatch out in the small intestine. They then burrow into the wall of the intestine and enter the tiny blood vessels or lymphatic vessels which drain into the blood system, so that the larvae

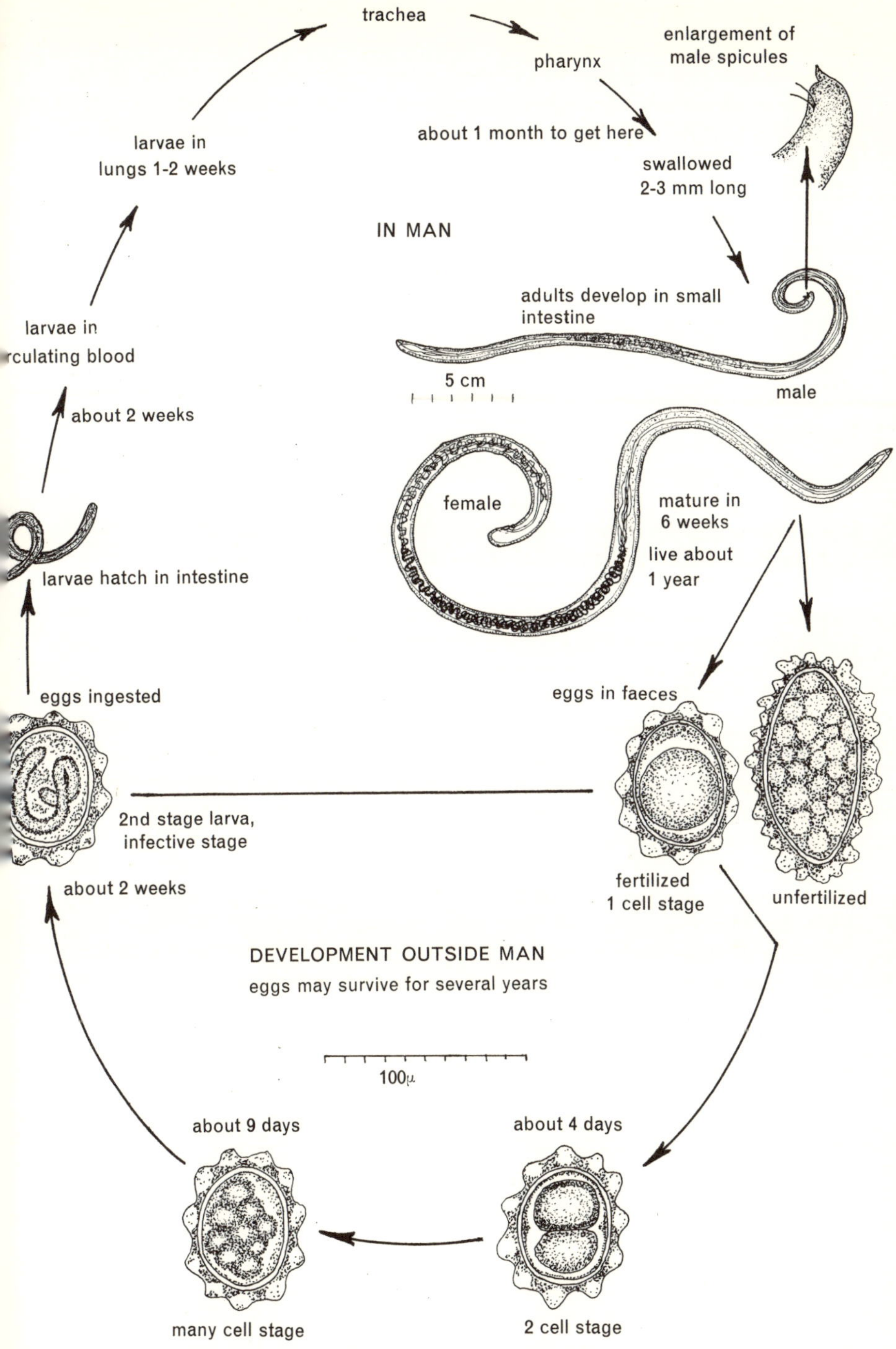

Fig. 4.3 Life cycle of *Ascaris*

are carried around in the blood as it circulates. Many are carried to the lungs, but some end up in other organs.

The lungs consist of tiny blood capillaries which branch and fuse, in between which are air spaces. The larvae break out of the tiny blood capillaries in the lungs and wriggle into the air spaces. They stay in the lungs for a week or two. During this time they moult twice. Moulting is a process where their skin comes off and they grow very quickly before their new skin becomes too tight.

After moulting, the larvae wriggle out of the lungs and up the windpipe to the back of the mouth. They are then swallowed into the stomach and pass into the intestine. At this stage they are 2 to 3 millimeters long, and they have taken about a month to hatch and migrate to the lungs and back to the intestine.

When they reach the intestine they start to grow. They are fully mature after another six weeks and the females then start to lay eggs. The adult worms die after about a year and they then pass out in the faeces.

Pathology

The larval stages of *Ascaris* affect man in different ways from the adult worms.

After they hatch from the eggs the larvae move into the bloodstream. About half of them end up in the lungs where they burrow out of the blood capillaries and migrate back to the intestine. However, some of them are carried in the blood to other organs such as the liver, lymph nodes, thymus, spleen and brain. Enlargement of the liver is sometimes a feature of such migrations. Perhaps the larvae are unable to tell the difference between these organs and the lungs; or perhaps they are caught in capillaries which are too small for them to pass through. About half of the larvae never reach the lungs but burrow out of the capillaries, into other organs, where they eventually die. However, before they die, they may wander about in these organs for some time damaging the cells about them. When they die they cause an allergic reaction. As the worms disintegrate (break up), their proteins leak into the surrounding tissues and this causes various white cells to gather in the area (refer to Section 2.2).

When it breaks out of the capillaries in the lungs, each larvae causes a little bit of bleeding. It also irritates the tissues of the lungs and causes coughing. If there are large numbers of worms, they may produce symptoms like those of pneumonia or asthma, with a fever, which may last for about two weeks. During this time larvae are often coughed up and may be seen moving in the sputum (or spit) when it is examined under a microscope.

Whenever large numbers of nematode larvae migrate through tissues of the body, there is an increase in the number of eosinophils in the blood. This is called eosinophilia. Eosinophils are one of the kinds of white blood cell (see Section 2.2). Usually from 1 to 4% of the white blood cells are eosinophils, but when the body is infected with certain helminths, they may make up 10 to 70% or more of the total number of white blood cells in the body. Eosinophilia is not a feature of true pneumonia.

Adult *Ascaris* worms do not cause high eosinophilia, because they are not in contact with the tissues.

When the lungs are damaged by the parasites other organisms such as bacteria, may enter the lungs and start to grow. This is known as a secondary infection; that is, a second organism becomes established after the first organism has damaged the tissues and provided a place for the second infection to enter.

Sometimes secondary infections are more serious than the first infection. For example, bacteria which cause true pneumonia may become established in the lungs.

The adult worms may cause symptoms in the intestine. These are not usually severe unless many worms are present. An infected person may sometimes get bad pains in the belly and often there is a loss of appetite, that is, the person does not feel much like eating.

If large numbers of worms are present they may block the inside of the gut so that the food cannot pass through. The gut may also become blocked when certain drugs are given to kill the worms.

Sometimes the worms burrow through the wall of the gut into the body cavity. Occasionally they may block the appendix or the ducts leading from the gall bladder and pancreas into the duodenum. These things can cause serious damage.

As many as 500 worms have been found in one person, but most people who are infected only have a few worms.

Ascaris is common in pigs in Niugini. Experiments on human volunteers in other countries have shown that pig *Ascaris* eggs will hatch in man and cause the usual symptoms when the larvae migrate through the lungs. Most of these experiments have shown that adult worms of pig *Ascaris* do not develop in man. However, in some experiments in Japan pig *Ascaris* eggs did produce adult worms in man.

In villages in Niugini where pigs live in close contact with the people, it is almost certain that the larvae from pig *Ascaris* frequently migrate through the lungs of man. They may even be able to develop into adult worms in man as they can do in Japan. Work should be done to find out whether cross infection can occur and how important it is. Pneumonia is responsible for about 10% of all hospital admissions in Niugini and, the part *Ascaris* larvae, from man and pigs, play in opening the way for secondary infection due to bacteria should be investigated.

One way to find out about this would be to do an experiment in several villages where the incidence of *Ascaris* was high. The following measurements should be made:

(i) The incidence and load of *Ascaris* in man.
(ii) The incidence and load of *Ascaris* in pigs.
(iii) The amount of 'pneumonia' due to *Ascaris* larvae and to bacterial infections.
(iv) The number of *Ascaris* eggs in the soil around the houses and gardens. (There is a special technique for getting eggs out of the soil.)

It would be best to make these measurements every month for a year, because they probably vary with the seasons. An all-out effort could then be made to control *Ascaris* infections, using drugs and an intensive education program. The measurements would be continued for at least another year. The control program could be designed to find out:

a) How often treatment must be given to control infection (people in one village could be treated once; in another, every six months; in another, every month, and so on).
b) How important pig *Ascaris* infections are in man (this could be done by treating the pigs in some villages, as well as the people).

The only way to find out if pig *Ascaris* can infect man in Niugini is to experimentally infect volunteers who do not have any *Ascaris* worms in their bodies, and see if they develop.

Adult *Ascaris* worms and many other gut nematodes, produce substances which neutralize the enzymes that digest proteins in the intestine. Presumably their function is to prevent the skin of the worm from being digested by the host in the same way that food is digested. These substances have the effect of reducing the amount of protein that is digested by the host. In areas where the people do not get enough protein to eat, the presence of many *Ascaris* worms may seriously increase the amount of malnutrition. Most protein comes from meat and most of the people in Niugini do not eat enough meat, so that what little they do get may not be efficiently used if they are heavily infected with nematodes. No one knows exactly how much protein digestion in man is reduced by nematode infections, and this should be found out.

Epidemiology

Most *Ascaris* infections are picked up in the house, under it, or near it, because the ground is polluted with faeces. Small children, who have not yet learnt to use toilets are probably the main source of infection for people of all ages. The eggs are not infective until they contain mature larvae which develop about two weeks after the eggs have passed out in the faeces.

Children and adults pick up the eggs from their dirty fingers. Small children may become infected by eating dirt which contain eggs. Eggs are sometimes present in dust and may be taken into the body in this way.

Occasionally, the water supply may be contaminated with faeces containing eggs and in countries where human faeces are used as garden fertilisers *Ascaris* eggs may sometimes be found on vegetables.

The eggs survive best in damp, shady places and in clay soils, which hold more water than sandy soils and thus do not dry out as quickly. Under ideal conditions the eggs may survive for several years. They are killed by exposure to direct sunlight, especially under dry conditions. Exposure to a temperature of about 45°C (113°F) kills all eggs in half an hour. The eggs are not killed by disinfectants that are commonly used to kill bacteria and they can survive in weak acid solutions.

Each female lays about 200,000 eggs each day and lives for about a year. For the species to survive, only two of these eggs (one to produce a male and one a female worm) need to enter man and mature. So we can see that many eggs can be wasted.

The incidence of *Ascaris* infection usually varies with age. In some surveys the incidence is higher in children, in others it is higher in adolescents, and in others it is highest in adults. There is no fixed pattern as there is with malaria. Most writers think that the incidence of infection depends only on the amount of exposure to infection. They think that differences in incidence in different age groups, are due to the amount of contact with infected eggs. However, situations where the incidence is highest in adolescents and lowest in older people might be due to the development of immunity to infection. The more a person has been infected in the past, the more resistant he would be to further infection. There is some evidence that pigs develop immunity to *Ascaris* infections as a result of previous infection.

TABLE 4.1 Incidence of *Ascaris* infections in different age groups, in various countries.

Age Group (years)	Iran Males	Iran Females	Taiwan	Western Samoan Village	British Honduras Village A	British Honduras Village B	Dutch New Guinea
0–10	58	62	45	43	30	46	79
10–20	70	80	30	50	30	28	57
>20	73	68	26	36	Not measured		46

TABLE 4.2 Incidence of *Ascaris* infection in different age groups in Niugini. (1966 — after Vines).

Age Group (years)	Highlands	North Coast Mainland	Niugini Islands
1–4	44	8	52
5–14	55	15	36
15–44	63	10	26
>45	65	5	20
All ages	59	10	31

The incidence of *Ascaris* infections in various age groups is given in Tables 4.1 and 4.2. In most localities, the incidence is lowest in the older age groups. This may indicate the development of immunity, or it may indicate that adults are less frequently exposed to infection.

Studies in some countries have shown that the incidence of *Ascaris* infection changes seasonally. The average number of worms per person also increases at certain times of the year. These fluctuations are probably connected with local climatic conditions. For example, during a hot dry season most eggs probably die before they become infective, but they would stay alive during the wet or cool season. In one study, the fluctuations were connected with seasonal changes in diet. In this case most infections occurred when certain vegetables, which were contaminated with *Ascaris* eggs, were eaten. These vegetables were only available during certain months.

One study of *Ascaris* infection in Germany, showed that the incidence of infection was related to the system of sewage disposal. In this study 3 per cent of the school children were infected in city areas where there was a sewage system connected to every house. In rural areas, where there was no sewage, 7 per cent of the children were infected and in one area where faeces flowed into a pool which overflowed into gardens, 14 per cent of the children were infected.

Sometimes the incidence of infection in males is different from that in females. For example, a survey in West Irian (Dutch New Guinea) in 1945, showed that the incidence in males was 49 per cent and in females, 61 per cent. Similar results are given for a survey in Iran in Table 4.1. No such sex differences have been reported for Niugini. Where such differences are found the incidence is usually higher in females than in males. This is probably because girls and women spend more time near the houses and gardens which are often polluted with faeces. They also look after small children who are a constant source of infection.

Ascaris in Niugini

Table 4.3 gives the incidence of *Ascaris* as recorded in various surveys in Niugini. It varies considerably from place to place and is highest in the Highlands and at Kavataria in the Trobriand Islands.

TABLE 4.3 Incidence of *Ascaris* in Niugini

		Incidence %
HIGHLANDS		
Highlands Survey	1966	59
Mogay (near Chimbu)	1940	90
Koruguru (near Chimbu)	1940	9
Kundiawa School	1947	66
Goroka	1949	26
Chauve	1947	50
Chimbu	1947	62
Chimbu—adult men	1957	65
Chimbu—children	1957	74
North Coast-Mainland		
North Coast Mainland Survey	1966	10
North Markham (7 localities)	1944	4
Busama (Huon Gulf)	1947	13
Kaiapit (Markham Valley)	1947	1
Patep No. 2 (near Lae)	1947	0
Territory of New Guinea (including Islands)	1921	11
Plantation workers		4
Village people		16
Niugini Islands		
Niugini Islands Survey	1966	31
Papua		
Coastal Papua	1920	12
Plantation Workers		6
Village people		17
Mission stations		8
Karavagi (Purari Delta)	1947	24
Kavataria (Trobriand Islands)	1947	84
Omu (Papuan Highlands)	1956	0

There is not enough information for us to be able to analyse the reasons for the wide variations in incidence from place to place.

Dr Vines found that in the Highlands the incidence was higher in villages built on flat muddy land (67 per cent) than in villages where the land was steep (50 per cent) and the water ran away quickly. Obviously, where the land was steep, the rain water was washing away many of the eggs.

In places where toilets are built over the sea or a big river a low incidence of infection might be expected, because many of the eggs would be washed away. However, no one knows how long eggs can survive if they are washed up onto the beach.

No one knows whether infection varies seasonally in Niugini. It probably does, in areas where there is little or no rain during the dry season.

The data in Table 4.3 suggests that there has not been any decrease in the incidence during the past fifty years. It would be interesting to re-survey some of the villages for which incidence data was collected many years ago.

Many patients in Niugini are admitted to hospital with vague belly pains. The only cause that is often found for their symptoms is *Ascaris* infection. These cases accounted for 12 per cent of all hospital admissions in the Highlands (1962-3) and 4 to 5 per cent of admissions in other areas (1963-4).

No deaths were recorded due to *Ascaris* infection during these periods although six deaths were recorded in 1964-5 as being due to helminths other than hookworm. They were probably due to *Ascaris*.

Other diseases of the gut (gastro-enteritis and colitis) account for about 5 per cent of hospital admissions and some of these are probably due to nematodes but are not diagnosed because of the lack of trained staff, or because no one has thought to examine the faeces for eggs.

Pneumonia accounts for about 11 per cent of all hospital admissions and some of these are probably due directly to migrating nematode larvae or to secondary infections resulting from them.

For every hospital admission which is diagnosed as an *Ascaris* infection, there must be many more people with belly pains due to *Ascaris* worms, who are never treated.

Very few measurements have been made of the 'load' of *Ascaris* infections in Niugini. In the Vines survey of the Niugini Islands, about 86 per cent of the people who were infected with *Ascaris* had egg counts of less than 1000 eggs per gram. This means most of the people had only one or two worms. About 7 per cent had egg counts of more than 5000 eggs per gram; about five female worms, or more.

Egg count data for *Ascaris* infections in adult males and children from Chimbu villages are given in Table 4.4. It shows that the load was higher in children than in adult males. About half the children that were infected, had a load of 15,000 eggs per gram, and about 9 per cent of all children had higher loads. However, only 3 per cent of the adult men had loads exceeding 5000 eggs per gram.

TABLE 4.4 Egg count data for *Ascaris* infections from Chimbu villages 1957. Showing the number of people with various loads (after Venkatachalam)

	Number	Egg counts (eggs per gram)		
	Uninfected	<1000	1–5000	>5000
Adult males	28	30	19	2
Children	30	28	46	10

Migrating *Ascaris* larvae cause eosinophilia. Measurements of the incidence and degree of eosinophilia give some indication of the importance of migrating nematode larvae. This is discussed, in relation to Niugini, in the section on larva migrans. (Section 6.3.)

4.3 Hookworm

Introduction

Hookworm is number-one nematode infection of man in the world. Like *Ascaris* it is common in Niugini and other tropical countries. It has been controlled in many European countries by the introduction of good sanitation and the use of drugs, but about a quarter of the world's population are still infected with hookworm.

Hookworms infect about 70 per cent of the people in Niugini, but it does not cause much obvious disease. It seldom makes people very sick. However, year after year it makes people a little less efficient. It makes them lazy. They are less able to work hard. It affects children so that they cannot think and learn as quickly as they would if they were free of this disease.

We will give just one example of the effect of hookworm. Many years ago the International Health Board conducted a great campaign against hookworm in many countries. Managers of tea plantations in Darjeeling, India, estimated that after hookworm was controlled, each labourer did from 25 to 50 per cent more work than he had done before.

Hookworms are nematodes that live attached to the lining of the small intestine. They suck blood, and if large numbers are present they cause anaemia due to the loss of blood.

Two species of hookworm occur in man in Niugini. They are *Necator americanus* and *Ancylostoma duodenale. Ancylostoma* is rare in man here. Two other species, *A. braziliense* and *A. caninum*, occur in cats and dogs and these sometimes enter the skin of man and cause a condition known as creeping eruption.

Morphology

Hookworms are pink or white and about one centimeter long. They have a large buccal capsule or mouth, which has cutting plates or teeth for holding on to the wall of the intestine. They have a strong oesophagus or pump which sucks fluid (mostly blood) from the wall of the intestine. Some of the characteristics of the two species in man are outlined below (see also Fig. 4.4).

	Necator americanus	*Ancylostoma duodenale*
Size: Males	5-9mm × 0.3mm	8-11mm × 0.4mm
Females	9-11mm × 0.4mm	10-13mm × 0.6mm
Shape:	Head hooked in opposite direction to curvature of body.	Body in one curve
Mouth:	One pair of cutting plates.	Two pairs of teeth
Size of eggs:	64-76μ × 35-40μ	56-60μ × 35–40μ

The males have an unusual tail. Instead of being pointed it is expanded into two flaps of skin, called a copulatory bursa. These flaps fold around the body of the female during copulation (or mating) so that sperm can be injected into the vagina of the female. Two worms are shown in copulation in Figure 4.4.

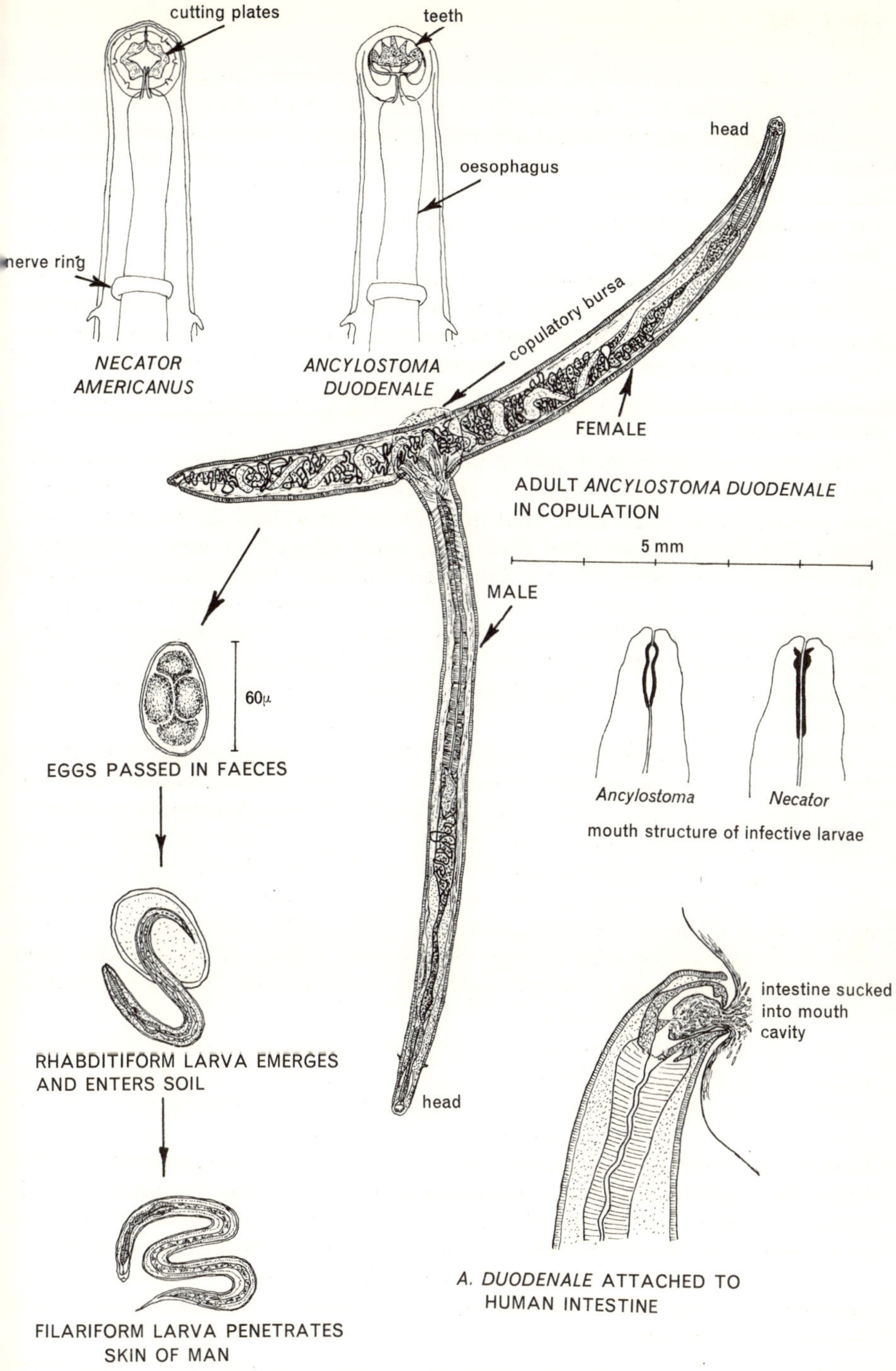

Fig. 4.4 Hookworm morphology and life cycle

Life Cycle

The adult worms live in the small intestine. The females lay many eggs which pass out in the faeces. Estimates of the number of eggs laid by each female per day have been made by several different scientists and they vary a great deal. Recent work in South America found that *Necator* lays an average of 3400 eggs per day. Most other estimates are higher than this (5000 to 10,000). Estimates for *Ancylostoma* are usually about twice as large as those for *Necator*. It seems that egg production depends partly on the number of worms present. It has been suggested that if only a few worms are present each female may produce up to 50,000 eggs per day. No estimates have been made of egg production in human hookworms in Niugini. This should be done because there may be considerable variation from one part of the world to another.

The adult worms have a life span of 5 to 17 years, but many of them probably die within two years.

The eggs pass out in the faeces, and a tiny worm or larva hatches from each egg about 24 to 48 hours afterwards. It is called a rhabditiform larva because it looks like a nematode called *Rhabditis*. These larvae feed on bacteria in the faeces. The larvae moult (shed their skin) twice before they become infective. The second moult occurs 5 to 8 days after hatching. The infective larvae (or filariform larvae) do not feed, but they remain active for several weeks under favourable conditions. They come to the top of the soil and may be found in clumps, standing on their tails, with their heads waving in the air. They attach themselves to any warm-blooded animal that touches them. They are about half a millimeter long.

Hookworm larvae are easily grown using the following method: mix about 5 grams of faeces from an infected person, with about the same quantity of wood charcoal that has been crushed to pieces about 2 to 4 millimeters in diameter. This can be done in a small jar or tin. Place this container in a larger jar, in which there is a little water to keep the humidity high. Cover the large jar with a lid or a plastic sheet that is tied on tightly. Keep in a shady place. Examine the culture under a dissecting microscope or a good hand lens after about six days and the larvae will be seen waving about on the surface of the mixture. Dog faeces may also be used, as most Niugini dogs are infected with hookworms that are very similar to those in man.

The infective (filariform) larvae of *Ancylostoma* and *Necator* are very much alike to look at, but they differ in the shape of the mouth. The mouth of the infective larvae of *Necator* is heavily thickened and has out-turned tips while in *Ancylostoma* it is hardly thickened at all and has inturned tips (see Fig. 4.4). In order to tell what species are present it is necessary to culture infective larvae from faeces and identify them. Hookworm larvae must also be distinguished from the larvae of another gut nematode of man, called *Strongyloides*. The eggs of *Strongyloides* usually hatch before they are passed in the faeces, while those of hookworms take 1 to 2 days to hatch. This means that if nematode larvae are found in fresh faeces, they are probably *Strongyloides* (see Fig. 4.9 and Section 4.5 for details).

Soil and plant nematodes are sometimes found in human faeces. They may have been ingested with vegetables or may contaminate faeces from the soil.

Hookworm larvae infect man by wriggling into the skin. This happens when our feet or hands, or other parts of our bodies come into contact with polluted soil.

During the last war, Australian soldiers often became infected while sitting on contaminated soil. This method of infection is probably not common in adult Niuginians because they generally prefer to squat down.

After they have penetrated the skin, the larvae enter the lymphatic vessels or blood capillaries and are carried in the blood stream to the lungs. Here they break out of the blood capillaries and move up the windpipe, to the back of the mouth, where they are swallowed into the gut. They moult for the third time and then grow into adults. The females start to lay eggs about five weeks after they have entered the skin.

Pathology

We will discuss the pathology of hookworm under three headings:

(i) Ground itch
(ii) Damage caused by larvae
(iii) Effects due to the adult worms.

(i) Ground Itch

When hookworm larvae penetrate the skin, they cause intense itching and the skin where they enter becomes red and a little bit swollen. This is called ground itch because it comes when a person has been in contact with polluted ground. It does not last long. The larvae soon find their way into the lymphatics or blood vessels and stop irritating the skin.

Creeping Eruption

The larvae of hookworms of cats and dogs may penetrate the skin of man. The cells a millimeter or two below the surface of the skin of man are tightly packed, and these foreign parasites seem to be unable to penetrate through them, into the deeper layers where the blood and lymph vessels are found. They just wander about in the surface layers of the skin. This condition is called creeping eruption.

The larvae tunnel a few millimeters or centimetres each day, leaving an irregular winding track in the skin (see Fig. 4.5). This causes severe itching. It may persist for several weeks or months. Creeping eruption occurs in Niugini where people sit on soil that is polluted by cats and dogs. Sometimes it occurs on beaches. It can be treated by freezing the skin at the end of the track, with dry ice (solid carbon dioxide) or ethyl chloride sprays.

(ii) Damage caused by migrating Larvae

While most larvae are carried in the blood to the lungs, some are caught in other organs and produce various symptoms before and after they die. When they break out of the blood capillaries in the lungs, they cause tiny haemorrhages (bleeding). However, it is unusual for migrating hookworm larvae to produce any pronounced symptoms unless they occur in very large numbers at the one time. They can then cause symptoms like pneumonia, as does *Ascaris*. The larvae may sometimes open the way for secondary lung infections.

Sometimes there is eosinophilia, but this seems to be less frequent and less severe than that associated with migrating *Ascaris* larvae.

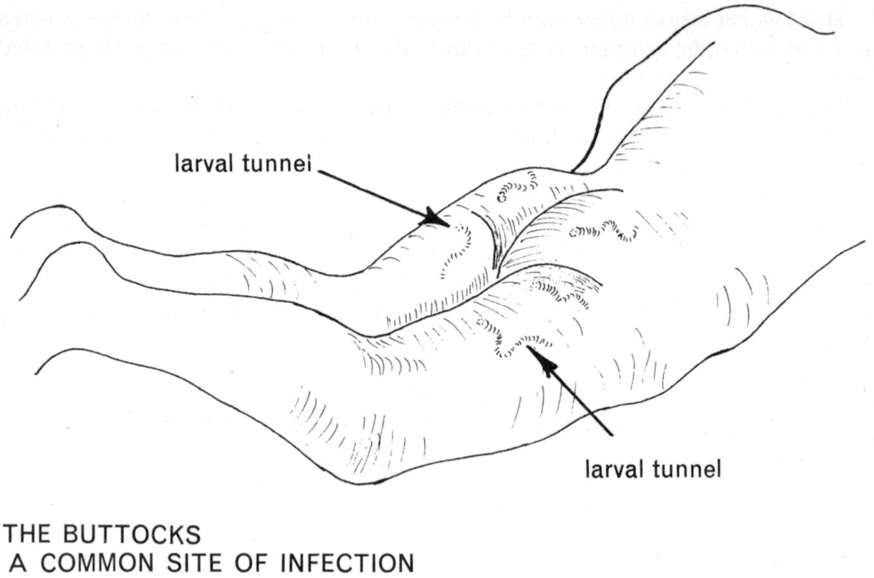

Fig. 4.5 Creeping eruption. Tracks caused by larvae of the hookworms of cats and dogs

(iii) *Effects due to Adult Worms*

Adult hookworms damage the host in a number of ways. The most important is the removal of blood. If many worms are present the host may become anaemic. This anaemia seems to be caused mostly by the loss of blood, but other factors may also be involved. Much more research is needed before this host-parasite relationship is completely understood.

The symptoms caused by hookworms vary with factors such as the number of worms present, the age of the infection, the nutritional status of the host and the amount of reinfection going on.

People who have a good nutritional status get enough of the right kinds of things to eat. Our bodies need a certain amount of several kinds of substances in order to work properly. For example, we need foods which provide energy. Fats, carbohydrates and proteins all do this. Protein is also necessary for growth and the repair of worn out and damaged tissues. Small amounts of various elements such as sodium, calcium, iron and iodine are also necessary. We also need a whole range of substances called vitamins.

The nutritional status of Niuginians is generally good although Niuginians eat about half as much protein as Australians do. But in some areas, the diet of the people appears to be deficient in protein, iron, calcium and iodine.

The Symptoms of Hookworm Disease

Light infections cause vague belly pains and a general feeling of laziness. Heavy infections produce more severe symptoms including:

(a) diarrhoea, alternating with constipation
(b) stunted growth and delayed sexual maturity

(c) mental retardation
(d) anaemia.

Children are particularly affected by heavy infections. They may develop a swollen belly and a puffy face, although other diseases cause these symptoms. Most adults with heavy infections do not look undernourished even though they may be anaemic.

Anaemia

Recent research work shows that each *Necator* sucks about 0.03 milliliters of blood per day. The figure for *Ancylostoma* is about 0.08 milliliters per day. There is some evidence that when only a few worms are present, each worm sucks more blood. Most textbooks give figures about 10 times greater than these values, but they are based on old measurements that were made using poor techniques.

Working on a blood loss of 0.05 milliliters per worm per day, an infection of 100 worms would cause a loss of about 2 liters of blood in a year. This is not enough to cause anaemia in a healthy person. Blood donors often give a liter of blood every 2 to 3 months without any ill effects.

Loads of about 500 worms often cause anaemia in adults and smaller loads can cause anaemia in children.

Blood loss seems to be the most important cause of hookworm anaemia. Anaemia develops when iron, which is an essential part of haemoglobin, is lost from the body faster than it is replaced from the food. This does not happen quickly because the body usually contains some reserves of iron. Hookworm anaemia appears to be almost entirely due to iron deficiency. People who eat little or no meat may not absorb enough iron from their food. They may also lose some iron when they sweat.

Other factors may be involved in hookworm anaemia. The relative importance of these factors is not known at present. Further research work may show that some of them are not important at all. Some of these factors are:

(a) Toxic substances secreted by the worms may inhibit red blood cell production in the bone marrow. This idea is supported by experiments where waste products from hookworms grown in test tubes, caused anaemia when injected into rabbits.

(b) There is some evidence that the amount of hydrochloric acid secreted into the stomach, is less than normal when a person is infected with hookworm. This seems to result in a decrease in the amount of digestion of protein, iron, vitamin A and folic acid. Folic acid is one of the 'B' vitamins which is necessary for the manufacture of red blood cells.

(c) Many gut helminths, including hookworm and *Ascaris*, produce anti-enzymes which neutralise the enzymes secreted into the gut by the host. These anti-enzymes reduce the amount of food that is digested and thus available to the body. Hookworm infection does not appear to be important in this regard. Most adults with heavy hookworm infections do not appear to be suffering from malnutrition, although this can happen. However, in places where there is constant protein deficiency, any interference with the utilization of available food may be important, especially in growing children.

(d) The life of red blood cells is reduced in patients suffering from heavy hookworm infection. This is usually corrected when iron tablets are given.

Hookworms damage the lining of the gut where they attach themselves. As they move about from place to place within the gut, they leave behind damaged areas

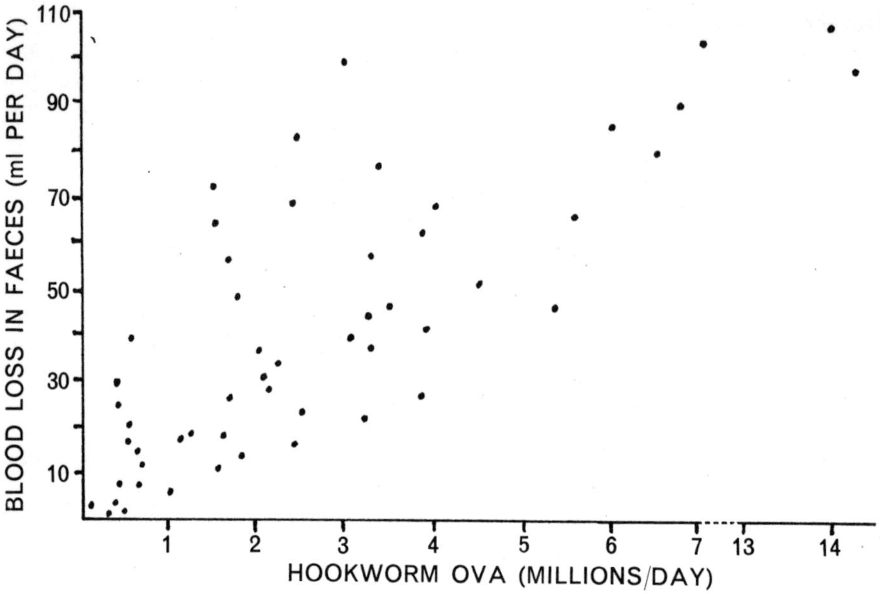

Fig. 4.6 Relationship between blood loss in faeces and hookworm eggs per day (after Martinez-Torres, Ojeda, Roche and Layusse, 1967)

which probably continue to bleed for a time. However, these areas appear to heal fairly quickly and the gut soon returns to normal if the worms are removed by drugs.

Several drugs are used to kill hookworms. They are listed in Section 4.11. If a person is anaemic due to hookworms, it is often necessary, not only to kill the worms, but to give iron tablets and extra protein in the diet to make them well again.

The Relationship between Load and Hookworm disease

Most studies of hookworm disease have shown that the level of anaemia (as measured by the amount of haemoglobin in the blood) is related to the number of worms present, and thus to the egg count. When egg counts are low there is no anaemia. In *Necator* infections, the haemoglobin level usually starts to drop below normal when the egg count reaches 5000 eggs per gram in adult men and 2000 per gram in women and children. Women are more susceptible to hookworm anaemia because they lose some blood every month when they menstruate. This does not happen while they are pregnant, nor does it happen in very old women.

As the egg count becomes higher the number of worms that are present is greater, the anaemia is more severe, and the haemoglobin levels tend to be lower. The haemoglobin values are said to be inversely related to the worm load. They become smaller as the load increases.

Fig. 4.6 is a graph showing the relationship between the amount of blood lost in the faeces (in milliliters per day) and the total egg production per person per day. It is taken from a study done in South America on 53 people who had severe *Necator* infections. Each dot represents the data from one person. It shows quite clearly that the blood loss increases with the total egg count. It is interesting to note that there is considerable variation in the amount of blood lost from different individuals which have about the same load. This provides some support for the idea that, at least in some individuals, anaemia depends on other factors besides the worm load.

As many as 6000 hookworms have been removed from one person. Large numbers of worms cause severe anaemia. For example, haemoglobin values as low as 4 grams of haemoglobin per 100 milliliters of blood (or 4 grams percent) have been recorded. The normal values for healthy Europeans are about 16 grams per cent for men and 14 grams per cent for women.

A few studies of hookworm disease have not shown any relationship between haemoglobin levels and anaemia. They are exceptions. One of these studies was done in Niugini and we do not know why these results were different from most other studies. We would like to make two suggestions about this which could easily be tested.

The first is that the nutritional status of people here is high, and they get enough iron in their food to make haemoglobin as fast as it is lost. The second is that egg counts may give a false indication of the worm load. We have already mentioned that if only a small number of worms are present, the egg production per worm might be very much higher that it is when the load is very high. For example, one worker has estimated that when only about ten females were present per person, each *Necator* produced nearly 30,000 eggs per day. That is about 300,000 eggs were passed in the faeces per day. Another worker has shown that when the load was about 600 worms (that is about 300 females) each female produced only about 3000 eggs per day. This would amount to 900,000 eggs per person, per day. In other words, when the load was increased thirty times the egg output only increased three times. We do not know whether this sort of thing really happens or not, and there is no data available for Niugini which can be used to test it.

Epidemiology

Hookworm exists mainly because people walk in their own and other people's faeces. Infective larvae take about 6 days to develop after the faeces have been passed. They can live for several weeks. Infection occurs when the skin comes in contact with infective larvae. In this country most infection probably occurs through the feet. In countries where faeces are used as garden fertilizer infection may take place through the hands when people are working with contaminated soil. As far as we can find out, faeces are not used as garden fertilizer in Niugini. Niuginians seldom defaecate in the gardens, but go into the bush, which is usually not far away.

Under ideal conditions, the infective larvae can remain alive for up to six or seven weeks. The survival of infective larvae is affected by many factors. They survive best in sandy soils with plenty of humus (decaying plant material) and in shady areas, so that the soil does not become too hot and dry. The best temperature for survival is between 24 and 32°C (75 to 90°F). Any shady place in Niugini is within these limits most of the time.

Heavy hookworm infections are rare in places with less than about 40 inches of rain per year, because the larvae need moist conditions. Transmission probably does not take place during the dry season in parts of Niugini, such as the southern coast around Port Moresby, where there is a long dry season with very little rain.

In places where there is rain almost every day, many larvae are washed away, especially where the ground is steep, and many others cannot survive because the soils are waterlogged. Heavy clay soils become water logged more easily than other kinds of soil.

Human habits play a most important part in infection. Many people who work away from toilets, for example, in gardens or among plantation crops, have their own little place which they regularly use for defaecation. This becomes a site for constant reinfection. Such people should be encouraged to dig small pits and to use them. A hole about as big as a large post hole, serves one person for quite a while. Where this is not practical, people should be told why they should use a different place each time and why it is important not to defaecate near tracks through the bush.

Everybody should be encouraged to wear sandals or thongs. This would not only prevent infection by hookworm, but it would also prevent many foot injuries. Shoes and long socks are becoming fashionable, but we think that the continual use of long socks in a hot humid climate, may encourage the growth of various fungi which often live between the toes.

One text book suggests that the habit of chewing betel nut, may be important in reducing the hookworm load. Because this habit causes spitting, many of the hookworm and *Ascaris* larvae which are migrating from the lungs to the mouth, are probably removed from the body before they reach the gut. Betel nuts also contain a chemical called Arecoline, which is an alkaloid that is used to treat tapeworms.

Europeans appear to be much more susceptible to hookworm disease than American Negroes. There have been many cases of hookworm disease among Europeans living in Niugini. It is possible that Niuginians are less likely to be seriously affected than Europeans are.

In areas where there is a lot of hookworm infection, the incidence and load of infection is usually highest in adolescents and young adults. Old people are said to develop immunity or resistance as a result of constant infection. The chance of becoming infected increases with time. As a child gets older he is more likely to become infected, and, because the worms live a .long time, the worm load will gradually increase. As a result of constant infection, it appears that immune mechanisms develop, so that the larvae are killed before they grow into adults.

It is absolutely essential to determine not only the incidence, but also the load, in all hookworm surveys. The amount of hookworm disease depends on the number of worms that are present. Light infections do not usually cause any severe symptoms.

Hookworm in Niugini

Most hookworm infection in Niugini is due to *Necator*. Only about 0.05 per cent to 1 per cent of the infections are due to *Ancylostoma*.

There are usually more female worms present than males. Lambert found about 60 per cent of all the worms passed after drug treatment were females.

The incidence of hookworm infection in Niugini is given in Table 4.5. Lambert's reports give the incidence for individual villages, missions and plantations, but they are too numerous to be listed here. With a few exceptions the incidence is high: generally between 70 and 90 per cent.

No hookworm was found in the Mortlock or Tasman Islands in 1921, or in

Table 4.5. The incidence of hookworm in Niugini

			Incidence %
Papua			
Plantation workers	1917		74
Papua	1920		59
Plantations		66	
Villages		55	
Mission Stations		60	
Motu area—dry coastal		5	
Wet coastal		66	
Dry Inland		30	
Wet Inland		73	
Purari—Delta District		72	
Kavataria (Trobriand Is.)	1947		90
Koravagi (Purari Delta)	1947		94
Omu (Papuan Highlands)	1956		0
Territory of New Guinea			
(Including Islands)	1921		74
Plantations		79	
Villages and Missions		71	
Europeans		46	
Niugini Islands			
Rabaul District	1921		84
Kieta District	1921		76
Kavieng District	1921		72
Manus District	1921		77
Mortlocks and Tasman Is.	1921		0
Niugini Islands Survey	1966		78
Highlands			
Wahgi Valley	1940		94
Koruguru (Near Chimbu)	1940		100
Highlands Survey	1966		74
Chimbu Adult Men	1957		64
Chimbu children	1957		61
Chimbu	1947		92
Chuave	1947		75
Kundiawa School	1947		74
Goroka	1949		75
North Coast-Mainland			
Madang District	1921		82
Aitape District	1921		93
Morobe District	1921		86
North Markham Area	1944		80
Busama (Huon Gulf)	1947		93
Kaiapit (Markham River)	1947		87
Patep No. 2 (near Lae)	1947		89
North Coast-Mainland Survey	1966		92

the village of Omu in the Papuan Highlands in 1956. It would be interesting to know if it has spread there in recent times.

Lambert's data for Papua is interesting. The incidence in Motu villages was only 5 per cent. He thought this was due to a combination of two factors. This area has a long dry season with very little rain, during which time hookworm larvae could probably not survive. And in addition, most of the Motu people then lived on the coast and had toilets over the sea. Hookworm larvae cannot live in sea water. Dry inland areas in Papua had an incidence of about half that of wet inland areas.

Another interesting feature is the higher incidence of hookworm among plantation workers than among villagers.

The Vines survey showed that within the Highlands, the incidence varied with altitude. At high altitudes it averaged 65%, while in lower parts, it averaged 84%.

Table 4.6 gives data on the incidence of hookworm in people of different ages. It shows that about one third of all the children are infected before they are one year old. Almost everybody is infected, or has been infected, before the age of 10 years. Figures for the Highlands suggest that the incidence of hookworm is lower in old people than young adults. However, these figures are based on a small sample and a much larger sample would be necessary to show this sort of change in incidence with age.

TABLE 4.6 Incidence (%) of hookworm infections in different age groups in Niugini (after Vines, and Bearup and Lawrence, pooled data)

Age Groups in years	Highlands	North Coast-Mainland	Niugini Islands	Five Villages
<1	33	33	38	23
1–4	48	77	72	83
5–14	78	98	82	98–100
15–55	81	94	76	
45+	65	95	95	
All ages	74	92	78	90

Hookworm survey data is of little value unless it includes measurements of the worm load as well as the incidence. Most workers in other parts of the world have shown that when the egg count exceeds 5000 per gram, hookworm disease that is, anaemia, is likely to develop. We will come back to this point again, but first let us look at the relationship between the age of the host and hookworm load in Niugini.

Table 4.7 shows the percentage of people of various ages with egg counts of more than 5000 per gram. We have compiled this table from the figures of Bearup and Lawrence, by pooling their data from five villages, as if they were one. All villages had a similar incidence of hookworm, and most of them had similar worm loads. (See Table 4.8.) We should probably not pool data from different areas in this way, but it is the only way that we can get a sample that is large enough for analysis. It is the best we can do. But we must be careful in how we interpret the data. The data may suggest that certain things are happening, but these would need to be tested by carrying out a specially designed survey before we could be sure

Table 4.7 shows several things:

(i) Children under one year do not have very big worm loads. They were all less than 5000 eggs per gram.

(ii) Egg counts of 5-10,000 eggs per gram were found most often in children in the 6 to 15 age group.

(iii) Counts exceeding 10,000 were found most often in people older than 11 years.
(iv) About 16 per cent of the population had egg counts exceeding 5000 eggs per gram, and about 5 per cent of these had counts exceeding 10,000 eggs per gram.

TABLE 4.7 Percentage of people with hookworm egg counts of 5-10,000 and more then 10,000 per gram. (Pooled data for 656 people from five villages surveyed by Bearup and Lawrence)

Age in	% with egg counts as under		
years	5–10,000	> 10,000	> 5000
0–1	0	0	0
2–5	8	4	12
6–10	20	3	23
11–15	16	6	22
16–20	5	4	9
> 21	10	6	16
Total	11	5	16

Table 4.7 also suggests that some immunity develops to infection. If people of all ages were equally likely to pick up infective larvae we would expect the load to keep on increasing with age, until the number of old worms that were dying was balanced by the number of new worms coming into the host. The table shows that the number of people with egg counts exceeding 10,000 does not increase much with age. If there was no immunity developing, we think the number of people with high loads would increase much more than it does.

Table 4.9 gives data on hookworm loads in the Niugini Islands. Only 2 per cent of the people under 15 years had egg counts exceeding 5,000 eggs per gram, and only 6 per cent of older people had such loads.

TABLE 4.8 Percentage of people in villages surveyed by Bearup and Lawrence, having egg counts of 5-10,000 and more than 10,000 per gram

	Percentage with egg counts as under		
Village	5–10,000	> 10,000	> 5000
Busama	11	10	21
Kaiapit	12	3	15
Patep No. 2	20	4	24
Kavataria	3	3	6
Koravagi	6	6	12

Heydon did egg counts on 26 specimens from the Mogay area of the Highlands. The counts averaged 220 eggs per gram and the highest was only 650.

Venkatachalam reported hookworm loads in the Chimbu people. Most infections were very light, being less than 1000 eggs per gram. No counts exceeded 5000 per gram in adult men, or 3000 eggs per gram in children.

Table 4.8 gives the percentage of people in the various villages surveyed by Bearup and Lawrence having high egg counts. Between 6 and 24 per cent had

TABLE 4.9 Percentage of people infected with hookworm, having egg counts of given ranges. (From Vines, Niugini Island Survey)

Age Group	Percentage with eggs as under			Total No of egg counts
	< 1000	1–5000	> 5000	
< 15 years	70	28	2	74
> 15 years	55	39	6	82

counts exceeding 5000 and 3 to 10 per cent had counts exceeding 10,000 eggs per gram.

Lambert recorded the number of worms passed by people after treatment with drugs. His figures are probably underestimates because only one treatment was usually given and some worms probably remained in the host. Furthermore, the people picking out the worms probably missed some, especially when they were working under field conditions. In Papua, under field conditions, an average of 31 worms were recovered from each person. In another series, which were better supervised the average was 87. In the Purari Delta area, the average was 157 worms per person. In the Territory of New Guinea (including the Islands) the figure was 96 worms per person.

The highest number of worms recovered from one person was about 2000.

We must now consider the importance of hookworm disease in Niugini.

Lambert, in 1920, suggested that about 5 to 10 per cent of the population had hookworm disease. While more than 60 per cent were infected with hookworm, only a small proportion were actually sick because of hookworm parasites.

We have seen that the most obvious sign of hookworm sickness is an abnormally low value of haemoglobin in the blood (anaemia). People who have heavy loads usually have low haemoglobin values. However, Bearup and Lawrence found that the haemoglobin values of Niuginians with egg counts of more than 5000 per gram were no lower than people who were not infected with hookworm, or than people who had low egg counts. This kind of result is unusual. Most workers in other countries have found that people with egg counts exceeding 5000 eggs per gram often have anaemia.

Dr Vines found the same thing as Bearup and Lawrence, but his data included only five people with egg counts exceeding 5000. This is too small a sample to prove anything.

Tables 4.7, 4.8 and 4.9 suggest that at least 5 to 10 per cent of the people in this country might have hookworm disease because they have egg counts exceeding 5000 eggs per gram.

Data on hospital admissions for hookworm disease are difficult to interpret. Annual reports for the Territory of New Guinea, and the Territory of Papua, are published each year. They contain tables listing the number of patients admitted to Administration hospitals for various reasons. (Mission hospitals have nearly as many patients as do Administration hospitals, so the figures given represent about half the total for Niugini). These tables include admissions for hookworm disease ('Ankylostomiasis') 'Disease due to other helminths', and 'All other diseases classified as infective and parasitic'.

There are also separate entries for 'Malaria', 'Dysentery, all forms', and 'Filariasis'.

The number of patients admitted for the Territory of New Guinea for 1967-8 for the categories of interest are as follows. (The number of deaths are given in brackets).

Ankylostomiasis	509	(1)
Other diseases due to helminths	690	(0)
All other diseases classified as infective and parasitic	2682	(15)

Now, hookworm and *Ascaris* are the only two important gut helminths in Niugini. The others that occur here, are discussed later in this chapter. None are of major importance. We cannot tell what diseases are noted under the heading 'All other diseases classified as infective and parasitic', Amoebic dysentery, Filariasis (*Wuchereria*) and Malaria should all be accounted for, under separate categories. There are no other important parasitic diseases that occur in Niugini that cannot be accounted for under another heading. Perhaps this category is used as a 'waste paper basket' for admissions of unknown cause, or for helminth diseases (including hookworm and *Ascaris*) which should be included in their specific category. If no other cause of illness can be found, it is very likely that helminth ova will be present in the faeces, thus providing a diagnosis which can be entered on the appropriate form.

The fact remains that about 500 admissions, or more, a year are diagnosed as hookworm disease. In 1962-3 about 8 per cent of all hospital admissions in the Highlands were diagnosed as hookworm disease.

Presumably these diagnoses have been made because hookworm eggs were found in the faeces and the patient was anaemic. It is doubtful if egg counts are done as a routine in any hospital. They should be. It may be that in many of these cases, the anaemia was not due to hookworm at all, but to some other disease. If we accept that the hospital data truly represents the amount of hook worm disease, then hookworm is an important disease in Niugini. It would appear that more research is needed into the importance of hookworm disease in Niugini. There is also a great need to keep accurate hospital statistics on helminth diseases.

Hookworm and malaria are probably the most important causes of anaemia in Niugini, although there are others. It is difficult to tell how much anaemia is caused by each. One way of doing this, would be to select a group of people with heavy hookworm loads, and give them drugs to stop them getting malaria. Measurements should then be made of their haemoglobin levels before and after treatment to kill the hookworms. In other countries it has sometimes been found that the haemoglobin levels do not rise quickly after treatment unless iron tablets and extra protein are added to the diet. This could also be tested under local conditions.

The haemoglobin values of most Niuginians are lower than those of Europeans. In some cases they are considerably lower. Haemoglobin values vary with age and sex and the altitude where a person lives. Table 4.10 gives some haemoglobin values of Niuginians. The mean (or average) values for adult Europeans is about 16.0 for males and 14.0 for females. It is higher in people living at altitudes greater than 5000 feet.

Table 4.11 gives the average haemoglobin values of males and females of various ages from the North Coast-Mainland and Niugini Islands surveys, for people who did not have enlarged spleens. (That is, people who showed no long term signs of malaria infection). The average values for people in Europe of the same age groups are also given.

These tables show that the haemoglobin values of Niuginians are often 20 per cent below those for Europeans living in Europe or Australia.

The North Coast-Mainland seems to be the worst area and the Highlands the best. A great many words have been published giving various theories for the low haemoglobin values found in Niugini.

We suggest that the best way to find out the real causes is to conduct carefully planned field trials and surveys, to test some of these theories. We still suspect that hookworm may be one important cause of anaemia in Niugini (see also Section 3.7 for the relationship between malaria and anaemia).

TABLE 4.10 Average Haemoglobin values for Niuginians. (In grams of haemoglobin per 100 millilitres of blood)

	Adult Females	Adult Males
Sepik	12.4	14.1
Bougainville	12.0	13.1
Busama (age 11–60 years)	9.9	10.6
Kaiapit (age 11–60 years)	10.0	11.1
Patep No 2	11.5	12.0
Kavataria	10.1	11.0
Koravagi	11.8	12.5
Altitude above *5000* feet		
Minj		15.8
Telefomin	13.4	14.2
Chimbu (age 20–30 years)	14.8	16.1
Chimbu (age > 30 years)	14.2	16.0
(Comparable figure for Australia)		16.8

TABLE 4.11 Average haemoglobin values from the North Coast-Mainland and Niugini Islands surveys for people who did not have enlarged spleens. (after Vines).

Age Group (years)	Europeans living in Europe	Mainland		Islands	
		males	females	males	females
1–4	11.7	10.2	10.6	12.7	12.6
5–9	12.9	10.6	11.4	13.4	13.1
10–14	13.4	11.4	11.0	13.7	13.3
15–29		13.5	11.3	15.5	13.4
30–44		12.2	10.9	15.2	13.8
>45		11.3	11.8	14.7	13.8

Average values for European adults are about 16.0 for males and 14.0 for females.

4.4 Trichuris

Trichuris trichiura (Tri-cure-iss tricky-ura) is another nematode of man. It lives in the large intestine, but seldom causes any symptoms unless very large numbers of worms are present. Closely related species are found in many domestic animals.

Morphology

The adult worms are about 3 to 5 centimeters long. The front end of the body is very long and thin. It penetrates into the wall of the gut of the host.

The tail end of the male is tightly curled. It has a copulatory spicule which can be extended from a sheath which is covered with tiny spines. The tail of the female is straight. The eggs are oval with plugs at each end. (See Fig. 4.7.)

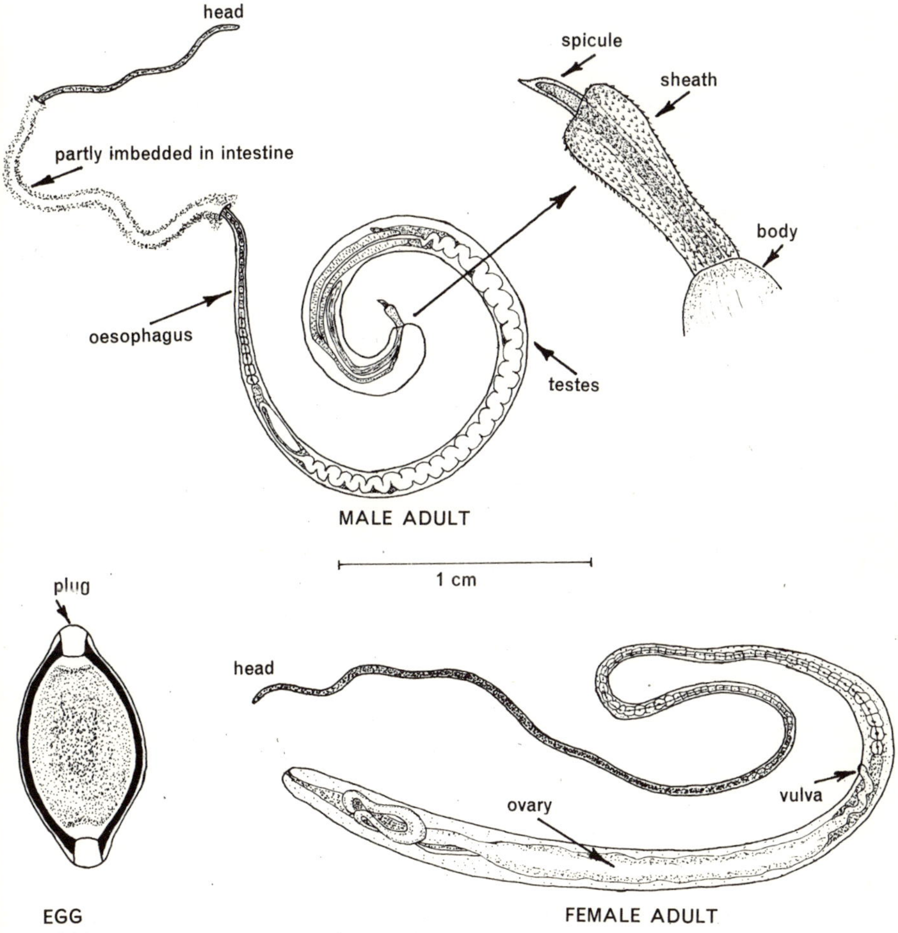

Fig. 4.7 *Trichuris trichiura*

The Life Cycle

The life cycle is very simple. The eggs pass out in the faeces of the host. Each female lays about 6000 eggs per day. The eggs are not infective until a larvae has developed within them. This takes about four weeks. The larvae hatch from the eggs soon after they are eaten by man. They take about two months to mature and live for several years in the large intestine.

Pathology

Trichuris causes no obvious symptoms unless large numbers of worms are present. Heavy infections are rare, and most infected people have no symptoms.

However, if several hundred worms are present, as indicated by an egg count exceeding 30,000 eggs per gram, the infected person usually has some symptoms. These include nervousness, loss of appetite, insomnia (does not sleep well) and vague belly pains. There may be constipation and occasionally diarrhoea. The number of eosinophils is usually increased. In very severe cases, the patient may be anaemic and lose weight.

Epidemiology

The eggs survive best in warm, shaded, moist soil. They are killed more easily than the eggs of *Ascaris*. They cannot survive in direct sunlight or in dry conditions. *Trichuris* infections are usually due to pollution around the house.

Trichuris in Niugini

The incidence of *Trichuris* in Niugini is given in Table 4.12. It varies considerably from place to place.

Only two samples of egg count data are available. Both suggest that infections in Niugini are very light. The Vines Survey (1967) of the Niugini Islands reported

TABLE 4.12 The incidence of *Trichuris* in Niugini.

		Incidence %
Papua	1920	13
Karavagi (Purari Delta)	1947	82
Kavataria (Trobriand Islands)	1947	88
Highlands		
Mogay (Salt Nomani Area)	1935	85
Koruguru (near Chimbu)	1935	2
Chimbu	1957	45
Highlands Survey	1966	37
Kundiawa School	1947	44
Chuave	1947	27
Chimbu	1947	24
Goroka	1949	16
North Coast-Mainland		
North Markham	1944	10
Busama (Huon Gulf)	1947	76
Kaiapit (Markham River)	1947	5
Patep No. 2 (near Lae)	1947	10
North Coast Mainland Survey	1966	16
Niugini Islands Survey	1966	36

that all egg counts were less than 1000 eggs per gram. Dr Venkatachalam (1957) found that less than 2 per cent of the Chimbu people examined, had counts of more than 1000 eggs per gram, and none exceeded 2000.

Trichuris infections are not listed separately in hospital admission figures, but we think that they are unimportant as a cause of sickness in Niugini.

4.5 Strongyloides

Strongyloides stercoralis (Stron-gil-oid-es ster-cor-alis) is a rare nematode of man in Niugini. Bearup and Lawrence found only two infections during the nutrition survey in 1947, and Lambert found about 0.2 per cent of the people were infected in 1920-1. *Strongyloides* is an interesting nematode because it can live as a parasite in man, or for a few generations as a free-living organism in faeces. The parasitic and free-living forms are different in size, shape and other characteristics.

The Life Cycle

All the adult parasites are females. They are parthenogenetic, that is, they lay eggs which develop normally without being fertilized by sperm from a male. There are no parasitic males.

The parasitic females are about two millimeters long and live in the surface tissues of the gut (see Fig. 4.8). Each female lays about 50 eggs per day. The eggs are almost the same as hookworm eggs, but are a few microns smaller. Most eggs usually hatch before being passed in the faeces, and all hatch within about 24 hours. The first larvae are called rhabditiform larvae. The eggs of hookworms take about 1 to 2 days to hatch, so that if nematode larvae are found in fresh faeces, they are usually *Strongyloides*. It is necessary to culture the faeces to tell if both *Strongyloides* and hookworm are present. (See Section 4.3, p. 152.)

An unusual feature of *Strongyloides* is that sometimes some larvae grow into adults without leaving the human host. Thus the load can increase without new infective larvae entering the host.

Many larvae and eggs are passed in the faeces. These larvae moult and grow into adult males and females which give rise to a free living cycle, which can go on for several generations, until conditions become unfavourable, or the food supply runs out.

A few of the larvae from each free-living generation usually develop into a larval stage which is infective to man. And when conditions start to become unfavourable, most of the larvae are of this kind.

Man becomes infected with *Strongyloides* when he comes in contact with infective (filariform) larvae. They penetrate the skin and migrate, by way of the circulatory system to the lungs, in the same way as hookworm larvae do. They then burrow out of the lungs and wriggle their way up the windpipe to the back of the mouth, where they are swallowed into the gut.

The rhabditiform and filariform larvae of *Strongyloides* and hookworms are very similar. They are illustrated in Fig. 4.9. In rhabditiform larvae of hookworms, the mouth is about as long as the width of the body: in *Strongyloides* it is very short. In the filariform (infective) larvae of hookworms the oesophagus is about one quarter of the total length of the body and the tail is bluntly pointed.

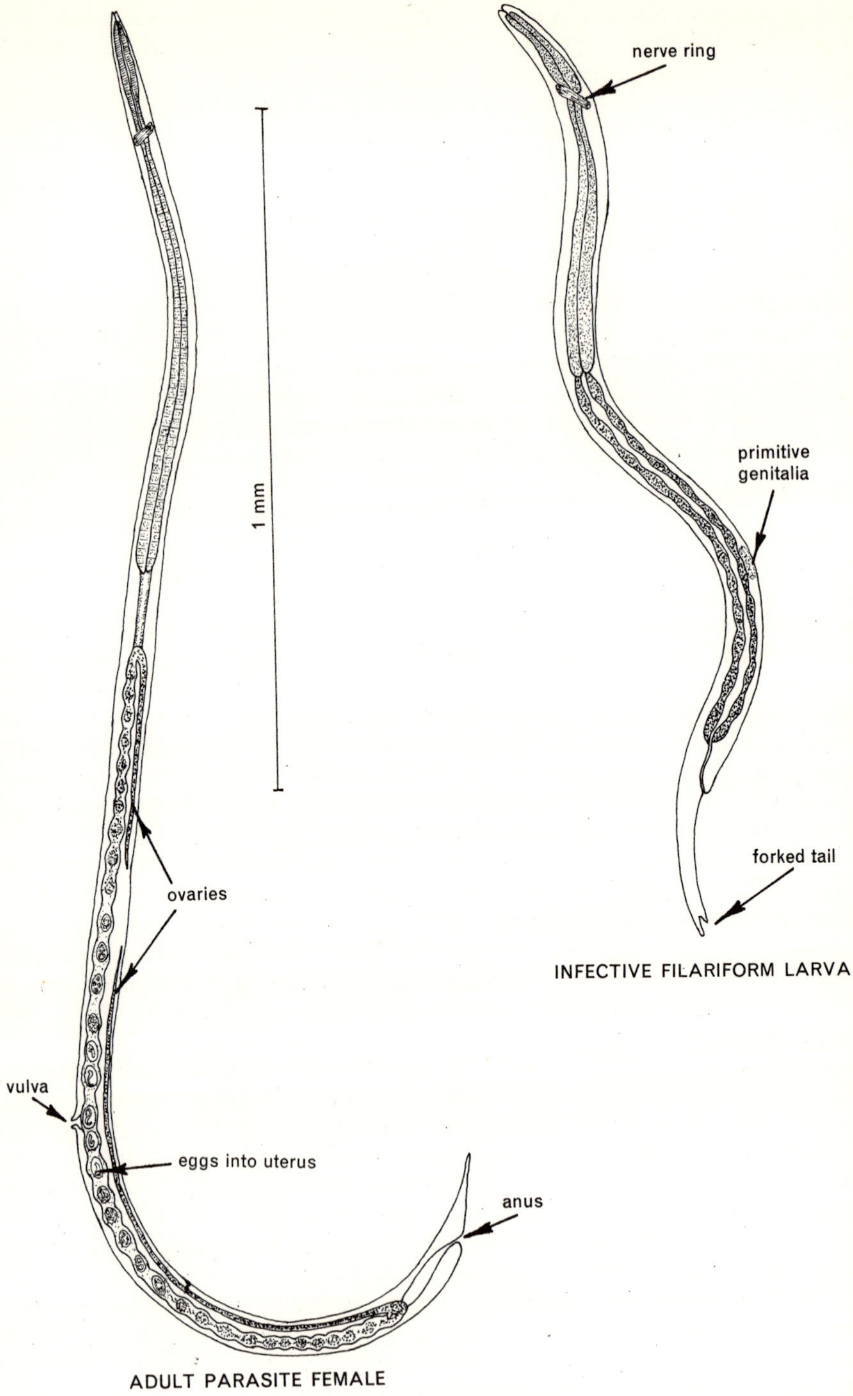

Fig. 4.8 *Strongyloides stercoralis*

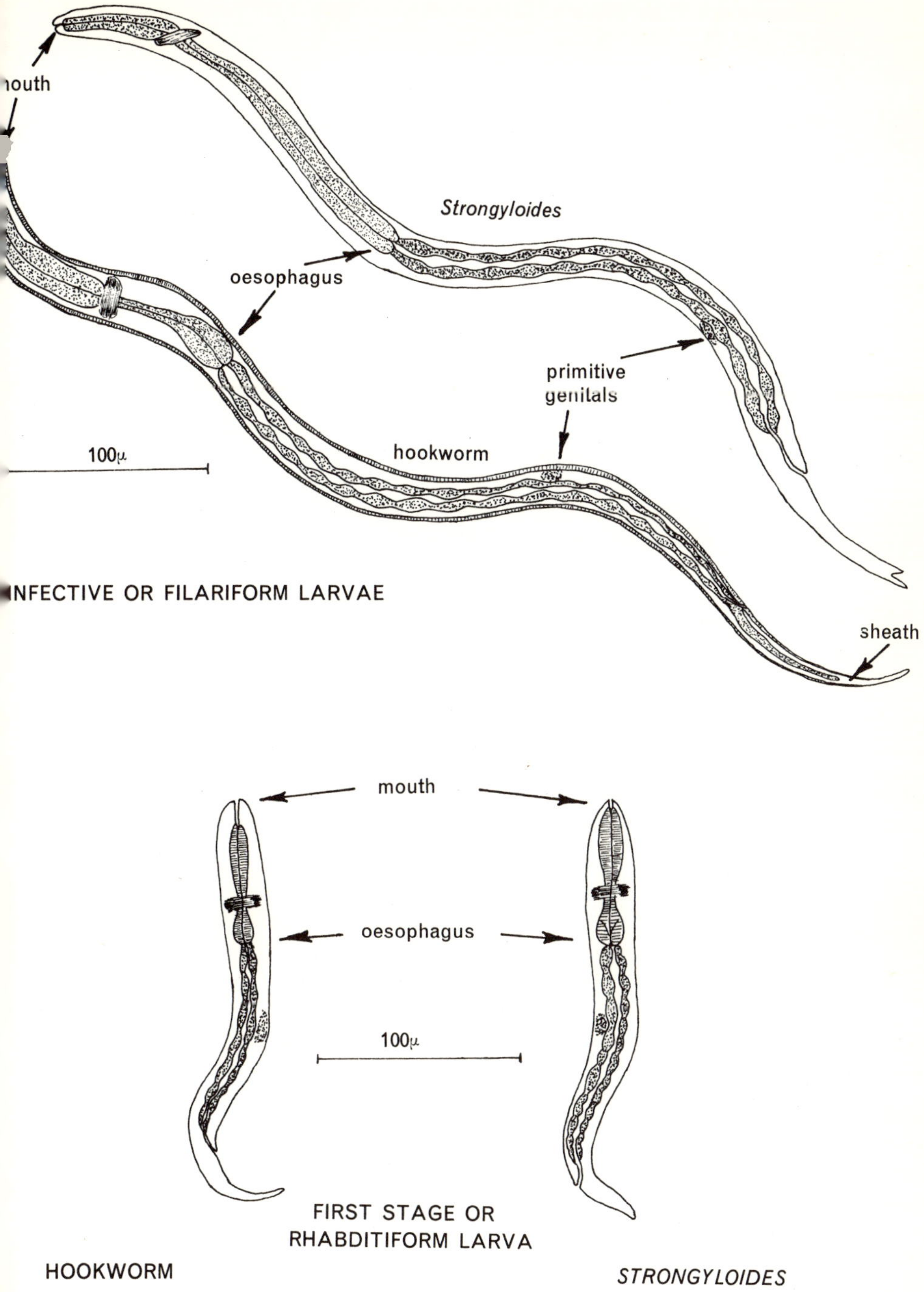

Fig. 4.9 Comparison of hookworm and *Strongyloides* rhabditiform and infective (filariform) larvae

In *Strongyloides* the oesophagus accounts for about 40 per cent of the total length, and the tail is forked.

Pathology

The migrating larvae produce symptoms like those of hookworm, but these are seldom noticed, unless large numbers of larvae are migrating at the one time. A high eosinophilia (40 per cent) is a feature of such migrations and a slight eosinophilia persists as long as adult worms are present, because they are in constant contact with the tissues of the gut.

Most infections seem to be light and without symptoms. In heavier infections there are vague belly pains and occasional diarrhoea.

Very heavy infections may build up over a long period because the life cycle can be completed within the host. In such cases, anaemia may develop, blood may be found in the faeces and the faeces often contain a lot of undigested food. When large numbers of worms are present, digestion is probably inhibited by anti-enzymes in the same way as it is with other nematodes.

We cannot find any reports of clinical cases due to *Strongyloides* in Niugini. .(A clinical case is where a person is sick.) However, they may occur here sometimes, and be diagnosed as hookworm disease. If there are nematode larvae in fresh faeces, it is usually a sign of *Strongyloides* infection, although sometimes soil and plant nematodes, which are not parasitic in man, are found in the faeces. They have been accidentally eaten and have passed through the gut without being digested. The larval stages must be cultured, using the same method as previously described for hookworm, (Section 4.3) to be certain that a double infection of hookworm and *Strongyloides* is not present.

Strongyloides is much more common in some countries than it is in Niugini. The factors governing survival of infective larvae and the infection of man are probably similar to those already discussed for hookworm.

4.6 Enterobius

Enterobius vermicularis (Enter-o-bee-us) is often called 'threadworm' because it looks like a small piece of white thread; or seat worm, because it causes itching around the seat, or the anus.

Enterobius infects about a third of all children under five in Niugini. Almost everybody has been infected by the time they become adults. *Enterobius* does not cause serious sickness, but heavy infections make children very miserable. There is often a surprising improvement in their behavior after they have been treated with drugs to kill the worms.

The eggs of *Enterobius* are seldom found in routine examinations of the faeces, although adult worms are sometimes seen wriggling in fresh faeces. The diagnosis of *Enterobius* depends on a special technique which is described later.

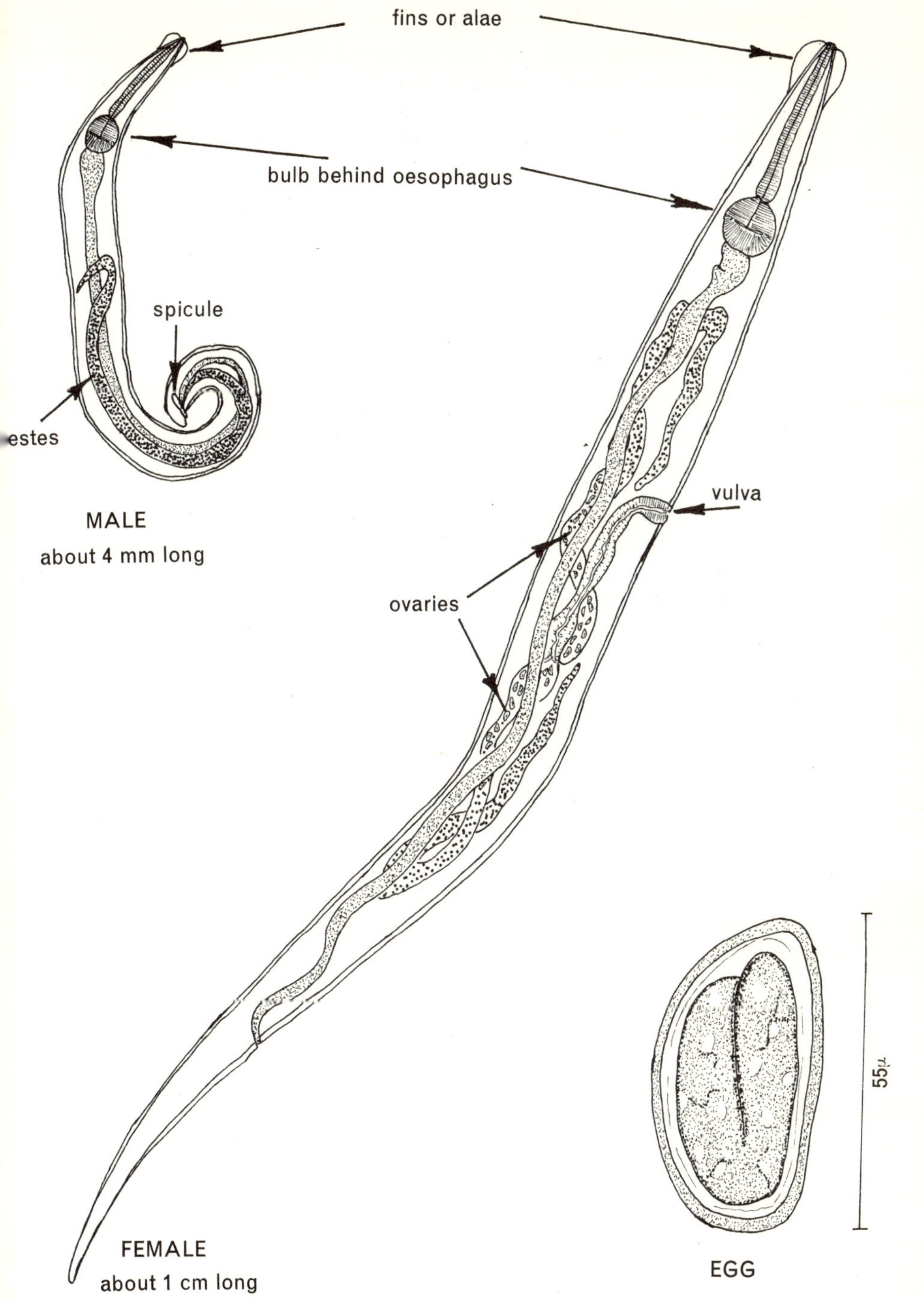

Fig. 4.10 *Enterobius vermicularis*

Morphology

The adult worms are white and they live in the large intestine. The cuticle, or skin around the mouth is expanded into two small fins or alae. At the end of the oesophagus there is a large circular bulb (see Fig. 4.10).

The males are 2 to 5 millimeters long, and have a curled tail with one spicule. They are seldom found, probably because they die soon after fertilizing the females.

The females are larger, about one centimeter long, and have a long pointed tail. They lay about 10,000 eggs, which are flattened on one side. The eggs can live for about 2 weeks, under ideal conditions.

The Life Cycle

Infection takes place when infective eggs are eaten. Larvae hatch from the eggs lin the duodenum and become established in the large intestine. The females start to lay eggs 2 to 3 weeks after infection takes place.

Enterobius has evolved several interesting features which increase the chances that an egg will infect man. They are adaptations to a parasitic way of life which make *Enterobius* a more efficient parasite.

The females migrate out of the anus of the host at night and lay their eggs on the moist, warm skin around the anus. The larvae develop within the egg and become infective in a very short time—about six hours. When the females move out of the anus, they irritate the skin and cause intense itching. The host reacts by scratching. When children scratch their anus, it is often a sign that they are infected. Now the eggs are very sticky and stick to the fingers. They are transferred from fingers to mouths, and onto walls, door handles, toys, crockery and cutlery, where they are easily picked up by other people. Some eggs hatch while they are on the skin near the anus and reinfect the host by migrating into the anus.

The combination of; (i) quick development of the larvae, (ii) irritation of the skin around the anus and (iii) sticky eggs, greatly increases the chance of infection.

Pathology

The adult worms seldom cause any symptoms while they are in the gut. Sometimes however, they gather in the appendix and cause inflammation or appendicitis. This is a serious condition because an inflamed appendix may burst and release the gut contents into the body cavity. Bacteria in the gut contents then start to grow and this can cause a serious condition called peritonitis. In order to prevent this from happening, it is usual for a doctor to operate and remove the appendix. Although *Enterobius* is unlikely to cause the appendix to burst, a doctor cannot easily tell the difference between appendicitis due to *Enterobius* or appendicitis due to other causes.

Children who are heavily infected with *Enterobius* often have a variety of nervous symptoms that appear to result from the irritation caused by the females when they are laying eggs. These symptoms include insomnia, bad dreams, loss of appetite and bad behaviour. They are generally miserable, uncooperative and inattentive at lessons. Children with heavy infections may lose weight.

The skin around the anus may become infected with bacteria as a result of scratching.

The adult worms sometimes wander into the vagina, uterus and Fallopian tubes, and from there they can enter the body cavity. They can cause irritation and a discharge from the female reproductive organs.

Over 5000 worms were recovered from one person after treatment with drugs to clean out the worms. However, most infections are very light and consist of only a few worms. It is not unusual for the load in small children to slowly increase as they continually reinfect themselves.

Diagnosis

Because the female worms generally come out of the anus to lay their eggs, *Enterobius* eggs are seldom found in the faeces of infected people. The best method of diagnosis is to press the sticky side of a piece of transparent cellulose (Scotch or Durex) tape against the skin, in the area around the anus. The cellulose tape is then stuck, sticky side down, onto a microscope slide, and examined under a microscope. The eggs stick to the tape and can be seen under the microscope. It is best to do this first thing in the morning before the person has washed.

Epidemiology

Enterobius is a common infection among children, although adults sometimes have heavy infections as well.

Infection occurs when people eat mature eggs. These can survive for several days under shady humid conditions but are killed by sunlight.

Eggs are frequently taken from the anal region of infected persons and left on toys, table tops, walls, door handles, plates, cups and spoons and other objects. Other people pick up the sticky eggs when they touch these objects. Over 5000 eggs were once recovered from one square foot of toilet wall. Eggs are often present in bed clothes and house dust and may be accidentally swallowed.

Enterobius is often a family and school infection. If one member of the family is infected, others are likely to be infected also. Infection frequently occurs in schools and institutions such as orphanages and mental homes.

Infected people frequently have eggs under their fingernails. The nails of infected children should be cut short.

Treatment with piperazine (Antepar) or Vanquin, will kill the adults, but several treatments, 2 to 3 weeks apart, are often necessary to eradicate all the parasites. It is usually necessary to treat all adults and children living in the same house at the same time. Bedding and toys should be warmed in an oven, or put in the sun, to kill the eggs. Toilet walls and children's furniture should be washed with warm soapy water to remove eggs.

In America, white people have a much higher incidence of infection than Negroes. This is probably due to genetic differences in susceptibility.

Enterobius in Niugini

As far as we can determine, there has only been one survey to determine the incidence of *Enterobius* in Niugini. This was done by Dr Vines (1967), using the cellulose tape method. He found that 35 per cent of the children under two years, and nearly 60 per cent of the children between 2 and 5 years, were infected. His data is given in Table 4.13 together with similar data from one survey in America.

Dr Vines also recorded the incidence of *Enterobius* eggs in the faeces survey.

It has already been explained that *Enterobius* eggs are seldom found in the faeces. Most textbooks consider they are only found in the faeces in about 5 per cent of the infections. However, in the Highlands survey, 8 per cent of the people had *Enterobius* eggs in the faeces. The incidence was about the same in adults and children. In the other regions only about 1 per cent of the faeces contained eggs.

TABLE 4.13 Incidence (per cent) of *Enterobius vermicularis* in young children.

Age	Niugini	United States of America
0-2	35	2
2-5	58	42
5-6		44

4.7 The Amoebas

The parasitic amoebas are protozoa with blunt pseudopods. They all belong to family Endamoebidae. Members of this family reproduce by binary fission and usually form cysts. Binary fission is a method of reproduction where the nucleus divides to form two identical nuclei and then the rest of the cell splits to form two new organisms, each with one nucleus. Cysts are resting stages which have a thick wall for protection.

The following amoebas live in man, they are all parasites of the large intestine except *Entamoeba gingivalis* which lives in the mouth.

Entamoeba histolytica (Enta-me-bar his-toe-lit-ikar)
E. coli (E. ko-lie)
E. hartmanni (E. hart-man-i)
E. polecki (E. po-lecky)
E. gingivalis (E. ging-iv-alis)
Endolimax nana (Endo-lie-max nar-nar)
Iodamoeba butschlii (Eye-o-dar-me-bar bootch-lee-eye)
Dientamoeba fragilis (Dye-en-tar-mee-bar fraj-ill-is)

The only pathogenic (harmful) species is *E. histolytica* and then only sometimes. Most of the time it lives without harming the host.

The Life Cycle

Most of the parasitic amoebas have similar life cycles. They usually consist of two stages: the trophozoite or feeding stage, and the cyst or resting stage.

The trophozoites feed on bacteria and other particles of food in the gut, in the same was as other amoebae do (see Section 1.2). They move by amoeboid movement. They are continually reproducing by binary fission. Many trophozoites are carried along in the food stream and pass out in the host's faeces. In preparation for their entry into the external environment, the trophozoites of most species change into a cyst or resting stage as they approach the end of the gut. The cyst has a protective cell wall and does not feed.

Dientamoeba fragilis does not form cysts and cysts have only been recorded once for *E. gingivalis*. In most species, the nucleus of the trophozoite divides several times so that each cyst has 4, 8 or more nuclei. The number of nuclei is characteristic of the species. Infection occurs when the cyst is eaten by a new host. The cyst hatches in the small intestine and the multinucleate amoeba emerges from a hole in the cyst wall. It is likely that enzymes from the host, and an alkaline environment, stimulate the cyst to hatch. The newly emerged amoeba divides to form as many individuals as there are nuclei. These are carried in the food stream to the large intestine where they normally live.

The large intestine does not have villi like the small intestine, but its surface is covered with tiny round pits called crypts. (It is as if someone had poked a tiny finger into a sheet of soft clay, leaving millions of tiny holes or crypts.) Most of the amoebae live in the crypts where they continually reproduce by binary fission. Those that wander from the crypts are likely to be carried out in the food stream.

THE DIAGNOSIS OF ENTAMOEBA INFECTIONS

We will discuss the characteristics used in the identification of the human species of *Entamoeba*, because some of this information is not readily available in Niugini, and because most of the surveys done here so far, have been uncertain about the identification of many of the specimens found.

The diagnosis of the amoebas depends on the examination of stained specimens under a microscope. Iodine solution is the most commonly used stain (1% potassium iodide in water, saturated with iodine crystals and then filtered). A tiny bit of fresh faeces is mixed with a drop of iodine, covered with a coverslip, and examined. This technique is used for all cysts and eggs that occur in faeces. Sometimes haematoxylin is used to stain gut protozoa. This takes several hours, but it does provide well stained material and the preparations are permanent and can be used for teaching.

Most species of gut protozoa can be grown in test tubes cultures. This provides plenty of parasites for identification, which is important when only a few cysts are present in the faeces. Culture techniques are used as a routine in some hospital laboratories.

The characteristics used in the identification of *Entamoeba* are as follows:

(*i*) *Size*

Size is a characteristic of most living things. The cysts and trophozoites of different species have different sizes. Of course, there is variation between one individual and another, but the average size of a number of specimens and the range of sizes found, are characteristic of the species. There is some overlap in the size ranges of the species of *Entamoeba* found in man, so other characteristics must also be used. Small parasites can be measured under a microscope with a micrometer eye piece.

(*ii*) *The number of nuclei in a mature cyst*

The trophozoites of all species have only one nucleus. As they become cysts, the number of nuclei usually increases. Mature cysts of *E. histolytica* and *E. hartmanni* usually have four nuclei; occasionally they have eight. Mature cysts of *E. coli* usually have eight nuclei, but sometimes they may have 16 or even 32. Mature cysts of *E. polecki* have only one nucleus. Some times the nuclei in a cyst cannot be clearly seen because they are hidden by other structures. In such cases, another cyst must be found and examined. Sometimes only trophozoites and uni-nucleated cysts are found in the faeces. This is often the case with faeces from people who have diarrhoea. If the faeces are allowed to stand for a few hours, mature multi-nucleate cysts may develop. It is not unusual for two or more species of *Entamoeba* to be present in the one host. Because of this, a number of cysts must be carefully examined and identified from each specimen.

(*iii*) *The characteristics of the nucleus*

Inside the nucleus there is a dark staining dot, which is called the karyosome. It is central in some species and eccentric (that is, off centre) in others. In *E. polecki*, it is large, and consists of several small granules.

Around the edge or rim of the nucleus there is dark staining material. This is called the peripheral chromatin. It may consist of a number of granules of the same size, or of granules of different size. Usually these characteristics of the nucleus can be seen in specimens stained with iodine, but they are easier to see in specimens stained with haematoxylin. The nucleus of cysts and trophozoites of the same species usually have the same characteristics. In *E. polecki*, the karysome of trophozoites is usually small but it is large in most cysts.

(*iv*) *Chromatoid bodies*

These bodies stain black with haematoxylin. They are largest in immature cysts, and usually small in mature cysts and often absent from trophozoites. They are sometimes clearly visible in unstained cysts and look like pieces of glass. The size and shape of chromatoid bodies are characteristic of the species

(*v*) *Glycogen masses*

Immature cysts usually contain masses of glycogen. These stain brown with iodine. They do not stain with haematoxylin but leave an unstained space. Glycogen is a food reserve.

Cysts of *E. polecki* have an 'inclusion mass' which is not glycogen. It does not stain with iodine. It may occupy about one quarter of the area of the cyst.

The characteristics of the species of *Entamoeba* which live in the gut of man are given in Table 4.14. The cysts of each species are also illustrated in Fig. 4.11. The cysts of *E. polecki* are seldom diagnosed with certainty, unless the preparations are stained with haematoxylin.

There has been a lot of doubt about the identity of uni-nucleate cysts found in man in Niugini. They are usually identified as *E. histolytica* or *E. hartmanni.* Some of them may be *E. polecki. Entamoeba debliecki* and *E. suis* are two other uni-nucleated species that have been described from pigs. They are not well known. They may also occur in man here.

If trophozoites or uni-nucleated cysts are found in a specimen and are difficult to identify, another specimen collected a day or two later may contain multi-nucleate cysts that can be identified more easily.

THE OTHER AMOEBAS OF MAN (see Fig. 4.12)

(*i*) *Endolimax nana*

The cysts are usually oval in shape and have four nuclei, with large eccentric karyosomes. They usually measure about 4 × 10 microns.

(*ii*) *Iodamoeba butschlii*

The cysts are usually irregular in shape and contain a very large glycogen mass. which stains deeply with iodine. The karyosome of the nucleus is large, and touches the outer edge of the nucleus. The cysts are about 6 to 15 microns long.

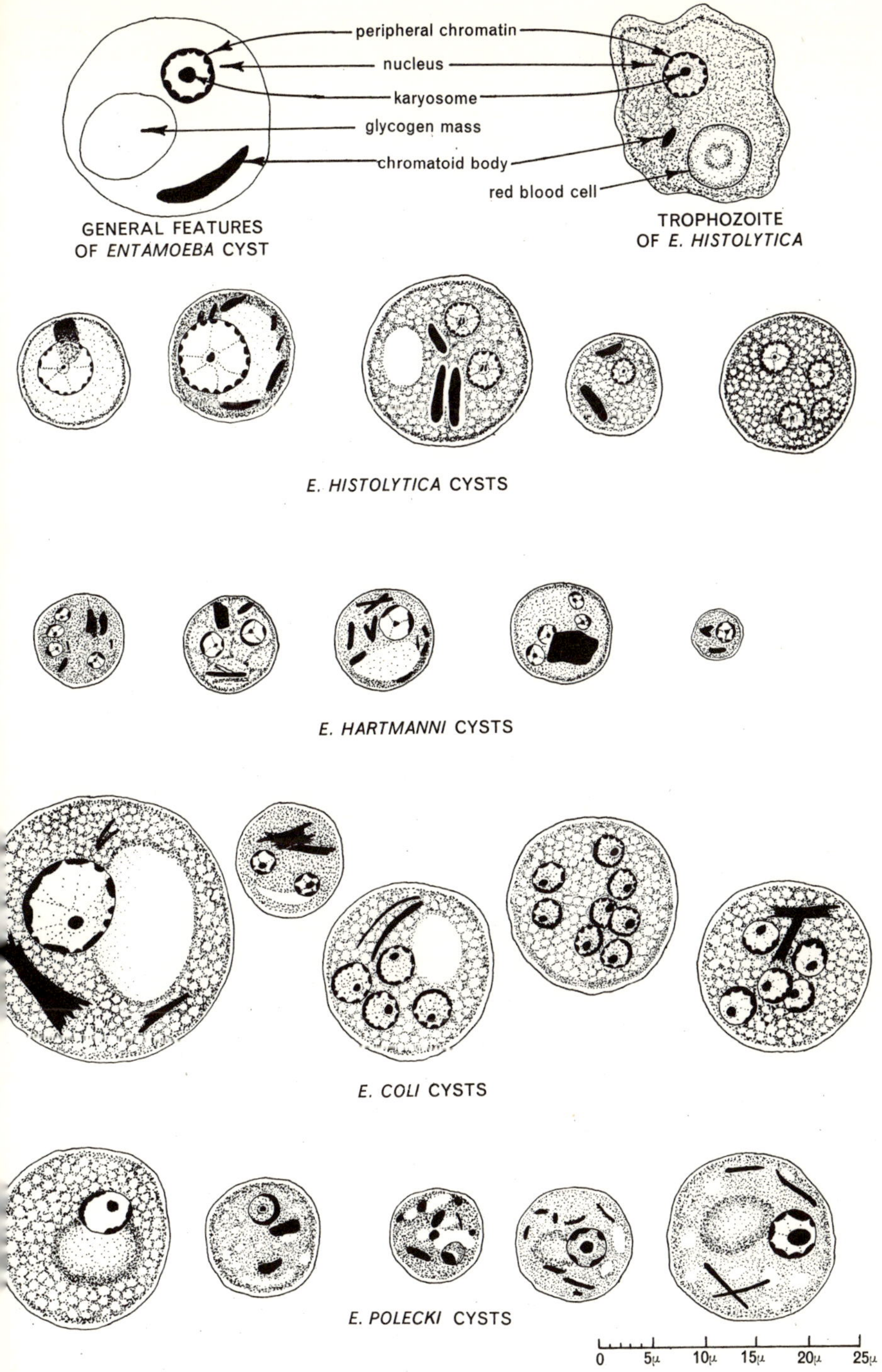

Fig. 4.11 Cysts of *Entamoeba* species (all drawn to one scale) (after Burrows and others)

	E. histolytica	*E. hartmanni*	*E. coli*	*E. polecki*
Size range of trophozoites	10–40	3–20	15–50	8–15
cysts. avg. size size range	11–13 9–16	7–9 4–11	17 10–33	12–14 9–18
Numbers of nuclei in mature cyst.	usually 4 rarely 8	4	usually 8 rarely 16 or 32	one rarely 2
Karyosome	usually small and central, sometimes eccentric	eccentric or central	eccentric and large	generally eccentric, large, of several granules or very small
Peripheral chromatin	usually many small evenly distributed dots of equal size	varies but usually a few large granules, irregularly distributed	a few large granules usually evenly distributed	usually irregularly distributed granules
Size of nucleus in uninucleated cyst	>3 40–60% of cyst diameter	2–3 25–40% of cyst diameter		2.4–4.3 25–30% of cyst diameter
Chromatoid bodies	bars with rounded ends, often absent	usually several present, but vary in size, shape and number	usually splinter-like with sharp ends	usually small and numerous (6-15 in each cyst) round, oval or with pointed ends
Glycogen mass	one large mass, usually present	usually absent, when present, several small masses	one mass, with indistinct border	sometimes present, often absent. Inclusion mass (not glycogen) often present.

TABLE 4.14 Characteristics of the *Entamoeba* species of man (all sizes in microns)

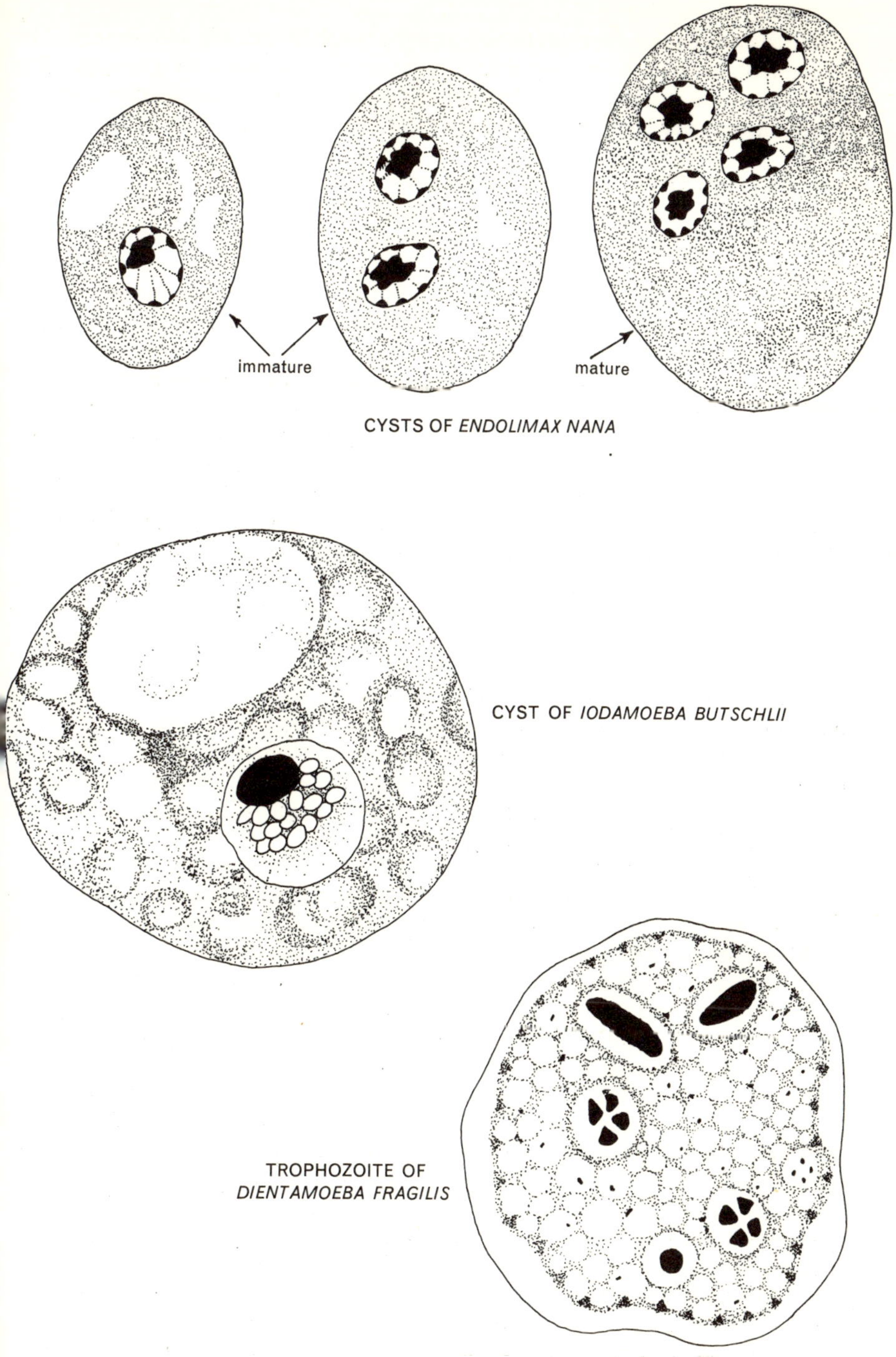

Fig. 4.12 *Endolimax nana*, *Iodamoeba butschlii*, *Dientamoeba fragilis* (after Belding)

(*iii*) *Dientamoeba fragilis*

No cysts are known for this species, and the trophozoites are seldom seen. The mature trophozoites have two nuclei, containing a number of granules. The trophozoites are usually 5 to 12 microns in diameter. So far as we can find this species has not yet been recorded from Niugini.

THE PATHOLOGY OF ENTAMOEBA

It is generally thought that *E. histolytica* is the only pathogenic species of amoeba in man. It causes amoebic dysentery and sometimes other symptoms which we will discuss presently.

Until recently it was thought that *E. histolytica* consisted of two races or forms, one large and one small. It was generally thought that the small race was harmless, although there have been rare cases where it has caused disease. This small race is now considered to be a separate species and is called *E. hartmanni*. It has a number of characteristics which separate it from all other species (see Table 4.14). Now that the characteristics of *E. hartmanni* are well known, people will be looking to see if it is pathogenic or not. The size range of the cysts of the two species overlap and it is necessary to use characteristics other than size, in the identification of these species. Both species are pathogenic in laboratory animals.

In most cases *E. histolytica* does not invade tissues. It normally lives on bacteria and other particles in the gut of the host. Only about 5 to 10 per cent of the people that are infected show any symptoms, and these are mostly mild. Occasionally *E. histolytica* invades tissues and causes ulcers or sores in the wall of the gut.

No one is sure why the trophozoites suddenly invade tissue. Some people think that it may be caused by changes in diet, or in the kinds of bacteria that are present in the gut. Attacks of amoebic dysentery sometimes follow other diseases or surgery. There are probably many different strains (or varieties) of *E. histolytica* and the strains in some parts of the world seem to be more pathogenic than others. Some strains may always be harmless.

The sores or ulcers formed by *E. histolytica* usually occur in the large intestine. They are commonest near the appendix and the rectum; the region in-between is less frequently affected. If there are only a few small ulcers, the symptoms are usually not severe. They consist of occasional diarrohea which starts with severe (acute) cramps in the belly and the person must rush quickly to the toilet. There may be no symptoms for weeks at a time, except that the infected person is easily tired and may pass a lot of wind (wind belong pekpek, or farting). There may be frequent constipation and pains in the belly.

If an ulcer wears away the wall of a blood vessel, there will be blood in the faeces. It often happens that this also brings on diarrhoea (faeces with lots of water in it). Bloody diarrhoea is called dysentery. If the ulcer gets big enough, it will wear right through the wall of the gut and let the gut contents leak into the abdominal cavity; this causes peritonitis which is very serious.

Other complications may follow if an ulcer wears through a blood vessel or a lymphatic vessel. Parasites can then be carried in the blood or lymph, to other parts of the body where they start to reproduce, and form new ulcers or abscesses. The liver, lungs and brain are the most common organs affected. These are secondary infections and follow a primary infection in the gut. The skin, penis, vagina, kidneys and other organs may also be affected.

A number of cases of amoebic infections of the penis have been recorded from Niugini. They probably occurred because of the practice of sodomy (copulation between males).

When *E. histolytica* invades tissues, the trophozoites keep on dividing without forming cysts. Cysts are not found in ulcers, and are seldom found in runny faeces.

Cysts may be found in the urine of people with infections of the kidneys or genital organs.

EPIDEMIOLOGY OF THE GUT AMOEBAS

The amoebas that infect the gut of man occur throughout the world. The incidence varies with the species and with the level of sanitation. In countries with a good standard of living, the incidence is seldom higher than 10 per cent, and is usually much less. In tropical countries, where there is poor sanitation, the incidence of *E. histolytica* is sometimes as high as 65 per cent and it is usually more than 10 per cent.

Infections are almost always due to the ingestion or eating of cysts. Many people do not wash their hands properly after going to the toilet and they pass cysts on to others.

We shall talk about the epidemiology of *E. histolytica* because it is the most important species.

Several studies have shown that if one person in the family is infected with *E. histolytica* it is very likely that others in the same family are also infected.

In one study in Nigeria, 62 per cent of all patients admitted to hospital with acute amoebic dysentery, were children under 10 years old. Most of them were under five. The study showed that 96 per cent of their mothers were infected, but none of them had diarrhoea or dysentery. They were infected, but they were not sick; but their children became sick when they became infected.

Other studies have shown that infected people who prepare food, often pass on the infection to people who eat the food that they prepare. In one study, all the food preparers were examined and those that were infected were treated to kill the *E. histolytica* in their bodies. After this was done, the incidence of infection in people who ate the food that they prepared, fell by 90 per cent over a three year period.

Sometimes epidemics of amoebic dysentery occur. An epidemic is a sudden outbreak of sickness in many people, who live in the one place. It seems that these epidemics are caused by a strain of the parasite that is particularly pathogenic, that is, for some unknown reason it causes sickness when it infects people, while other strains usually do not cause sickness. Some of these epidemics have been due to a polluted water supply. In several cases the water supply in a big hotel has been connected to the sewers. When this happened, people drank water that was polluted with faeces containing cysts of *E. histolytica*. In other cases a water well had been contaminated with faeces from a nearby toilet.

The cysts of *E. histolytica* have been found on the hands of infected people after they have washed them. They are probably often left behind on towels that are used by everybody and other people become infected. Paper towels that are thrown away after they have been used should be put in all public toilets.

Everybody should keep their fingernails cut short so that dirt (faeces) does not gather under them.

The cysts of *E. histolytica* can live in faeces for two weeks at a temperature of 70 to 80 °F. They can stay alive for two weeks in water. However, they are killed by heat and low humidity (dry conditions).

Until recently 1 to 2 per cent of people who were admitted to hospital with amoebic dysentery or amoebic ulcers in other organs, died. If the infection is diagnosed early and treated with modern drugs, very few people die. The faeces of all people admitted to hospital with dysentery or diarrhoea, should be examined to see if parasites are present.

The incidence of *E. histolytica* often increases with age. The infection usually lasts for many years and as a person gets older, he is more likely to become infected. Few people develop acute symptoms, but people who are infected, may become sick if they drink too much alcohol or become very tired or are very upset.

In most cases *E. histolytica* does not cause serious sickness. However, most of the people who are infected with it are a little bit sick sometimes. Every now and again they suffer from diarrhoea and constipation and cannot work as efficiently as they would if they were not infected. The total effect of *E. histolytica* is difficult to measure, as it is with most other parasitic infections.

THE GUT AMOEBAS IN NIUGINI

Table 4.15 gives the incidence of the various gut amoebas in Niugini. There has not yet been a survey done in Niugini in which the various species of *Entamoeba* have been identified with certainty.

TABLE 4.15 Incidence of the human gut Amoebas in Niugini.

	E. histolytica	*E. coli*	*E. hartmanni*	*E. polecki*	*Endolimax*	*Iodamoeba*
Mt Hagen 1933	18*	6	–	–	0	0
West Irian 1945	2	9	–	–	4	0
Busamba 1947	12*	14	–	–	0	0
Kaiapit 1947	16*	2	–	–	8	0
Patep No 2 1947	26*	25	–	–	12	3
Kavataria 1947	9*	24	–	–	3	0
Koravagi 1947	9	3	–	–	0	0
Highlands Survey 1966	16†	25	2	2	5	9
North Coast-Mainland 1966	14†	22	3	–	3	3
Niugini Islands 1966	11†	19	5	–	0	0
Koruguru 1933	50*	–	–	–	–	–
Samarai 1946	1	–	–	–	–	–
Finschhafen 1937	10	–	–	–	–	–
Kundiawa School 1947	20*	22				

*mostly uni- or bi-nucleated cysts
†'*E. histolytica* like'

Disease due to *Entamoeba histolytica* does not appear to be common in most parts of Niugini. Only four or five deaths a year are recorded as being due to this parasite. But until every hospital has properly trained technicians who can diagnose this infection, and until the faeces of all patients with diarrhoea and dysentery are examined, we will not know exactly how important *E. histolytica* is.

Over a period of 15 months (1964-5), 46 patients were admitted to a hospital at Baiyer River with disease caused by *E. histolytica.* Of these 14 had symptoms in organs other than the gut. They included disease of the genital organs or urinary system, skin ulcers, liver ulcers and pneumonia. Of the 32 patients with gut symptoms many were small children. Some of them had bad diarrhoea but no dysentery.

In a recent analysis of 2000 admissions to the medical wards at the Port Moresby General Hospital, there were only four cases of amoebic dysentery and three cases of liver abscesses due to *E. histolytica.*

4.8 The Gut Flagellates

The following species of Flagellates occur in the gut of man (see Fig. 4.13):
Retortamonas intestinalis (Re-tort-amoan-ass intestin-alis)
R. sinensis (sine-en-sis)
Enteromonas hominis (Entero-moan-ass hom-in-is)
Chilomastix mesnili (Kilo-mas-tix mess-nil-i)
Giardia lamblia (Gee-ard-ia lamb-lia)
Trichomonas tenax (Trike-o-moan-ass ten-axe)
T. hominis.

All gut flagellates reproduce by binary fission. Some of them form cysts with protective walls. Man becomes infected when he eats cysts or trophozoites which occur in faeces.

Retortamonas intestinalis and R. sinensis

Neither of these species have been recorded from Niugini. They are very rare; in most countries only one person in several thousand is infected. The trophozoite is 4 to 9 microns long and has two flagella. It lives in the large intestine. The cyst has a pointed end in *R. intestinalis* and is oval in *R. sinensis*. It is about 5 to 7 microns long. The cysts have one nucleus. *R. intestinalis* is thought to be non-pathogenic but *R. sinensis* has been found in people suffering from diarrhoea.

Enteromonas hominis

Heydon found specimens that he thought belonged to this species in people in the Highlands. But he did not positively identify them and they have not been recorded in Niugini since then.

The trophozoites live in the large intestine. They have four flagella, one of which is joined to the surface of the body for most of its length. The cysts contain two or four nuclei and are 6 to 8 microns long. *E. hominis* is thought to be harmless. In one survey, in the Philippines, it was found in 3 to 15 per cent of the people.

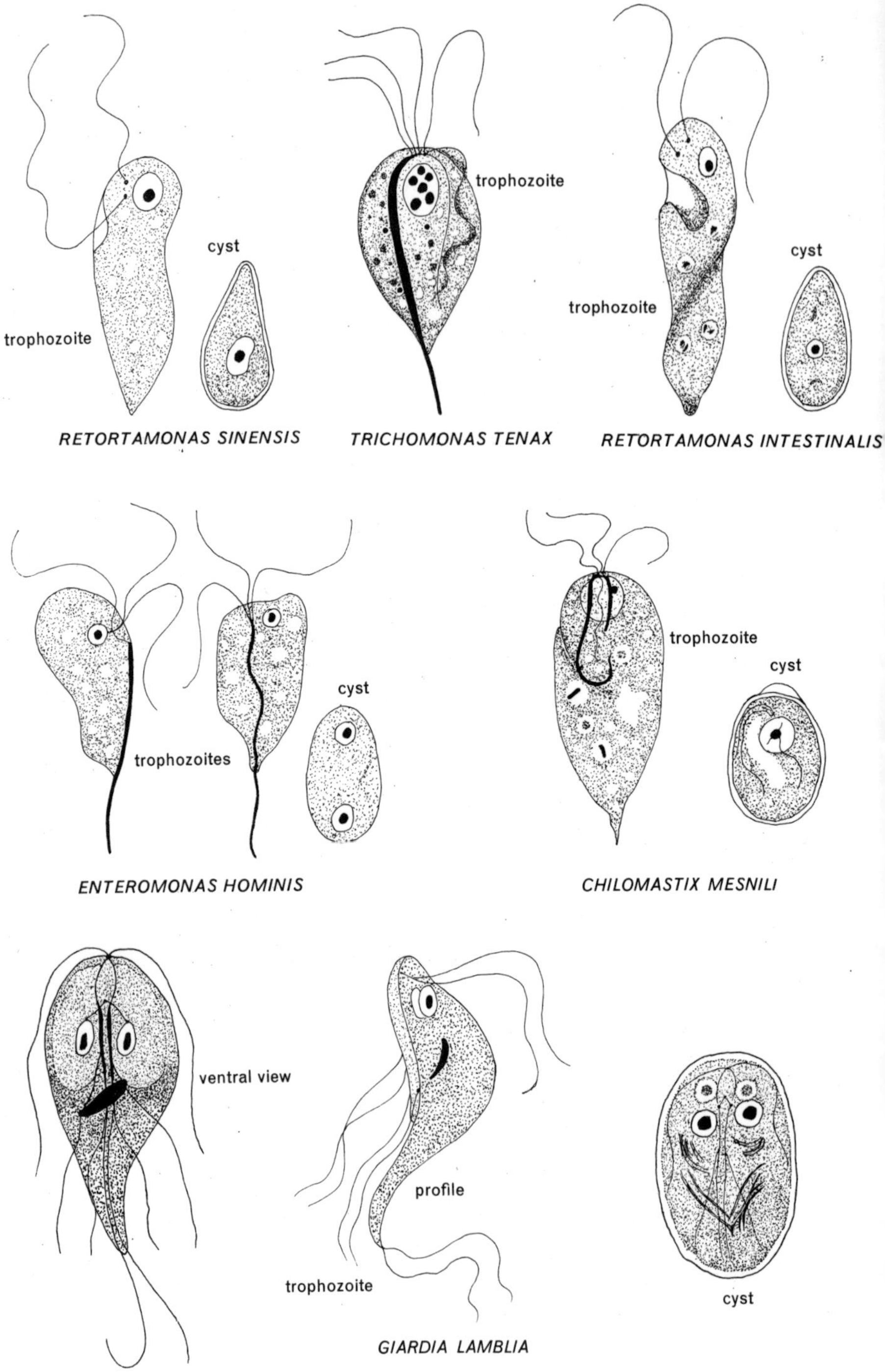

Fig. 4.13 Gut flagellates, trophozoites and cysts (after Craig & Faust and others)

Chilomastix mesnili

This species also lives in the large intestine. The trophozoites have three free flagella, which are about as long as the body. A fourth flagellum lies within the body. The trophozoites are 6 to 20 microns long. The cysts are shaped like a sipora or muli and have a bump on one side. They are 7 to 10 microns long and contain only one nucleus. This species does not cause disease.

Giardia lamblia

This species lives in the small intestine of man. In most countries, it is the most common flagellate found in the gut of man. The trophozoites are 10 to 20 microns long, and are shaped like half a pear. They have four pairs of flagella and two nuclei. They look a bit like a face (see Fig. 4.13). One surface of the animal forms a sucking disc. This is obviously an adaptation for hanging on to the wall of the gut.

The cysts have two or four nuclei and are 8 to 12 microns long. Within the cyst several double strands of material can be seen. They are the remains of the flagella.

Giardia sometimes causes disease. It irritates the lining of the gut causing dull pains and recurrent attacks of diarrhoea. The faeces contain lots of undigested fat and mucus. The infection seems to stop the digestion and absorption of fat from the food. The parasites also cause the cells of the gut to secrete large amounts of mucus. Flatulence (farting) is also common. The gall bladder may become infected and the bile duct may become blocked, but this is very rare.

The infection is commonest in children, although adults are also infected. In most cases the cysts are very numerous in the faeces. Sometimes cysts may not be found in the faeces for a few days, but then they return again.

Trichomonas species

Three species of *Trichomonas* infect man. They are all very similar in appearance. The trophozoites usually have four free flagella and a fifth that is attached to the surface of the body. In all species, a stiff rod, called the costa, runs from the front of the body and pokes out of the other end. *Trichomonas* does not form cysts.

T. tenax

This lives in the mouth of man. It is more often found in people who have sores in their gums, but there is no evidence that this parasite causes the sores. Up to 25 per cent of some populations are infected. It has not been recorded from Niugini, but probably occurs here. The parasites are about seven microns long.

T. hominis

This species lives in the large intestine of man. No one is sure whether it causes disease or not. In some areas it is often found in the faeces of people with persistant diarrhoea, but this does not necessarily mean that it causes the diarrhoea. It may be that it thrives best in the gut of people who happen to have diarrhoea. It is not a very common parasite, but it is seldom recognised except in soft or liquid faeces when the trophozoites can be seen moving about. In some countries about 10 per cent of the people are infected. The parasites are about seven microns long.

T. vaginalis is found in the genital organs of men and women. It cannot live in the gut. We talk about this species in Section 6.1.

TABLE 4.16 Incidence (%) of gut flagellates in man in Niugini

	Chilomastix	*Giardia*	*Trichomonas hominis*
Mt Hagen 1933	2	0.4	10
West Irian, 1945	4	0.6	5
Busama, 1947	0	6	2
Kaiapit, 1947	2	2	2
Patep No 1, 1947	1	0	0
Kavataria, 1947	0	0	1
Koravagi, 1947	0	4	0
Highlands Survey, 1966	1	5	0
North Coast-Mainland, 1967	0	2	0
Niugini Islands	0	4	0

4.9 Other Gut Protozoa

There are two other gut protozoa of man. One is *Balantidium coli* (Bal-an-tid-i-um coli) which is a ciliate. The other is *Isospora* (I-sos-pora) which is a sporozoan. Both cause disease in man (see Fig. 4.14).

Balantidium coli

This species lives in the large intestine of man, monkeys and pigs. It is the largest of the human protozoa. The trophozoites are 50 to 200 microns long and the cysts are about 40 to 60 microns long. It is said that specimens from pigs are larger than those from man.

The trophozoites are covered with short cilia. They have a funnel shaped mouth that is lined with cilia. There are two nuclei, one large and one very small.

The cysts are almost round and have a thick cyst wall. They are about the same size as hookworm eggs. *Balantidium* sometimes causes dysentery in man. It may attack the lining of the gut and cause ulcers which are like those caused by *Entamoeba histolytica*. However, many people who are infected do not get dysentery. As far as we know, no cases of dysentery or diarrhoea have been recorded due to this parasite in Niugini. The symptoms are diarrhoea or dysentery, and occasional constipation, with belly pains and sometimes vomiting.

In 1933, Heydon found about 11 per cent of the Mt Hagen people were infected. The incidence was highest in women, presumably because they look after the pigs. Vines found the incidence was 2 per cent in the Highlands in 1966, but did not report it from the North Coast or the Niugini Islands. In one part of the Highlands of West Irian, 28 per cent of the people were infected.

ISOSPORA SPECIES

Isospora is a sporozoan parasite which belongs to the subclass Coccidia. It has a life cycle in which reproduction occurs by schizogony and sporogony. These

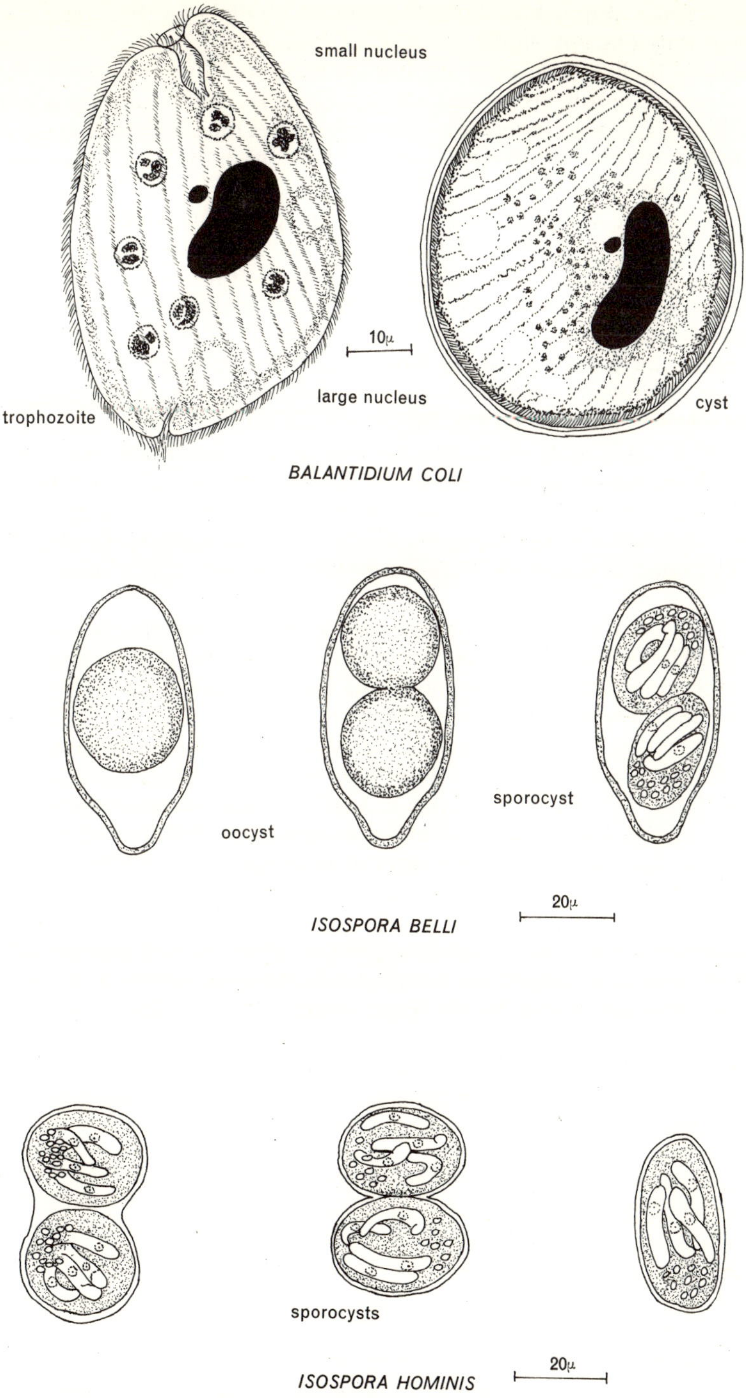

Fig. 4.14 *Balantidium coli, Isospora belli, Isospora hominis*

terms have already been explained in the chapter on malaria, (Section 3.1 and 3.2). There is only one host in the life cycle of *Isospora*.

The trophozoites live in the cells that line the small intestine and grow to form schizonts. There are probably several schizogony cycles. Eventually some of the trophozoites give rise to gametocytes, which also live in the cells that line the gut. Oocysts are formed after fertilization and these pass out in the faeces; the oocysts usually have a protective wall.

The nucleus of the oocyst divides to form two sporocysts which also have a thick protective wall. Inside each sporocyst four sporozoites develop. They are packed tightly within the sporocyst like small fish in a tin.

Man becomes infected when he eats oocysts or sporocysts which contain mature sporozoites.

Three species have been recorded from man. Two of them, *I. belli* and *I. hominis* occur in Niugini. *I. belli* has cysts (oocysts) about 20 to 33 microns long by 10 to 19 microns wide. When the oocysts are passed, the sporocysts have not yet developed. It takes up to three days before the cysts are mature and contain sporozoites. They can be cultured (grown) in the laboratory by covering the cysts with a 3 per cent solution of potassium dichromate. This stops the growth of bacteria and other micro-organisms, but allows the sporozoites to develop. The cysts are illustrated in Fig. 4.14. They are almost oval but have a bump at one end.

I. hominis has cysts in which the sporozoites develop before the cysts pass out in the faeces. The wall of the oocyst is not very strong and often breaks away, leaving two separate sporocysts, or sporocysts that are stuck together by a little bit of oocyst wall. The sporocysts are about 16 × 10 microns (see Fig. 4.14).

Isospora causes diarrhoea or dysentery. As the schizonts burst, they destroy the cells lining the small intestine. The faeces are usually light coloured and contain a lot of fat. Other symptoms are pains in the belly, and sometimes fever and eosinophilia. The symptoms develop about a week after infection occurs, and cysts appear in the faeces about ten days after infection. They are often not found until a few days after the diarrhoea starts. After a month the cysts disappear.

Many other vertebrate animals are infected by a closely related parasite called *Eimeria*. This sometimes causes serious disease in chickens, rabbits and cows.

Heydon found 5 per cent of the Mt Hagen people infected with *I. hominis* and Bearup and Lawrence found an incidence of 2 per cent at Busama in 1947. Dutch workers, in 1962, found both *I. hominis* and *I. belli* in people on the south coast of West Irian near the border with Niugini. Each species was present in about 1 per cent of the people.

4.10 Hymenolepis

The only tapeworms recorded from man in Niugini, are two species of the genus *Hymenolepis* (Hy-men-o-leap-is), *H. nana* and *H. diminuta* (dim-in-you-ta). Both species are fairly rare (see Fig. 4.15).

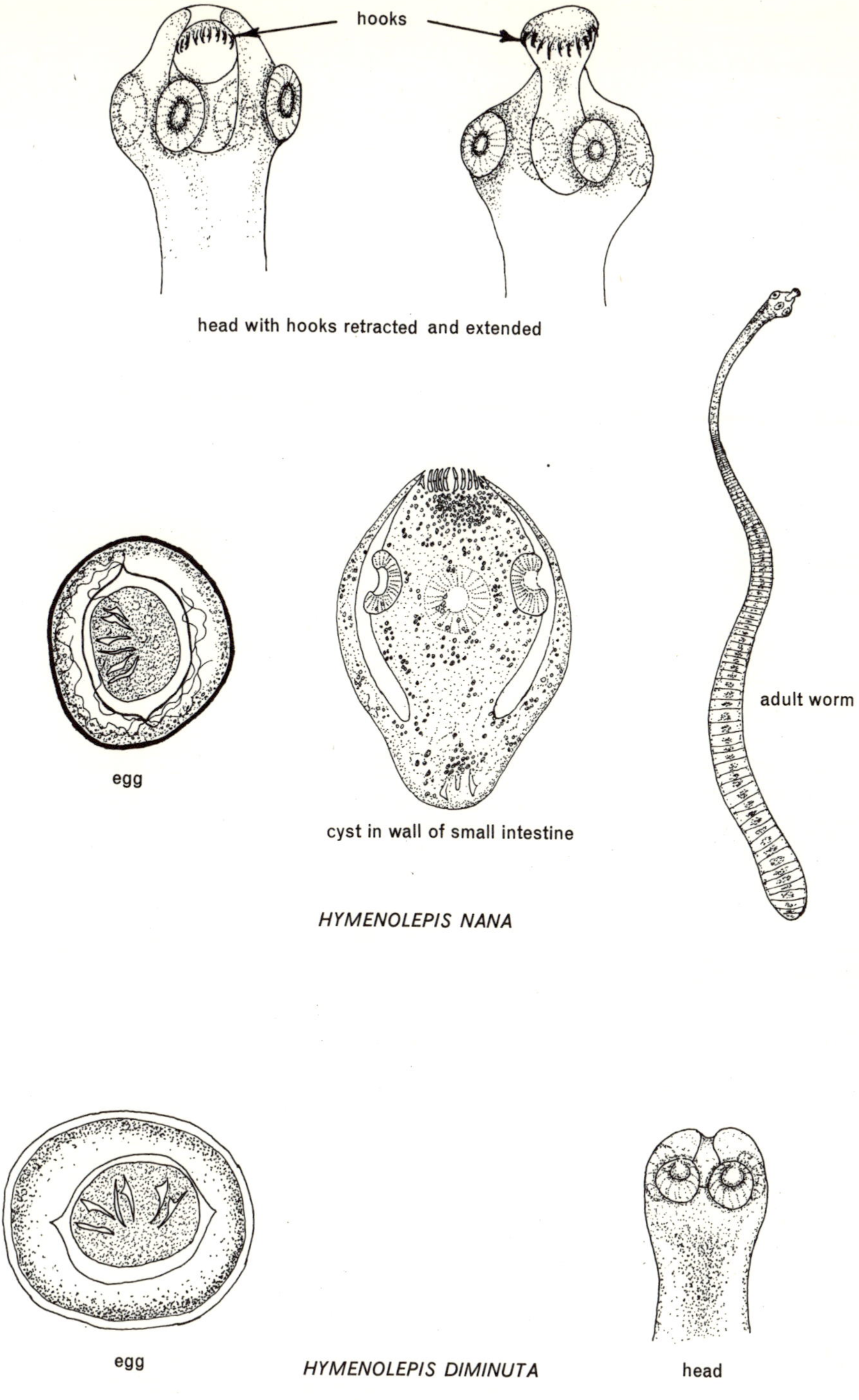

Fig. 4.15 *Hymenolepis nana*, *Hymenolepis diminuta*

Hymenolepis nana

The adult worms are usually from 1 to 10 centimeters long. The length of the worms depends on the number of worms present in the host. In heavy infections the worms are small, but if only a few worms are present, they are large. There are four suckers around the head and at the top of the head, there is a circle of 20 to 30 hooks which can be pushed in between the villi of the gut of the host or they can be withdrawn into the head of the tapeworm. They are said to be retractile.

New segments grow from behind the head and old segments are continually dropping off the tail end. The mature segments are about one millimeter wide and about 0.1 to 0.3 millimeters long. They are full of eggs.

The eggs are oval shaped and measure 30 to 47 microns in diameter. They consist of a central body within which there are six tiny hooks, and an outer case in which there are 4 to 8 long filaments which run around the egg and which join together at each end of the egg.

The life cycle of this tapeworm does not need an intermediate host, although it can have one. This is most unusual.

If the eggs are eaten by man or mice, they hatch out and the larvae burrow into the wall of the small intestine and develop into a cyst stage there. After about two weeks, the cyst becomes a small adult and lives with its head attached at the base of the villi and its tail hanging free in the gut.

If the eggs are eaten by certain fleas, or beetle larvae, they hatch out and develop into the cyst stage within these intermediate hosts. Man and mice can become infected if they eat an infected intermediate host. The cyst stage then becomes an adult.

Because an intermediate host is not necessary, large numbers of adults can build up within the one host. This can happen as the host reinfects himself with eggs from his own faeces, or when eggs hatch within the gut before they are passed out in the faeces. This is known as internal auto-infection (auto=self). The eggs hatch and start a new infection without ever leaving the host.

If only a few worms are present they do not produce many symptoms. However, because auto-infection occurs, it is not uncommon for large numbers of worms to develop within the one host.

The symptoms of infection are: vague pains in the belly, diarrhoea or constipation, hunger or sometimes a loss of appetite, headache, tiredness and sometimes insomnia, and dizziness. Most infected people are irritable. About one third of all patients have an eosinophilia of more than 5 per cent.

Infections are usually more common in children than in adults. Perhaps people develop an immunity to reinfection. Experiments have shown that this happens in mice. Mice that do not eat enough protein are more easily infected than mice that have a proper diet. Perhaps this applies in people too.

H. nana is fairly rare in Niugini. Lambert recorded six cases in 18,000 people examined from Papua in 1920. He found 0.5 per cent of the people from the Territory of New Guinea (including the Islands) infected.

Hymenolepis diminuta

This parasite is very similar to *H. nana*. The adults are usually smaller, being 2 to 6 centimeters long. They have no hooks on their head. Their eggs are much larger than those of *H. nana.* They are 60 to 80 microns in diameter and do not have any filaments in the outer case around the central body.

There is always an intermediate host in the life cycle. A wide range of insects can act as intermediate host, for example, moth larvae, fleas, beetle larvae and cockroaches. Rats and mice are the usual final hosts, but man is infected sometimes. The final host becomes infected when it eats an infected intermediate host. This sometimes happens when man eats uncooked cereals in which larval insects are living and which are contaminated by rats or mice. The symptoms of infection are similar to those listed for *H. nana.*

H. diminuta is usually less common in man than *H. nana.* It has been found in man in the Niugini Highlands. Vines recorded one case in 241 people, and Mr S. H. Christian found 2 per cent of the Goroka people infected in 1949.

4.11 The Control of Gut Parasites

Almost all the gut parasites of man exist because we are not as clean as we should be. In most cases infection occurs because we unknowingly eat faeces or walk in them. We could not become infected with most of the gut parasites unless this happened. The infective stages of most of the gut parasites of man are very small, which is an advantage for them, because they can infect man when he comes in contact with tiny particles of faeces.

We can expect the incidence of gut parasites to slowly decrease as people get to know about the importance of clean habits, and as the standard of living rises. If people have good toilets and water taps in their houses, so that they can wash their hands easily, they will have fewer gut parasites.

It is never too early to start to control these parasites. We need not wait until there are water taps and toilets in every house. Campaigns to control gut parasites must attack the problem three ways:

(1) mass treatment with drugs
(2) improved sanitation
(3) education.

(1) Mass Treatment with drugs

There are many drugs available that kill the gut parasites of man. Most of them are only effective against one or two species of parasite. Some of them produce toxic side effects in a few people, and they can only be given to people in hospital or when a trained person is constantly watching to see that nothing goes wrong. Other drugs do not cause serious side effects and they can be given to everybody. This is known as mass treatment. All the people are treated for a particular infection at once. They may become reinfected again soon afterwards, and several lots of treatment may be necessary to control the infection.

Mass treatment has been used in Japan to control *Ascaris* and Hookworm. At the end of World War II, the incidence of *Ascaris* in Japan was about 55 to 60 per cent. In 1952, mass treatment programs, using drugs called santonin and kainic acid, were started. By 1957, the incidence had been reduced to 18 per cent and to 5 per cent in the towns and cities. The incidence of hookworm in Japan was similarly reduced from 20 to 30 per cent (1950) to 3 per cent in 1959, by mass treatment using a drug called bromo naphthol.

(2) *Improved Sanitation*

Better toilets and sewage systems decrease the incidence of gut parasites. We cannot come in contact with faeces if they are put in a deep hole in the ground or if they are washed away into a sewage works. There are also many bacterial infections that cause diarrhoea and other sickness that decrease when there is good sanitation. There should be a water tap with soap and a towel, in every toilet, and everyone should wash their hands before they leave the toilet, and also before eating.

(3) *Education*

Everybody should be told why it is important to be clean. People cannot be expected to be clean unless they understand why this is a good thing. This should be the work of properly trained health workers that visit all schools, villages and local governent councils.

We would like to give one example of a campaign to control *Ascaris* infections which did not work very well, because the people did not use toilets which were provided for them.

The Effect of Sanitation on Ascaris Infections

About 850 people live in Shahabad village in Iran. A survey in 1963 showed that most of them (67 per cent) were infected with *Ascaris*. At that time, the village had no toilets at all. The people left their faeces inside the village, along the walls of their gardens. Their water supply came from an open canal. People also washed themselves and their clothes in this water. In 1963-4, the village was provided with a good water tank with taps to several parts of the village, a public bath house and laundry, and most importantly, 114 toilets. The village people did much of the work in this project.

In 1965, a second survey was carried out to see if the incidence of infection had changed. Table 4.17 gives the incidence of *Ascaris* infection before and after the toilets were built. It can be seen that the incidence dropped by 11 per cent. While this is an improvement, it is certainly not a very great change. However, the average load of worms per person dropped considerably, and this is just as important as a slight decrease in the incidence.

After the toilets were installed, the average number of female worms per person dropped from about three to one. About $\frac{2}{3}$ of the faeces were being deposited in the toilets instead of on the ground.

TABLE 4.17 Incidence of *Ascaris* in different age groups in Iran, before and after the construction of toilets.

Age Group (years)	Male		Female		Total	
	Before	After	Before	After	Before	After
< 10	58	48	62	62	60	54
11–20	70	60	80	65	77	63
> 20	66	58	70	55	72	56
Total	66	55	69	59	67	57

While there was a small drop in the incidence of infection and a large drop in the average number of worms per person, the building of toilets did not really control *Ascaris* infection because many people did not use them.

All control programs should start with a survey to measure the incidence of infection before treatment is started. In the case of helminth infections, the parasite load should also be measured by doing egg counts. It is necessary to assess the program by repeating these surveys after treatment.

When the incidence of infection is high and the drug being used for mass treatment is not toxic, everybody can be treated. But when the drugs sometimes cause side effects, or when the parasite is only present in a few people, it is better to treat only those people who are infected. This means that everybody must be examined to see if they are infected. In infections such as *Entamoeba histolytica*, where adults seldom show any signs of disease, but where children may become very sick when they are infected, it is necessary to treat the adults as well as the children. One infected person can be a constant source of infection to the other people who live or eat with him. Such people are called 'carriers' and they must be found and treated if infection is to be controlled.

Drugs

Table 4.18 lists some of the drugs that have been used against helminth infections in man. We also list the drugs that are used against *Entamoeba histolytica*, *Giardia* and *Balantidium*.

We do not give any details of the dosages required or the number of treatments that are necessary. Some drugs are toxic, some produce minor side effects such as nausea and vomiting in some people. Most drugs should only be given under the supervision of a doctor, nurse or medical assistant. The drug companies usually give detailed instructions with each lot of drugs. These should be followed. Further information on the use of drugs is given in the latest edition of *Extra—Pharmacopoeia—Martindale* edited by R. G. Todd and published by the Pharmaceutical Press.

Drugs for Entamoeba histolytica

Acetarsol, Amodiaquine, Arsthinol, Bialamicol hydrochloride, Bismuth glycolly-larsanilate, Carbarsone, Chiniofon, Chiniofon sodium, Chlorbetamide, Chloroquine, Clefamide, Diodoquin, Emetine hydrochloride, Emetine bismuth iodide.

Drugs for Giardia intestinalis

Amodiaquine, Atabrin, Di-iodohydroxyquinoline, Hydroxychloroquine sulphate, Mepacrine.

Drugs for Balantidium coli

Aureomycin, Carbarsone, Di-iodohydroxyquinoline, Diodoquin, Terramycin.

TABLE 4.18 Drugs used for helminth infections in man (√ = used; * = recommended).

	Ascaris	Hookworm	*Trichuris*	*Enterobius*	*Strongyloides*	*Hymenolepis*
Arecoline (from betel nut)						√
Ascaridole	√					
Bephenium hydroxynaphthoate (Alcopar)	√	√	√			
Bromonaphthol		√				
Chenopodium oil	√	√				
Chlorbetamide		√				
Dichlorophen						√
Dithiazine iodide	√		*	√	*	
Hetrazan	√					
Hexylresorcinol	√	√	√			
Mepacrine						√
Niclosamide						√
Nyxolan (Aloxyn)				√		
Piperazine (Antepar)	*			*		
Santonin	√					
Tetrachloroethylene	√	*				
Thiabendazole	√	√		√	√	
Triclofenol Piperazine	√	√				
Viprynium Embonate (Vanquin)				*		

Chapter 5

WUCHERERIA BANCROFTI

5.1 Introduction

Wuchereria bancrofti (Wook-er-ear-ia ban-croft-eye) is a roundworm or nematode that lives in the lymphatic vessels of man. It may block the lymphatic vessels and cause the enlargement of the legs or the sex organs. This is called elephantiasis (el-eff-an-tie-as-is). *Wuchereria* is fairly common in Niugini and it infects about 200 million people throughout the world.

Wuchereria belongs to the superfamily Filarioidea (Fil-ari-oid-ea); the filarial (fil-air-i-al) worms. These all live in the tissues or body cavities of vertebrates. Certain species of frogs, reptiles, birds and mammals in Niugini, are infected with various species of filarial worms.

The females of this family produce many tiny larvae which are called microfilariae. These larvae are found in the blood or skin of the vertebrate host. They are sucked up by certain species of blood sucking insects and mites, which act as the intermediate hosts or vectors. The larvae develop into a new stage within the insect host and can infect another vertebrate host when the insect later takes blood.

Several filarial worms infect man, but only *Wuchereria* occurs in Niugini.

Before we discuss the life cycle of *Wuchereria* in detail, we will tell the story of its discovery.

Sir Patrick Manson was born in 1844 and died in 1922. He was a Scottish doctor who went to work in a Mission hospital in Taiwan (Formosa) in 1866. He moved to the mainland of China in 1871. We have already told the story of the part he played in encouraging Ross to work out the life cycle of Malaria. He also played an important part in working out the life cycle of *Wuchereria*, as we shall see presently.

Manson has been called the father of tropical medicine. He returned to England in 1889 and became a specialist in tropical diseases. Ten years later, he established the London School of Tropical Medicine, which is now part of the University of London. This school has trained hundreds of doctors in tropical medicine and its teachers and students have made many important discoveries about human parasites. Much of the success of this institution is due to the fine traditions of scholarship and research that were started by Manson.

When Manson was in China, many of the people he saw had elephantiasis. This is a condition where the legs and genital organs become greatly enlarged and deformed. Manson did a large number of operations where he cut away much of the excess tissue so that these people could walk more easily. In one year he did 237 such operations and removed an average of about 50 pounds of tissue from each person.

When he went back to England on leave, in 1875, he went to the library at the British Museum to learn all he could about elephantiasis. He read about tiny worms that had been found in the blood of patients in India.

When he went back to China, he took a new microscope and soon found these worms for himself. He found that they only appeared in the blood at night. Very few worms were found in the blood during the day. He counted the number of worms in a drop of blood, taken every 3 hours, day and night, for 14 days, and made a graph showing how the numbers of worms in the blood, varied with time. A similar graph is reproduced in Fig. 5.1.

Fig. 5.1 Changes in the number of microfilariae of *Wuchereria bancrofti* in the peripheral blood of a Niuginian at different times of the day. The counts are based on the number of microfilariae in 40 cubic millimetres of blood. Samples were taken every two hours (data from Dr J. J. Saave)

He also saw that the worms were covered by a loose skin or sheath, and he found that the worms wriggled free from this when the blood was cooled. This made him think that the worms in the blood were larval stages which hatched from the sheath when they got out from the human host into a cold blooded animal. He also thought that a mosquito might be the most likely second host, because it was so common in the area, and it fed only at night, when the larvae were in the blood of man.

He tested these ideas by feeding mosquitoes on his gardener, who had elephantiasis. He later found larval stages in the mosquitoes. He tried several different species of mosquitoes but the larvae lived in only one species. He did not know much about mosquitoes, so he wrote to the British Museum for information about them. Six months later he had a reply saying that no book had ever been written about mosquitoes. He then set to work to find out about the anatomy and life cycle of the mosquitoes himself.

Manson found that the larvae from the blood of man, broke out of the sheath once they got inside the mosquito. The larvae then moved from the stomach of the mosquito into the muscles of its chest. Manson thought that man became infected when he drank water contaminated by larvae, which were freed from the mosquito, when it died in water after laying its eggs. This was found to be wrong. We now know that the larvae infect man, when man is bitten by the mosquito host.

Manson's work was tremendously important. He was the first person to show that a parasite of a vertebrate can spend part of its life cycle in a blood sucking invertebrate. This led many other workers to think that other diseases might be transmitted in the same way; by insects.

The adult worms of *Wuchereria* were found by an Australian doctor, Dr Joseph Bancroft, in 1876. He found several thin transparent worms, about two inches long, in a Chinese patient from Brisbane who had elephantiasis. When he examined the female worms under the microscope, he saw thousands of larvae within their bodies. He sent the worms to a parasitologist called Cobbold in London, who described the worms and called them *Filaria bancrofti*, after the man who sent them to him. In 1921, this species was put into the genus *Wuchereria*.

Manson did not learn of Bancroft's discovery for some years and he continued to search for the adult filarial worms. He eventually found them in 1880 in the tissues of a patient whom he was operating on.

For many years Manson wondered where the larvae went during the daytime. There were plenty in the blood at night, but very few during the day. One day he examined the body of an infected patient who had died suddenly. He found many larvae in the lungs. He thus showed that the larvae stop in the lungs during the daytime. This of course, raised two new questions—how do they know the difference between daytime and night time? and why is it important for them only to appear in the blood at night? These questions have now been partly answered by the work of Dr Hawking and his assistant who have been working in London.

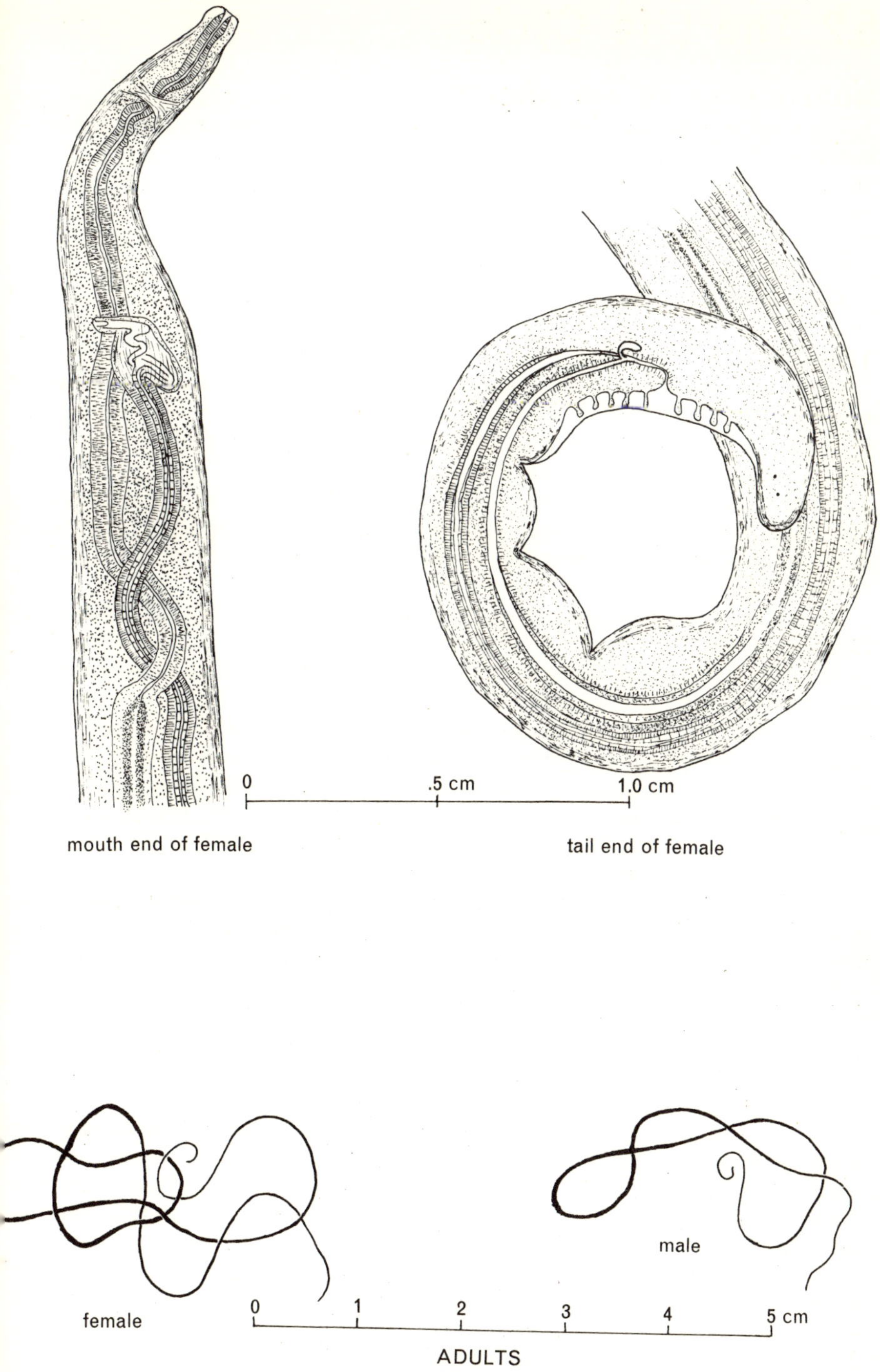

Fig. 5.2 Adults of *Wuchereria bancrofti*

5.2 The Life Cycle

MORPHOLOGY OF THE ADULTS

The adult worms are white and long and thin. The males are about 4 centimetres long and only 0.1 millimeters in diameter. The females are larger. They are up to 10 centimeters long and about 0.3 millimeters in diameter. The males have a coiled tail with two copulatory spicules one of which is much longer than the other. Along the inside curve of the tail of the male, are many small bumps called papillae. (See Fig. 5.2.)

The adults live in man for about ten years, sometimes more.

THE LARVAE

The female worms produce thousands of microfilariae which are surrounded by a delicate membrane or sheath. This is the remains of the egg shell. The microfilariae stay alive in the blood of the host for many months.

It is not unusual to find 70 microfilariae in 25 cubic millimeters of blood. This means than an adult man would have about 14 million microfilariae in his body. We do not know how many microfilariae are produced by one adult female in a day, but it is probably several thousand. Most people that are infected with *Wuchereria* have several adult worms in their bodies.

The microfilariae are about 250 microns long and about 7 to 10 microns wide (See Fig. 5.3.)

There are two races of *Wuchereria bancrofti* which have different geographical distributions. (See Fig. 5.4.) The race that occurs in Niugini, the Solomon Islands and the New Hebrides and places north and west, is called the periodic race, because it appears in the peripheral blood (blood of the skin) periodically; every night. Very few larvae are found in the peripheral blood during the daytime (see Fig. 5.1.) The other race is called the non-periodic race. About the same number of microfilariae are present in the peripheral blood throughout the day and night. This race occurs in New Caledonia, Fiji, the Ellice Islands and other Pacific Islands to the east of these. There is some evidence that the two races differ in the number of nuclei that occur in certain regions of the body, but too few samples have been examined to be sure of this.

The periodic race have vectors which bite man at night. The non-periodic race have vectors which bite man during the day. The phenomenon of periodicity raises two questions—why and how? The why question is partly answered by the different habits of the vectors. Each race has obviously become adapted to a life cycle involving vectors which feed on man at different times of the day. However, there must be a good reason why the micro-filariae of the periodic race disappear from the peripheral blood during the daytime.

Dr Hawking has worked on this problem for many years. He has confirmed the observations of Manson, that the microfilariae stay in the lungs during the daytime. It would appear that the lungs provide a better environment for them than other parts of the body. In other words, by spending most of the time in the lungs, they are able to live longer in man. Thus we have reasonable explanations for the 'why' question. But how is it that they can stay in the lungs during the daytime and not be swept around the body as the blood circulates? And how is it that they 'know' when to leave the lungs and circulate in the blood?

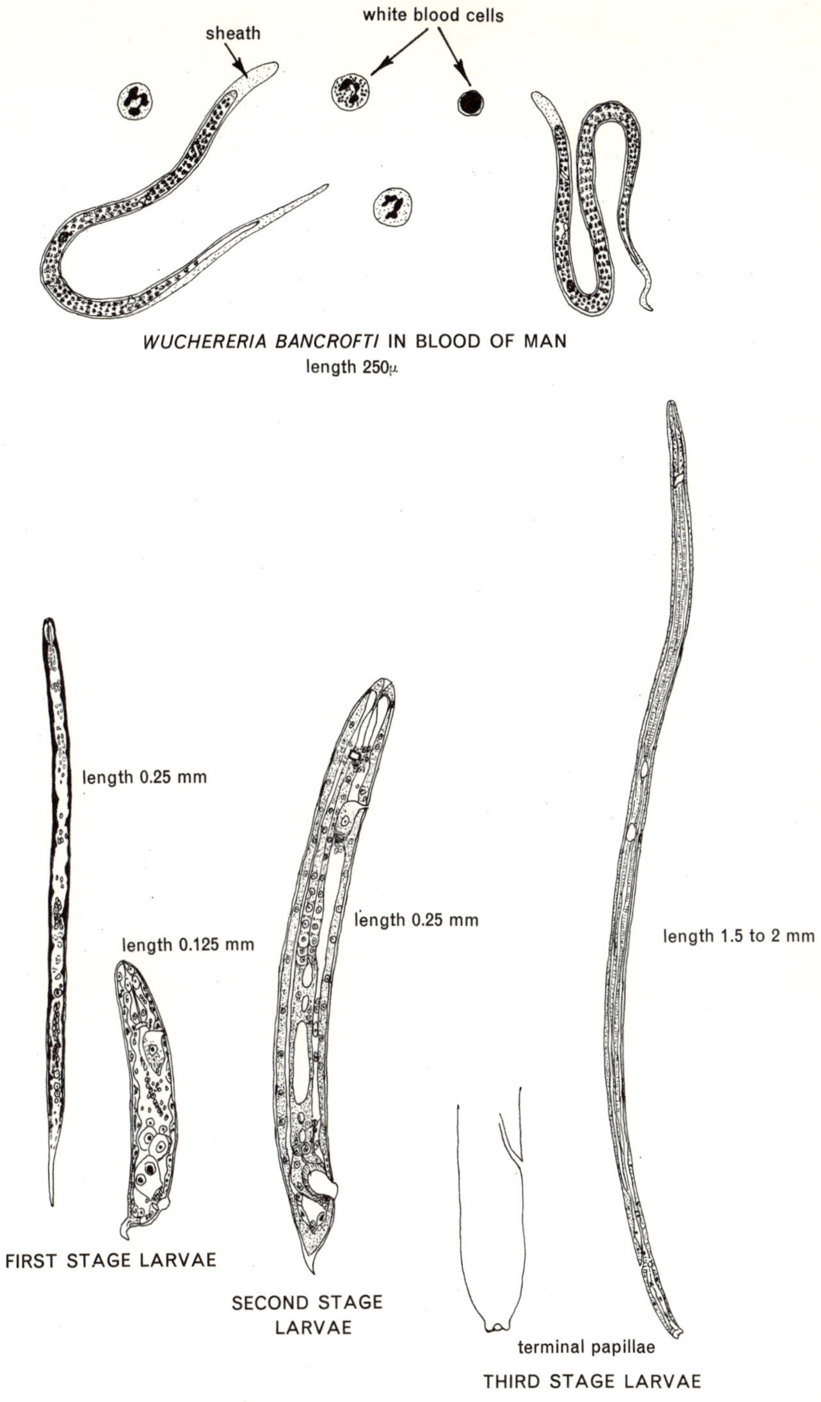

Fig. 5.3 Larvae of *Wuchereria bancrofti*

The work of Dr Hawking suggests that the microfilariae can stop in the lungs by swimming to maintain their position. They do this in the same way that a fish keeps its position in a river, by swimming against the current. During the daytime a few microfilariae are found in the peripheral circulation; these are the ones that have been swept away. They return again in less than a minute and then have another chance to stop in the lungs. During the daytime most of the microfilariae are in the lungs. A few are always leaving for a short time and then returning.

The microfilariae are able to detect differences in the amount of oxygen in the blood of the veins (which have less oxygen) and the arteries (which have more). As they are carried into the lungs, in the blood of the veins, the amount of oxygen starts to increase. This seems to provide the stimulus or signal for them to start swimming against the current so that they stay in the one place. During the daytime, the difference between the amount of oxygen in the veins and arteries is much greater than it is when the host is asleep at night. The difference during the night is not large enough to tell them to change behaviour when they enter the lungs. This means that their behaviour is largely governed by changes in the host's blood which occur periodically, as the host is active during the day, and asleep during the night.

If the host sleeps during the day and works during the night, the microfilariae change their periodicity. They appear in the blood during the daytime instead of the night. Now this change-over takes about a week to be completed. Each day the microfilariae occur in peak numbers at a slightly later time. This suggests that the microfilariae have some sort of clock mechanism which is adjusted to the activity of the host, but which cannot switch over immediately the host changes from being active during the day, to being active at night. Hawking has done a rather neat experiment which shows that this happens at least with the microfilariae of a parasite called *Edesonfilaria malayensis* which lives in monkeys. This species has a strong periodicity with the microfilariae appearing in the peripheral blood at night. An infected monkey was put under conditions of artificial light during the night and darkness during the day, so that the daily rhythm of the monkey and its microfilariae became inverted. They changed over so that they were common in the blood during the day, and were rare during the night. Blood containing microfilariae was taken from this monkey and put into the circulatory system of uninfected monkeys with a normal 24 hour rhythm (that is, awake during the day and asleep during the night). This was done by giving the monkeys a blood transfusion. The number of microfilariae in a given volume of peripheral blood was then counted every few hours for the next three days. For a time, the transfused microfilariae followed the periodic cycle that they had in their old host. But within 14 hours they changed over to follow the periodic cycle of their new host.

This experiment shows that the microfilariae have a periodic cycle of their own, which is independent of the behavior of the host. However, their own cycle is weak, and it soon changes if the behaviour of the host changes.

Thus the periodic race of *Wuchereria* has microfilariae which appear in the peripheral blood at night and which stay in the lungs during the daytime. They probably do this because they are transmitted by vectors which bite at night and because the lungs provide an environment which enables them to live longer.

They are able to stay in the lungs by swimming to maintain their position there.

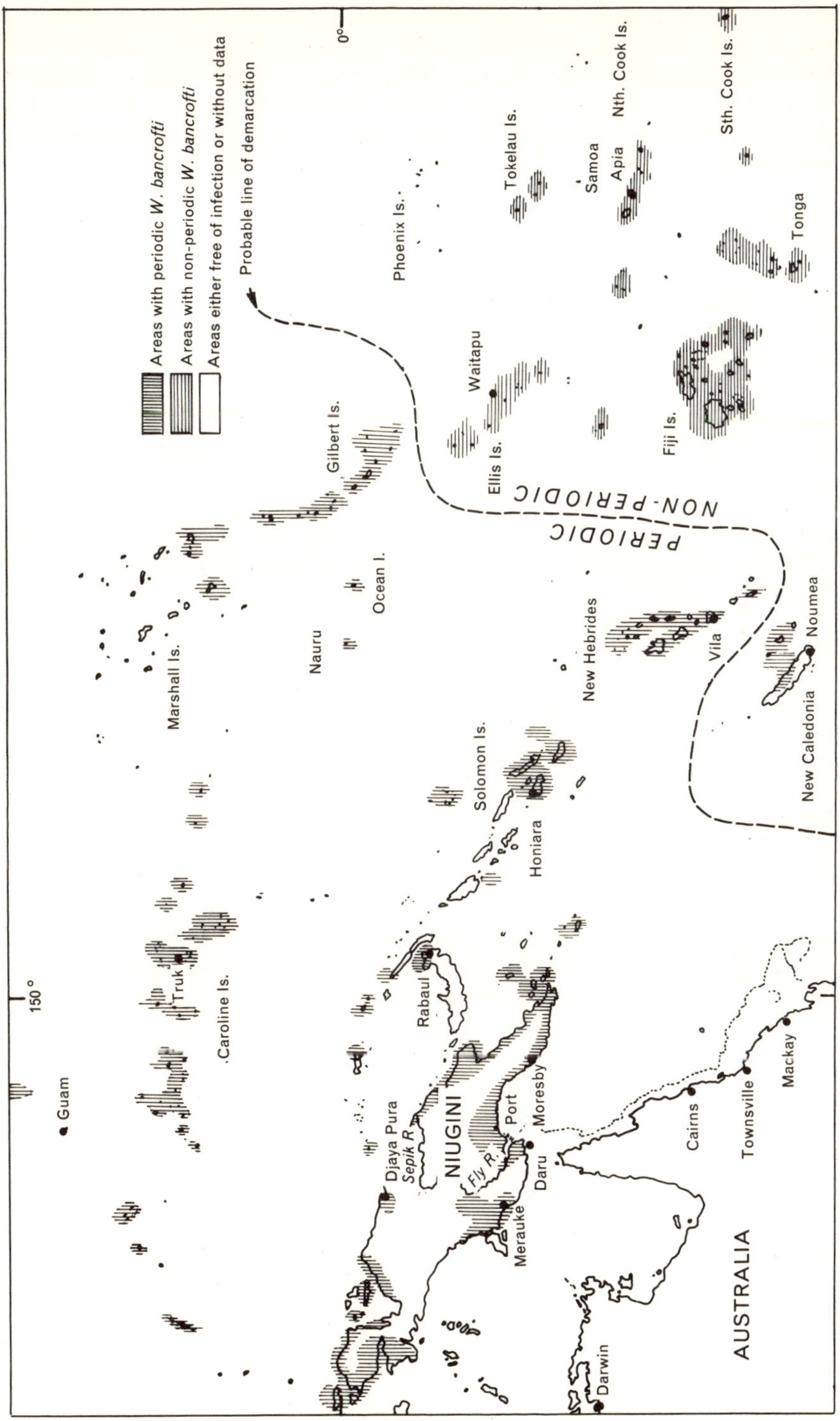

Fig. 5.4 Distribution of periodic and non-periodic forms of *Wuchereria bancrofti*

During the day they are able to detect changes in the amount of oxygen in the blood as they come into the lungs. This sudden change tells them to start swimming against the current. At night time, the amount of oxygen in the blood of the host does not change nearly so much, because the host is asleep. As the microfilariae come into the lungs at night, they do not get the message to start swimming, so they are carried around in the blood to all parts of the body. In addition to this, the microfilariae probably have a periodic cycle of their own, which makes them stay in the lungs by day and circulate in the blood at night. However, this is soon modified if the host's cycle is changed.

DEVELOPMENT WITHIN THE VECTOR

Many species of mosquitoes belonging to several different genera, can act as the intermediate host (or vector) of *Wuchereria bancrofti*. We shall discuss the vectors in Niugini presently.

In order to become infected, the mosquito must take blood containing microfilariae from man. These larvae move out of their sheaths almost immediately and wriggle through the wall of the gut and into the muscles of the chest of the mosquito. They then change in size and shape, so that they do not look very much like microfilariae at all. The body shortens to about half its original length and becomes fat, like a sausage. It is called the 'sausage stage'. It then moults and is known as a second larva, but its old skin does not come completely free. It grows in length and develops internal organs. It has a tiny tail. It moults again to become a third stage larva. This stage is found throughout the body of the mosquito. It is long and thin and has three little bumps or papillae around the tip of the tail. It is infective to man. The mature larvae are about 1.5 to 2.0 millimeters long. All stages are illustrated in Fig. 5.3.

The mature third stage larvae migrate to the head of the mosquito, and from there to the mouth parts or proboscis. Some larvae get lost and end up in other parts of the body.

The infective larvae take about 10 to 14 days to develop within most vector species. However, development is faster than this in a few vectors.

Experiments and observations on mosquitoes caught in the field, show that some species are better hosts than others. Some species can be experimentally infected, but mature larvae do not develop. The larvae die before they become infective. Other species develop mature larvae in experiments done in the laboratory, but are never found infected in nature. The percentage of mosquitoes that are infected with larvae of *Wuchereria* varies considerably with the species of vector.

Several hundred first stage larvae may be found in a single mosquito, but it is very seldom that more than ten infective larvae are found in a single host in nature. Most of the larvae die before they become infective and furthermore, many host mosquitoes die before the larvae have time to complete their development. Even in good vectors, only 1 to 2 per cent are found to be infected with mature larvae in nature. For example, in the Berau area of West Irian, de Rook found that 26% of all *Culex annulirostris* were infected with larvae of *Wuchereria bancrofti*, but only 1.6% of them contained infective larvae. About 20% of *Mansonia uniformis* and *Culex bitaeniorhynchus* were infected, but only 1 to 2% contained infective larvae. Experiments have shown that the concentration of microfilariae in the blood of man, influences the development within the vector.

If there are too many microfilariae in the blood (more than about 100 per 20 cubic millimetres) the mosquito will probably die. It cannot survive with large numbers of worms in its muscles. If the concentration of microfilariae is very low (for example less than one in 20 cubic millimetres of blood) the mosquito may not become infected.

Man becomes infected when infected larvae escape from the proboscis of the vector as it bites man. They enter the circulatory system by wriggling into the hole left by the mosquitoes mouth parts.

THE VECTORS IN NEW GUINEA

The species of mosquito that are vectors in Niugini are listed in Table 5.1. The incidence of larval stages of *Wuchereria*, in the species that have been found infected in nature is also given. A few species have been shown to be vectors under experimental conditions, but have not been found naturally infected. (They are marked E.) In addition, *Aedes vigilax* is a vector in New Caledonia but has not been reported to be a vector in Niugini, although it is common along our coast.

TABLE 5.1 Vectors of *Wuchereria bancrofti* in Niugini and West Irian, together with the incidence (per cent) of larval stages in wild mosquitoes. A few species have been infected experimentally, but have not been found to be infected in nature. They are marked 'E'.

Species	*Incidence %*
Anopheles farauti	2–46
A. punctulatus	4–15
A. koliensis	6–7
A. bancrofti	11
A. karwari	E
Aedes kochi	2–16
Armigeres obturbans	2
Culex annulirostris	2–26
C. fatigans	2–17
C. bitaeniorhynchus	2–20
C. squamosus	E
Mansonia papuensis	E
M. bonnewepsterae	E
M. crassipes	E
M. uniformis	1–20

There is little information available on the relative importance of the various vectors in Niugini, but there is much more information on the vectors in West Irian. It seems that certain species of vectors are more important in certain areas.

Three of the anopheline vectors are also vectors of malaria and their biology has already been discussed in Chapter 3.

Anopheles bancrofti does not seem to be very common in Niugini. It breeds in shaded swamps and pools containing plants. The adults attack man and they appear to be outdoor feeders.

Anopheles karwari is found only in a small area of West Irian.

Aedes kochi is common in the Markham valley and certain other coastal and island areas. It breeds in water which collects in the axils of plants such as bananas, pandanus and pineapples. (The axil of a plant is where a leaf joins the trunk.) It feeds in the late afternoon and night, and sometimes comes inside houses.

Armigeres obturbans is a common mosquito that breeds in pit toilets and dirty water.

Culex annulirostris breeds in a wide variety of temporary and permanent waters. It can breed in brackish water. It feeds on man at night both inside and outside houses. It is especially common in coastal areas.

Culex fatigans also breeds in a wide variety of temporary and permanent waters. It is often found breeding in tins, water drums and polluted (dirty, smelling) water. It often comes inside houses to feed, but also bites outside. It is common around towns.

Culex bitaeniorhynchus usually breeds in clean water. It usually bites at night outside the houses. It does not seem to be very common in Niugini.

Culex squamosus is rare in nature and is unlikely to be an important vector.

Mansonia uniformis breeds in swamps. It feeds during the day and night and often enters houses. It is often very common near large rivers (for example, the Sepik, Fly and Ramu).

A great deal more work should be done on the vectors of *Wuchereria* in Niugini.

DEVELOPMENT IN MAN

The infective larvae live in the lymphatic vessels and lymph nodes. It takes about a year before the females start to produce microfilariae. The females probably do not produce microfilariae unless they have first been fertilized by a male. Masses of adult worms have been found coiled up together in lymph vessels. This suggests that the young adults might wander about until they find a mate, but very little is known about their development or behaviour in man. The adult worms may live in man for up to ten years or more.

5.3 Pathology

Wuchereria bancrofti affects man in a number of ways. Not a great deal is known about certain aspects of the pathology of this infection. Different textbooks give different theories about what happens.

The first symptoms start about 3 to 12 months after infection. They probably start when several young adult worms gather together and block, or partly block, a lymphatic duct or lymph node. The symptoms will depend on how many worms are present, their location in the body, and the amount of immune reaction from the host against them. The lymphatic ducts and nodes in the area where the worms have come to rest become inflamed. This means that they become swollen (enlarged) and tender (sore to touch) and the skin covering them becomes red. Inflammation of the lymphatic system is called lymphangitis (lymph-an-jite-is). When this happens the infected person usually has other symptoms including fever, headache, backache, an itchy rash on his skin, a general feeling of being unwell (malaise), and unhappy (mental depression). These symptoms last a few days and come and go every few weeks, but they gradually become less severe until they stop completely. These first symptoms are characteristic of the acute stage of the disease.

In many cases, where only a few worms are present, the worms die within a couple of years. They are probably killed by cells of the lymphoid-macrophage system. Large numbers of white blood cells, especially eosinophils, move into the tissues near the worms. They are probably able to kill the worms by the action of their enzymes. Antibody reactions may also play a part. In many infections, all the symptoms gradually go away within a couple of years and there is little or no permanent damage. Severe symptoms, including fever, occur when the worms die due to the release of foreign proteins as the worms disintegrate. A high eosinophilia (up to 80 per cent) is often present during the acute stage.

In a small percentage of cases, elephantiasis and other permanent changes occur. The disease is then said to be chronic. Chronic cases of the disease are common in areas where the people are continually being reinfected. Usually less than 5 per cent of the infections become chronic, and chronic infections are seldom found in children or young adults. Chronic cases follow the complete blockage of the lymphatic vessels. It seems that this blockage is not entirely due to the worms themselves. The host reacts to the presence of the worms and this increases the amount of blockage. Special fiber cells grow along the walls of the lymphatic vessels in the area around the worms. They may completely surround the worms and as they do this, they block the lymphatic vessels. This means that lymph does not drain back into the circulatory system and the tissues become water-logged. Swelling occurs, and there may be an accumulation of fluid between certain membranes. The tissues become swollen and the skin is stretched tight. It may crack and secondary infections with bacteria and fungi may occur.

Several factors probably influence the outcome of infection. The number of worms that are present, the amount of reinfection and the location of the worms, will all obviously play a part in determining whether an infection becomes chronic or not. Even in areas where there is a lot of *Wuchereria*, only a small proportion of the infections become chronic. This suggests that a few people react to infection in such a way that chronic symptoms develop. But most people react in such a way, that this does not happen. We know very little about the mechanisms involved.

In Niugini, the legs and genital organs are most commonly affected in the chronic stage (see Fig. 5.5). The breasts and arms are affected in a very small number of cases.

Elephantiasis is most common in the legs, below the knee, although the upper leg may also be involved. The lymph nodes are enlarged and hard. The skin and tissues under the skin become greatly thickened and cracks may develop. The limb is swollen and may become so greatly enlarged that the person can hardly walk.

The testes and the ducts which carry sperm to the penis are often involved. They may become very swollen and sore. Fluid (lymph) may accumulate in the scrotum or in other parts of the genital organs. This is known as a hydrocoele. The fluid usually contains microfilariae. Sometimes these hydrocoeles burst into the urinary tract, and the urine then contains lymph.

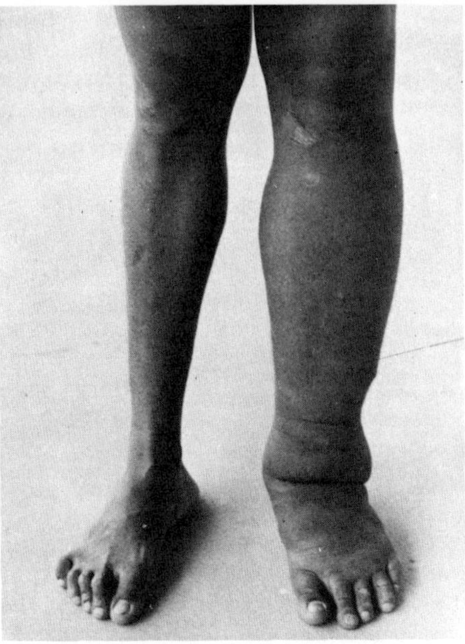

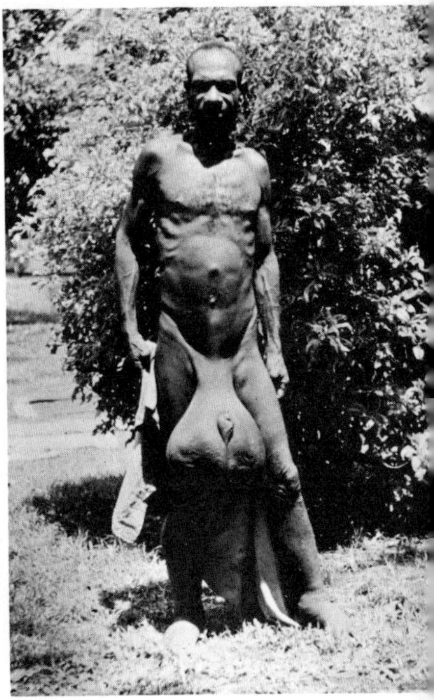

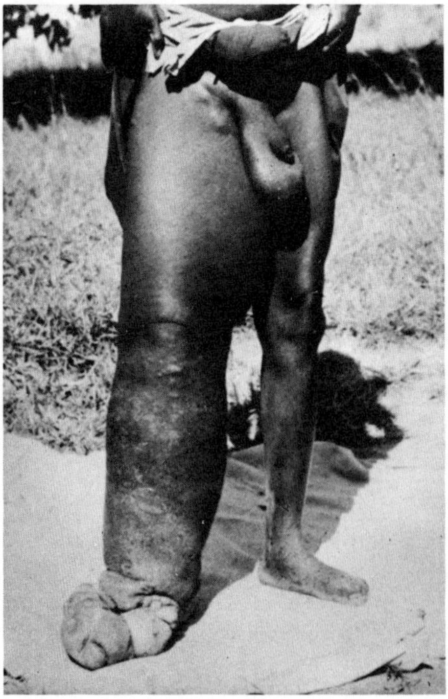

Fig. 5.5 Elephantiasis due to infection of *Wuchereria bancrofti*

The Diagnosis of Wuchereria

The diagnosis of *Wuchereria* depends on finding microfilariae or adult worms in the host.

Microfilariae of the periodic variety are seldom found in the peripheral blood during the daytime. Blood should be taken between 10 pm and 6 am, and stained with a Romanowsky stain or Haematoxylin. If fresh blood is examined the microfilariae can be seen moving about in it. Microfilariae are not present in the early stages of infection, and they may not be found in chronic cases. They may be found in the urine if it is milky (due to the presence of lymph), and they may be found in hydrocoele fluid. Adult worms may be found during surgery.

There are several other species of filarial worms of man that have adults and larvae that look like *Wuchereria*. None of them are known to occur in Niugini, but microfilariae of unusual appearance have been found in people from Kuyawa and Milne Bay. They were not identified by they may belong to another species closely related to *Wuchereria*.

In areas where *Wuchereria* is known to be present, the acute or chronic symptoms are suggestive of the infection. The drug di-ethyl-carbamazine (also known as heterozan and banocide) kills *Wuchereria* and as the worms die there is a characteristic reaction (see Section 5.5). If this reaction follows treatment, it is highly likely that *Wuchereria* was present.

5.4 Epidemiology

Wuchereria is a widespread disease. It occurs in tropical Africa and South America, and through Asia from Saudi Arabia through India to China and Japan. It also occurs on many of the Pacific Islands. About 200 million people throughout the world are infected with it.

Wuchereria infections are common in the coastal areas and islands of Niugini. They do not occur in the Highlands above 5000 feet. Infections appear to be very rare in Europeans.

Not a great deal has been published on the epidemiology of *Wuchereria* in Niugini, but there are several reports of work done in West Irian (Dutch New Guinea).

Three measurements are important in the epidemiology of *Wuchereria*. They are:

(i) the incidence of microfilariae in the blood
(ii) the density of microfilariae in the blood
(iii) the incidence of chronic disease due to *Wuchereria*.

The incidence of microfilariae is determined from blood surveys. The blood must be taken at night. It is usual to take 20 cubic millimeters of blood and spread it on a slide to allow it to dry. The slide is later stained and examined under a microscope for microfilariae.

It is usual to count the number of microfilariae in a fixed volume of blood (usually 20 cubic millimeters). This provides a measure of the density of the parasites.

The incidence of chronic disease is determined by examination of the legs, arms, genital organs and other parts, to see if elephantiasis or hydrocoeles are present.

Chronic symptoms are seldom found in children or adolescents, although they may have microfilariae in their blood. Some people who have been infected for many years, and who have elephantiasis, do not have microfilariae in their blood.

TABLE 5.2 Incidence (%) of microfilariae in different age groups from Inanwatan (West Irian) and the Maclay Coast (Niugini)

Age (years)	Inanwatan	Maclay Coast
0– 9	6	8
10–19	18	29
20–29	23	45
30–39	36	52
40–49	41	62
> 50	42	71
Total % positive	21	35

TABLE 5.3 Incidence (%) of microfilariae in the blood, and of chronic disease caused by *Wuchereria*, in different age groups, in the Berau Region of West Irian (after de Rook)

Age (years)	Incidence Microfilariae	Incidence Chronic Disease
1– 5	7	0
6–10	11	0
11–15	23	0
16–20	17	1
21–30	19	8
31–40	24	15
41–50	31	26
51–60	40	35
> 60	44	41

TABLE 5.4 Density of microfilariae in people of different ages, Inanwatan, 1959 (after Van Dijk)

Age (years)	Average Density (per 20 cubic millimeters) of Microfilariae in infected Persons
0– 4	19
5– 9	58
10–19	49
20–29	60
30–39	44
40–49	62
> 50	57
Total	54

Tables 5.2 and 5.3 show how the incidence of microfilariae and chronic symptoms increase with age, in areas where *Wuchereria* infections are common.

Very little data has been published on microfilarial densities in Niugini. Dr Van Dijk has published some data showing that at Inanwatan, the density of microfilariae in infected people, does not vary very much with age. This data is given in Table 5.4.

The Epidemiology in Niugini

Wuchereria is common in coastal and island areas of Niugini. It does not seem to be present in the Highlands in places above 5000 feet. No one knows why. The vectors are certainly present in the Highlands. Many highlanders must have been infected while working on the coast and must have taken the disease back to the Highlands when they returned home. *Wuchereria* occurs in the Bomai area of the Chimbu district. Most of this area is below 5000 feet in altitude.

Table 5.5 gives the incidence of microfilariae in various parts of Niugini. Some of this data has been supplied by Dr J. J. Saave and has not been published before.

The incidence of chronic symptoms (elephantiasis and hydrocoele) in Niugini, is seldom more than three per cent and often less than one per cent. In a few places, the situation is more serious; for example in 1959, ten per cent of the people on the Maclay (Rai coast, Madang District) coast had elephantiasis and 19 per cent of the males had hydrocoeles.

In the six years 1961-7, there were 1165 admissions to government hospitals with *Wuchereria* infections. About 80% of these patients were males. About 400 cases came from Papua, 450 from the North Coast-Mainland and 270 from the Niugini Islands.

Many cases of early acute infection, with fever as the most common symptom, are not diagnosed because microfilariae are not yet present in the blood. These cases are probably treated for malaria which is one of the commonest causes of fever.

TABLE 5.5 Incidence of microfilariae in various parts of Niugini

		Incidence per cent
North Coast Mainland		
Maclay Coast	1911	60
Lambringi	1912	71
Maprik	1959	33
Warup	1959	30
Vanimo	1967	21
Markham Valley	1966	26
North Markham	1946	30
Kaiapit (Markham)	1947	44
Busama (Huon Gulf)	1947	21
Patep (near Lae)	1947	0
West Coast	1924	71
Tami Is	1912	24
Sio	1959	36
Popondetta	1966	16
Samarai	1946	31
Milne Bay	1946	51

		Incidence per cent
South Coast		
Purari Delta	1947	30
Papua	1901	2
Highlands		
Bomai (altitude 4500 ft)	1966	30
Niugini Islands		
Papitalai (Manus)	1912	33
Manus	1923	19
Manus	1960	36
Pak Is (Manus)	1960	19
Lorengau (Manus)	1965	19
Admiralty Group	1924	39
Emirau Is (St Matthias Group)	1947	39
St Matthias Group	1924	21
St Matthias Group	1911–12	9
New Britain	1924	50–71
Rabaul	1924	15
New Britain	1930s	5–17
Rabuana (New Britain)	1912	50
Arawe (New Britain)	1912	50
Kokopo (New Britain)	1965	27
Gloucester (New Britain)	1965	20
Kaliai (New Britain)	1967	25
Makada (Duke of York Is)	1931	23
Kavataria (Trobriands)	1947	17
Kiriwina (Trobriands)	1965	17
D'Entrecasteaux Is	1968	8
Goodenough Is	1938	7
Kavieng (New Ireland)	1923	14
Matty Is (Western Is)	1930s	25
Buin (Bougainville)	1965	15

5.5 Treatment, Control and Eradication

Until recent times little could be done to control *Wuchereria* infections. Chronic cases of elephantiasis and hydrocoele have been treated surgically since Manson's time. In these operations, excess tissue is cut away and fluid drained out, so that the patient can walk more easily. These operations are very successful when the testes are involved, but are not so successful with elephantiasis of the limbs. Nowadays, surgery is seldom used for this. Instead, the limb is tightly bandaged. The bandages are changed each week until the swelling can be reduced no further. This squeezes the lymph from the tissues back into the circulatory system and the size of the limbs may be considerably reduced. Bandages work best in elephantiasis of young adults. Little can be done for people who have had severe elephantiasis of the legs for many years.

In recent years, two new developments have made it possible to control and eradicate *Wuchereria* infections. They are, the use of the drug di-ethyl-carbamazine, which sterilizes the adults so that microfilariae are no longer produced, and eventually kills them, and secondly, the use of residual insecticides such as DDT.

Di-ethyl-carbamazine was first used to treat *Wuchereria* in 1955. Microfilariae disappear from the blood a day or two after treatment with this drug.It eventually kills the adult worms as well.

The amount of drug taken by each person should vary with their weight, so that large adults take more of the drug than little children. Several types of treatment have been tried. The most successful seems to be 5 to 6 milligrams of drug per kilogram of body weight, taken every week or month for 6 to 12 doses. That is, relatively small amounts of the drug are taken at intervals over a long period. In other cases, one dose is given every day for a week, or a single very large dose is given once only. This is the easiest to administer to large numbers of people, but does not seem to work as well as smaller amounts taken at intervals over a long period.

The drug causes some minor toxic effects, which we shall talk about presently, but this drug is safe to use, and if given to everybody, interrupts the transmission of *Wuchereria*. It also kills the adult worms so that the symptoms of the disease disappear from many infected people. It cannot cure the symptoms of elephantiasis or hydrocoele found in long standing chronic infections.

Treatment with di-ethyl-carbamazine is often followed by side effects. In one drug trial in West Irian, about one quarter of the people had some side effects. About 1 to 3 days after taking the drug, these people had fever, headache and pains in the body, especially in the back, legs and arms. Sometimes there was swelling or pain in the groin or knees. These symptoms went away in a few days. They were not severe and were treated with aspirin. They probably resulted from the release of proteins from the microfilariae as they died. In a smaller proportion of people, further symptoms developed 2 to 4 weeks after treatment. They probably resulted from the death of the adult worms. In these cases, certain parts of the body such as the arm, thigh, knee or calf, became swollen. Occasionally these swellings developed into abscesses (sores).

Mass treatment with di-ethyl-carbamazine greatly reduces the incidence of microfilariae, the density of microfilariae in the blood, and the incidence of mature larvae in mosquitoes. Examples of this are given in Table 5.6.

In some cases the people have been given di-ethyl-carbamazine and their houses have been sprayed with residual insecticides. The effectiveness of residual insecticides will depend on the situation. Some vectors of *Wuchereria* do not come inside houses very much, (they bite outdoors) and where these vectors predominate, residual spraying will be of little use. Where the vectors regularly feed inside houses, residual insecticides may help to end transmission. Residual insecticides on their own, probably do not control transmission very much.

When the incidence of microfilariae reaches a few per-cent, every effort should be made to completely eradicate *Wuchereria*. This should be a much easier task than the eradication of malaria because a single course of treatment probably kills all the adult worms in most people. Those few people who remain infected after treatment, could easily be found by surveys in which the blood of every single person is examined. All infected people could then be given another course of drugs.

TABLE 5.6 Data for the incidence and density of microfilariae in man, and the incidence of larvae in vectors, before and after treatment with di-ethyl-carbamazine (from Kessel and Massal)

	Taiwan	West Irian	Maclay Coast Niugini	Tahiti	Fiji
Incidence (per cent) of microfilariae in man —					
Before	8	21	35	38	12
After	3	1	10	6	3
Density of microfilariae—(per 20 cubic millimeters).					
Before	–	67	55	35	4
After	–	13	1.5	0.6	0.4
Incidence (per cent) of larvae in vectors—					
Before	8	25	4.8	8.5	4.6
After	3	?	0	1.6	0.1

Chapter 6

OTHER PARASITES

6.1 Introduction

In this chapter we will deal with the remaining parasites of man in Niugini.

The first that we will consider is *Trichomonas vaginalis*. It lives in the vagina of women and the urethra of men. We have already talked about other species of this genus which live in the mouth and the gut.

We will then deal with a phenomenon called larva migrans. This is the name given to symptoms that arise when nematode larvae of foreign species infect man. These species develop normally in animals such as dogs, cats, pigs and rats. When they infect man, they do not develop normally and often wander about damaging various tissues before they die.

Finally we will deal with bed bugs, and mites, which are ectoparasites of man.

6.2 Trichomonas vaginalis

Trichomonas vaginalis lives in the vagina of women and the urethra and prostate gland of men. It is often pathogenic. The vagina is the female organ which receives the penis during mating or copulation. The prostate gland produces substances which form part of the semen. The semen also contains sperms which are formed in the testes. Semen passes out of the body through the urethra, which is a tube that runs to the end of the penis. Urine also passes out of the body through the urethra.

T. vaginalis is similar in appearance to *T. hominis* and *T. tenax*, which live in the gut and the mouth (see Section 4.8). They are all separate species. Each one is morphologically different from the others and each species can only live in its own place. That is, *T. vaginalis* will only live in the reproductive organs; *T. hominis* will only live in the gut; and *T. tenax* in the mouth.

T. vaginalis is bigger than the other species. It has an average length of 13 microns (range 7 to 23), which is about twice as big as the other species. It does not form cysts. It reproduces by binary fission, that is, the nucleus divides to form two nuclei and then the rest of the animal splits to produce two new individuals. During this process, new flagella are formed. *T. vaginalis* is illustrated in Fig. 6.1.

It can be kept alive in test tube cultures containing sugars and starch and other nutrients. It probably lives by absorbing food though its surface, although it can eat bacteria when it is kept in culture.

The vaginal solutions are normally acidic (pH about 4). *T. vaginalis* can survive only in solutions that are less acidic than this (pH about 5 to 7). It may only be able to infect women when the pH of the vagina is unusually high. The pH of the vagina rises to about 5 after each menstrual period. Once *Trichomonas* is established in the vagina, it seems to be able to keep the pH at this level.

T. vaginalis causes inflammation of the vagina and a frothy creamy discharge. This discharge causes severe itching and burning. Some infected people have no obvious symptoms.

Males are less often affected by the infection, but sometimes they develop severe symptoms. Often the patient urinates frequently or may have difficulty in

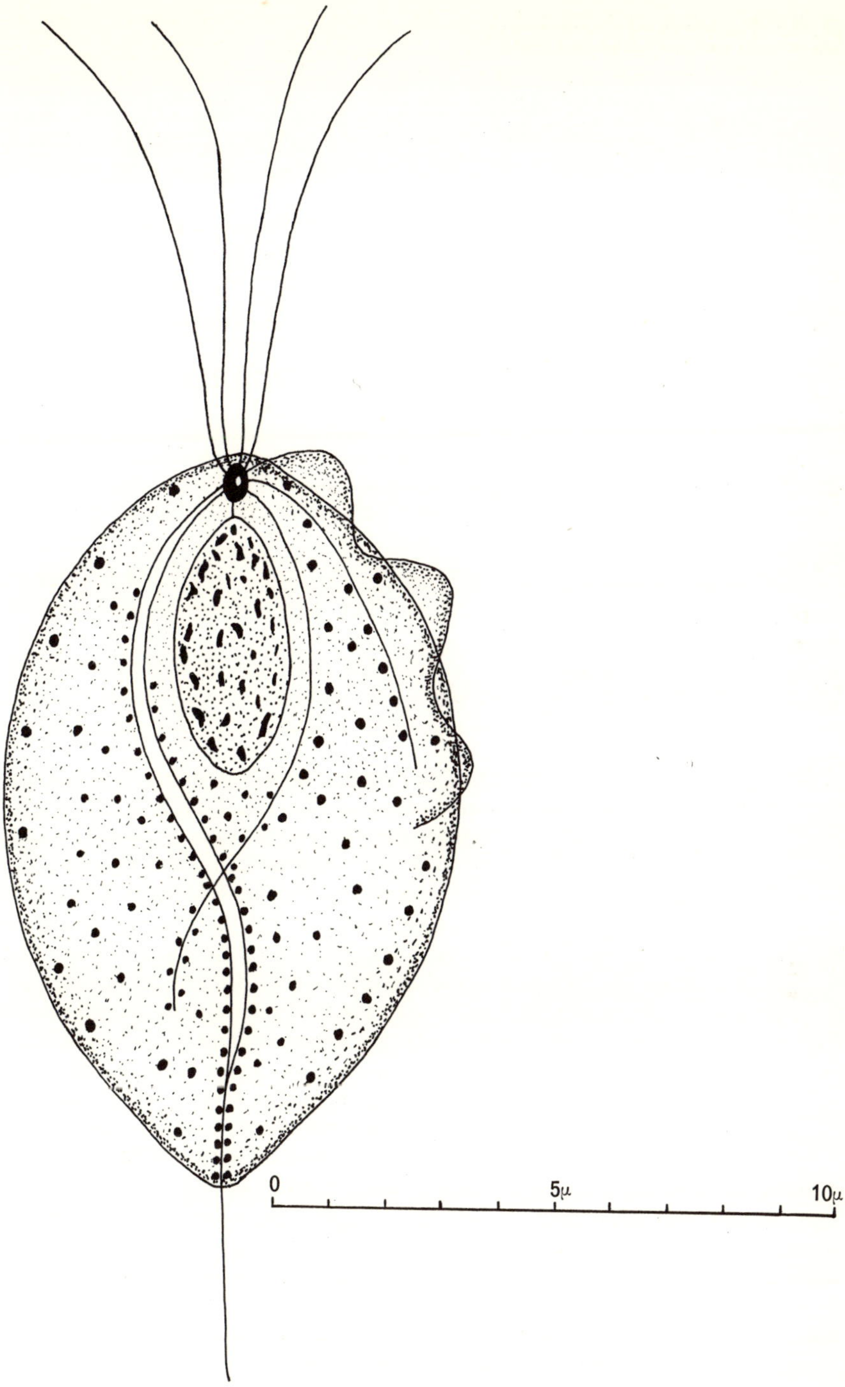

Fig. 6.1 *Triehomonas vaginalis*

urinating. There may be a dirty white discharge which contains a few parasites. The urethra may be itchy. The prostate gland may be enlarged and tender.

Diagnosis depends on finding the parasites in the discharge or in the sediment of a urine specimen. They remain active for several hours.

The parasites are transferred from person to person by sexual intercourse. Some people do not develop any symptoms and they may pass the infection on to their sexual partners.

Surveys done in Europe and America have found that 12 to 40 per cent of all women are infected. The incidence is much higher (up to 60 per cent) in women who come to venereal clinics. The incidence is much lower in men. In one survey, 9 per cent of men with urethral discharges were infected.

T. vaginalis occurs in Niugini, but there are, as yet, no published records of its incidence. It is not uncommon in patients coming to hospitals with venereal infections.

A number of drugs are used to treat *Trichomonas* infections. Some are taken by mouth and others are put inside the vagina. Sometimes several drugs must be tried in succession before the parasites are killed. It is necessary to treat the husband, who may have no symptoms, as well as his wife. Otherwise he will reinfect her again.

6.3 Larva Migrans

There are a number of species of parasites, the larval stages of which infect man. But man is not their normal host, and they are unable to develop into adults in man. They migrate through the skin or internal organs of man for weeks or months or even years, and as they migrate, they damage the tissues around them. Some of them produce waste products or secretions that are irritating. The symptoms that result from the migration of these larval parasites are known as 'larva migrans'. If they are moving in internal organs, it is known as visceral larva migrans. (Viscera is the collective name for the organs that lie in the body cavity.) Some species tend to occur in the brain and we should probably refer to this as cerebral larva migrans. If they are moving in the skin, it is known as cutaneous larva migrans. Most larva migrans is caused by nematodes.

When these nematodes are in their normal host they usually have a migration of the kind that occurs with *Ascaris* and hookworm larvae. In an abnormal host they seem to get lost and keep wandering until they die, or are killed by the host, or until they end up in an organ, such as the eye, from which they cannot escape. The normal host must provide stimuli or signs which direct their migrations, so that they end up in the appropriate organs. It must provide other stimuli that tell the larvae to stop wandering and start changing into adults. Abnormal hosts do not provide the correct stimuli and the larva wander into the 'wrong' kinds of places. They stay immature and do not develop into adults.

The symptoms produced in man will depend on the number of larvae that are present, the organs that they invade, their species, and the reaction of the host to their presence. In most cases the larvae eventually die and disintegrate, or they are surrounded by special host cells which form a wall around them while they are still alive.

In almost every case eosinophilia is a feature of the infection. The number of eosinophils in the blood increases and they gather in large numbers in the tissues around the worms. This is so much a feature of larva migrans that certain forms of larva migrans have been known as 'tropical eosinophilia'. This term has been applied to cases of larva migrans of unknown origin.

It is often difficult to diagnose larva migrans and it is usually even more difficult to determine the species of parasite responsible. The worms cannot be recovered without cutting the patient open and searching for the larvae. This is seldom done unless the patient is very sick or has died. The larval stages of many species of nematodes are also difficult to identify. Some progress has been made when volunteers have been experimentally infected, but this practice is full of unknown consequences. (The volunteers may become seriously sick!)

We shall now give some examples of parasites that cause larva migrans and the symptoms that result. Very little is known yet about larva migrans, and as time goes on, many more species of parasites will be found to cause this condition in man. At present we only know about the species that are commonly responsible for this condition.

Cutaneous Larva Migrans

We have already mentioned in Chapter 3 that the infective larvae of the hookworms of cats and dogs cause creeping eruption. This is a form of larva migrans. The infective larvae of human hookworms may sometimes cause this condition. We know of one experiment where three volunteers were exposed to the infective larvae of hookworm.

All the larvae had been cultured from the one batch of eggs. Two of the people infected developed normal hookworm infections in the gut, and the third developed a severe case of cutaneous larva migrans.

Visceral Larva Migrans

This is commonly caused by the larvae of *Toxocara canis* and *T. cati*. These parasites are closely related to *Ascaris* and normally infect dogs and cats. Children are more commonly infected than adults because they often play in areas polluted by their pets. Some small children become infected when they eat dirt containing infective eggs. The larvae have been recovered from the liver, muscles, lungs, brain and eye. The liver and spleen are frequently enlarged and if the lungs are infected there is often a bad cough. Larvae in the eye can cause blindness. Larvae in the brain may cause fever, convulsions (fits) or unconsciousness. There is no treatment for visceral larva migrans due to *Toxocara*. All pet dogs and cats should be treated with drugs each month, to kill their parasites. This will decrease the chances of human infection.

A nematode called *Gnathostoma spinigerum* (Nath-o-stome-a spin-ig-erum) causes larval migrans in man. It does not occur in Niugini, but a closely related species, *G. doloresi* (doll-or-easy) has been found in pigs in Rabaul and Goroka. It may be able to infect man, although there are no reports of human infections so far. People could become infected by eating uncooked meat containing larvae from animals such as fish, frogs, snakes, lizards, crabs or yabbies.

Filarial worms belonging to the genus *Dirofilaria* have been found in man a few times. They are normally parasites of dogs and cats. *Dirofilaria immitis* is very common in dogs in Niugini. It lives in the heart. One human case has been recorded from Niugini recently.

Angiostrongylus cantonensis (Ann-gee-o-stron-gillus can-ton-ensis) is another nematode that is known to cause larva migrans in man. The adults normally live in rats. They are found in the large blood vessels which run from the heart to the lungs. The larval stages are found in a wide variety of land and freshwater invertebrate hosts such as slugs, snails, prawns, crabs and planarians. Man becomes infected by eating these intermediate hosts without cooking them. Occasionally vegetables may be contaminated with the larvae. The giant land snail *Achatina fulica* is an important intermediate host. It occurs in the Solomon Islands and Niugini Islands and has recently been introduced into the Port Moresby area. Human cases of *Angiostrongylus* infection have been reported from many Pacific Islands and Taiwan, the Philippines, Thailand and Indonesia. So far, the parasites have not been found in rats or man in Niugini, but rats from Queensland are often infected.

Human infections are characterised by eosinophilia and meningitis (men-in-jite-is). Meningitis is the inflammation of the surface of the brain. The most common symptoms are headache, pains and stiffness in the neck, back and body, and fever. Vomiting sometimes occurs and sometimes paralysis of some of the face muscles. A large number of eosinophils are found in the spinal fluid. A small number of infected people die, but most recover in two or three weeks. The symptoms develop about three weeks after eating an infected intermediate host.

We do not know very much about larva migrans in Niugini. There have been several reports of people who have probably had larva migrans. Their sickness has involved eosinophilia and various other symptoms, but the parasites responsible are seldom found. Migrating hookworm and *Ascaris* larvae and *Wuchereria* infections all cause eosinophilia.

Table 6.1 gives the incidence of eosinophilia in people of different age groups in the Highlands, where there is no *Wuchereria*.

Some of these cases of eosinophilia are probably due to *Ascaris* and Hookworm and some are probably due to foreign parasites. Of these cases, 71 per cent had an eosinophilia of 10 to 15 per cent. that is 10 to 15 per cent of the white blood cells were esinophils (the normal figure is 1 to 4 per cent). Furthermore, 24 per cent of these cases had an eosinophilia higher than 15 per cent. Most of the cases were in children.

TABLE 6.1 Incidence of eosinophilia in different age groups in the Highlands (1967)

Age	Incidence of Eosinophilia (per cent)
<1	12
1–4	12
5–14	13
15–44	6
>45	5

6.4 Bedbugs

Bedbugs are insects which belong to the Order Hemiptera. This group has mouthparts that form a hollow needle. They feed by pushing this needle into plant or animal tissues and sucking out the tissue juices. Most species feed on plants but the bedbugs suck blood. In Niugini, bedbugs are known as musmus.

Two species of bedbugs live in houses and bite man. They are *Cimex hemipterus* (si-mex hem-ip-ter-us) and *Cimex lectularius*. As far as we know, only *C. hemipterus* has been found in Niugini. This species occurs in all tropical countries of the world while *C. lectularius* is generally confined to colder countries. Other species of this genus attack various birds and mammals.

Bedbugs live in cracks and crevices in wooden floors, beds, walls and furniture. They have a flattened body so they can easily fit into cracks and narrow spaces. They grow to about one centimeter long. They do not have any wings, but the adults have a pair of little pads which are the remains of the wings. The head and thorax are small and the abdomen is large and rounded. The body is covered with small hairs.

The two species of *Cimex* can be separated by a feature of the thorax. In *C. lectularius* the sides of the thorax are flattened into a thin narrow blade. The thorax of *C. hemipterus* is not like this (see Fig. 6.2).

At least two American textbooks state that a species of bedbug called *Leptocimex boueti* (Brumpt) bites man in Niugini. This is wrong. We have been unable to find any original report of this species in Niugini.

The females of *Cimex* lay one or two eggs each day. They are stuck to surfaces. She must return to the same place again and again to lay her eggs because they are found in clumps of ten to twenty with each one touching the egg next to it. Each female lays several hundred (up to 400) eggs during her lifetime.

The eggs are about one millimeter long and have a lid, or operculum, set at an angle at one end. (See Fig. 6.2.) This is pushed open when the young hatch. The eggs are white or grey and their surface is covered with tiny circles so that it looks as if it is covered with chicken wire. The eggs hatch in about 6 days. The immature stages are called nymphs. They moult five or six times; each time they shed their skin they become a little bigger. Provided the nymphs are well fed and the temperature is warm, the nymphs moult every three to five days. They become adults in about three weeks under tropical conditions. The adults live for about six months.

Bedbugs are usually nocturnal, that is, they feed at night. But they will feed during the daytime in darkened houses if people are asleep. All stages feed on blood from man. They can go without food for several months, but they usually feed every day or two. Feeding takes about 10 to 15 minutes. They bite exposed parts of the body, but do not burrow under clothing. They seldom cling to the host while feeding, but usually stay on the bed. After they have fed, they usually defaecate on the bedding. The faeces make large (2 to 4 millimeter) dark brown spots on the bed clothes.

The bites of bedbugs can be very irritating. Some people are more sensitive than others. The irritation is caused by the saliva which is injected before feeding starts. It probably contains an anticoagulant which prevents the blood from

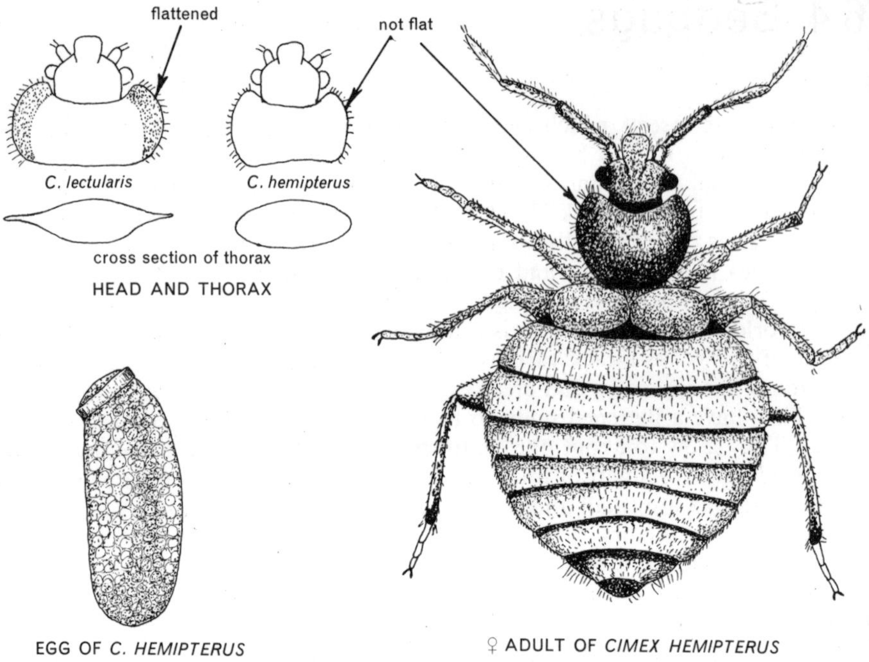

Fig. 6.2 Eggs and adult of *Cimex hemipterus*

clotting. The bites produce a red itchy lump. Often there are several bites in a line. If a person is continually bitten by a lot of bedbugs anaemia may develop, the victim becomes nervous and suffers from a lack of sleep. The bites of 50 to 100 adults cause influenza-like symptoms, palpitations of the heart, headache and eye disturbances. Bites on the lips may cause severe swelling.

Bedbugs are not known to transmit any human diseases. Experiments have shown that they can act as vectors of certain diseases under laboratory conditions, but they have not been shown to do this in nature.

A number of methods have been used to control bedbugs. Experiments need to be done to determine the best and cheapest methods to kill bedbugs in various kinds of houses in Niugini. We will list some of the methods that might be tried.

(i) Beds, furniture and walls can be sprayed with insecticide. A 5 per cent solution of DDT in kerosene has been used successfully in some places, but in certain countries including parts of Niugini, bedbugs have developed resistance to DDT. Gammexane (BHC) is also used, but in some areas bedbugs have become resistant to this too. Gammexane also has a bad smell. A mixture of 0.2 per cent pyrethrins and one per cent piperonyl butoxide is effective. Pyrethrins are also used on their own. They can be used in a water spray, as a powder, or an aerosol. Malathion is an effective insecticide, and it is relatively non-toxic to man. It is at present being used to control bedbugs in some parts of Niugini.

(ii) Sulphur fumes are effective. Sulphur is put in a bowl or dish and burnt (some methylated spirits should be added to make sure that it burns well.) All doors and windows must be shut and the house must be left closed for several hours. The smoke from burning sulphur is poisonous and everyone should stay out of the house until the smoke has cleared. About five pounds of sulphur are needed for every 1000 cubic feet of space. If sulphur is bought in bulk, it costs less than 5 cents a pound, so that most houses in Niugini could be treated for 25 to 50 cents each.

(iii) Bedbugs are killed by temperatures greater than 100°F. The temperature in the sun on a normal sunny day in Niugini is usually 110°F or more, so that bedbugs sheltering in movable beds will be killed if the beds are put outside in the sun for several hours. If the legs of the bed are then put in tins containing a little kerosene, bedbugs will not be able to reinfest the bed from other parts of the house.

(iv) A mixture of diesoline and creosote (in the ratio of 3:1) is also effective if sprayed on walls, beds, and other furniture. It is relatively inexpensive. In one trial in Niugini, the cost of labour and materials was less than $1 for a large house.

Bedbugs are widespread in Niugini and occur in the Highlands, coastal areas and the Niugini Islands. It is said that bedbugs have become a serious problem since houses have been sprayed with DDT to kill the mosquitoes that carry malaria. An entomologist is at present investigating this problem in Niugini.

We will give one example of the great numbers of bedbugs that can occur in one building. In January 1970, we organised a bedbug hunt in one of the wards of the Kundiawa Hospital. The patients had previously complained about the musmus biting them. Children, labourers and patients all took part. They chased the bugs from crevices and cracks in the beds and walls with dissecting needles, cotton wool soaked in chloroform, and pressure packs of insecticide. A total of 3,294 bedbugs were caught in 76 man hours (nearly one a minute)! Many could not be caught because they could not be pulled from the cracks. At a rough guess, less than half the bedbugs in the building were caught. This ward was occupied by 42 people and it is no wonder they complained about being bitten. The ward was subsequently sprayed with diesoline and creosote. Two weeks after spraying, 32 bedbugs were caught in seven man hours. The numbers gradually built up again and the building was sprayed with Dichlorovos (DDVP), an organo-phosphorus compound. A week later, 28 bedbugs were caught in 3 man hours. Other wards nearby were not sprayed and it seems that bedbugs from them, were soon entering the sprayed building.

6.5 Acarina

The Order Acarina, of the Phylum Arthropoda, contains the ticks and mites. It is a large order divided into 166 families. The general characteristics of ticks and mites have been given in Section 1.2.

Many species of ticks and mites are temporary parasites of man. They are often not very host specific and they occasionally infect man, but normally they infect other animals or are free living in dead organic material. Some of them are vectors of various diseases, others cause dermatitis (skin disease) and a few species of ticks cause paralysis when they feed on man. In Niugini the most important of these temporary parasites of man, are certain Trombiculid mites which act as vectors of scrub typhus. We will say more about them presently.

In addition to these temporary parasites, two species of mites live as permanent parasites in the skin of man. They are *Demodex folliculorum* and *Sarcoptes scabiei*. Both occur in Niugini.

DEMODEX

Demodex folliculorum lives in hair follicles and sebaceous glands. A hair follicle is a pit in the skin from which a hair grows. Sebaceous glands are small glands which produce an oily substance. They usually open into a hair follicle. *Demodex* is wormlike in shape and has four pairs of small stumpy legs. The females are almost half a millimeter long, and the males are a little smaller. (See Fig. 6.3.) They rarely cause any symptoms. Sometimes they are found in pimples on the face, called blackheads, and sometimes they cause a little irritation and redness of the skin. They usually occur on the face. Related species cause severe mange in dogs and some other animals.

SARCOPTES

Sarcoptes scabiei (Sar-cop-tease scabe-ee-eye) is more important than *Demodex*. It is called the itch-mite and causes a condition known in Niugini as *kaskas*. This species has many forms. One form infects man and others infect horses, dogs, pigs and other animals. All these forms look alike, but they can be distinguished by very tiny differences in shape and size. Each form is generally confined to its own host. Experiments have shown that the forms from animals will not live for more than a few days on man.

The adults are tiny. The females are about 0.3 to 0.45 millimeters long. The males are only about 0.2 millimeters long. The adults have four pairs of short stumpy legs. The females have a long hair trailing from each of the four hind legs, but the males only have hairs on the third pair of legs. On the fourth pair, they have suckers. The larvae have only three pairs of legs. (See Fig. 6.3.)

The males and females mate on the surface of the skin. The males die soon afterwards so that they are seldom found. After she has been fertilized, the female starts to burrow into the skin. She moves two or three millimeters a day. The burrow is often visible in people with fair skin, because it contains eggs and tiny dark specks of faeces, left by the female as she moves along. The burrows can be seen in people with dark skin, but it requires a lot of practice.

The female lays about three eggs a day, and she lives for about a month. Only about twenty eggs are found in one burrow. The eggs are transparent and oval and measure about 150 $\times$ 100 microns. The first larval stage hatches in about three days. The females moult three times, at intervals of three or four days and their life cycle is completed in about two weeks. The males moult only twice and their development takes about ten days.

These mites are most often found in the skin of the hands, wrists and forearms. Sometimes they are found in other places such as the feet and ankles, penis and scrotum and the buttocks. They are rarely found in other places.

Experimental infections in volunteers have shown that there is no reaction to the infection at first. The number of mites gradually builds up, and after about three months, itching starts and gradually gets worse. A red rash often appears

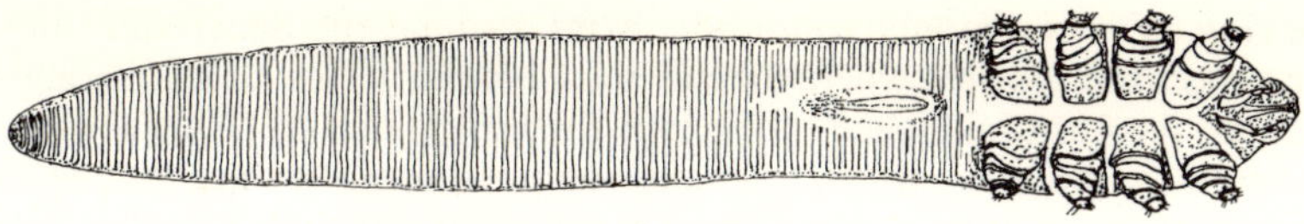

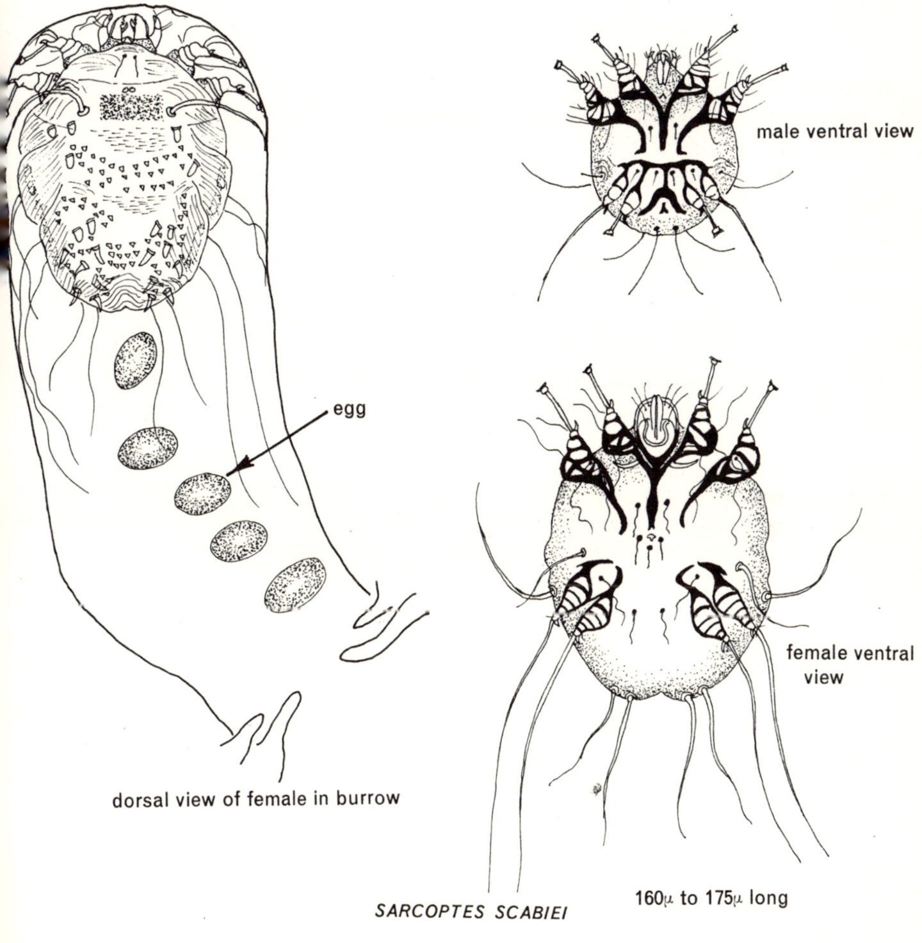

Fig. 6.3 *Demodex folliculorum* and *Sarcoptes scabiei* (after Belding and Mégnin)

in the skin. It is commonly seen under arms, around the belly, and thighs. (It does not occur in the area where the mites are present.) This rash is a reaction on the part of the host, and when it develops the host is said to be sensitised. After this occurs the number of mites usually declines considerably. As the itching gets worse, the host starts to scratch and many of the mites are probably removed from their burrows. A few mites may remain but in many cases they probably all die out. People who have been sensitised by a previous infection seldom become reinfected. Experiments have shown that reinfection causes an intense reaction and this appears to limit reproduction of the parasite. The new infection usually dies out within a few weeks. A few people develop very severe and long lasting infections. The skin may become greatly thickened, wrinkled and cracked. It may peel off in large flakes. It would appear that in these cases sensitisation does not occur. Dr Backhouse described cases like this, from New Britain, in 1929.

Infections are passed from person to person by body contact. Children may be infected when they sleep with their parents. Adults may be infected during sexual intercourse, or when holding hands.

The diagnosis of scabies depends on finding adults or larvae in the skin. People who have not become sensitised show no symptoms. Once they start to scratch, the mites can usually be found by digging them from their tunnels with a needle.

Scabies can be treated with various chemicals. The following are said to be the most reliable: Sulphur ointment (10% sulphur in vaseline); Tetmosol (Tetra-ethyl-thiuram monosulphide); Benzyl benzoate (equal parts of Benzyl Benzoate, soap and methylated spirits). All are applied to the skin each day for several days.

Scabies occurs in man in Niugini, but seems to be less common than it used to be. Dr Backhouse in 1926-7 found that 43 per cent of adult males in Rabaul were infected with scabies. The Vines survey (1966) reported that scabies was present only in the Highlands and rainforest areas of the North Coast Mainland of Niugini. His data is summarised in Table 6.2. The infections were much more common in children than adults.

TABLE 6.2 The incidence (per cent) of scabies in the Highlands and North Coast Mainland of Niugini.

Age (years)	Highlands	North Coast
<1	34	17
1–4	19	7
5–9	19	2
10–14	7	3
15–29	5	4
30–44	2	0.6
>45	0.7	0.7

TROMBICULID MITES

Several hundred species of mites belonging to the family Trombiculidae occur throughout the world. The larval stages of all species are temporary ectoparasites of vertebrates. They are found on snakes, lizards, turtles, birds and mammals. Over 40 species are known to attack man and domestic animals. Man

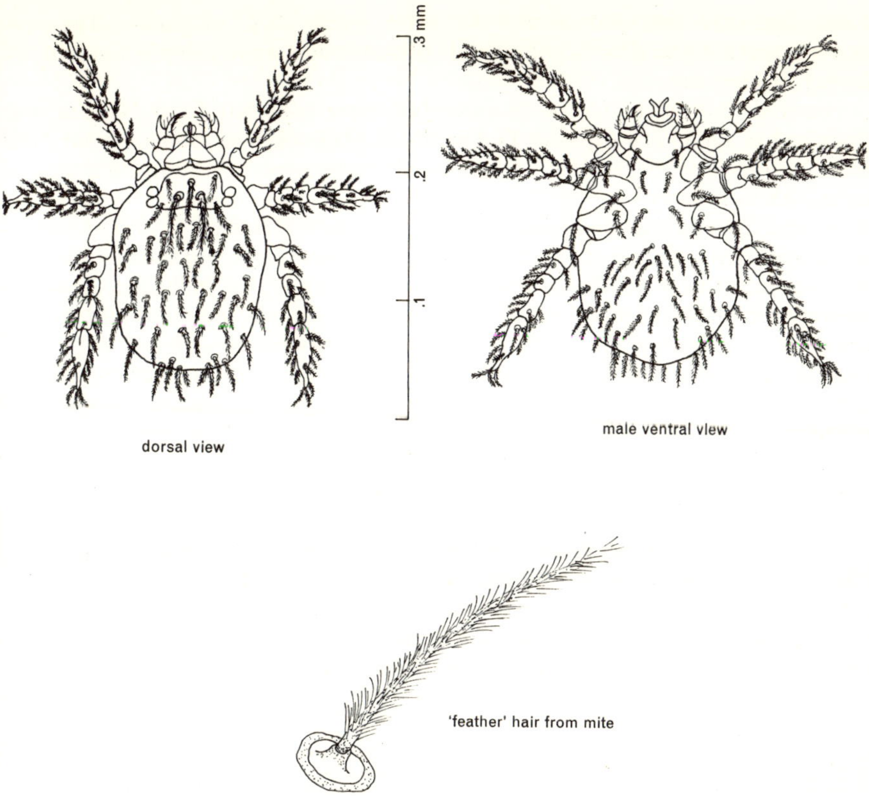

Fig. 6.4 Larvae of trombiculid mites

is not a normal host, but is occasionally infected by contact with other animals, or when walking in the bush.

Some species are the vectors of a disease called scrub typhus. Many species cause an intense itching reaction when they feed on man.

The larvae of trombiculid mites have six legs. They are usually orange or red in colour and are only about 0.2 millimeters long. The body is covered with tiny hairs which are branched like the feathers of a bird. They are sometimes called redbugs, chiggers or harvest mites. (See Fig. 6.4.)

The females lay eggs in the soil, which hatch in about a week. The larvae must feed on a vertebrate host before they can develop into the next stage. The larvae then moult once or twice before becoming adults. but do not feed on a host again. The life cycle takes about two months in some species that have been investigated.

The larvae usually stay on the host only a few days. They use their mouthparts to pierce the skin, and then inject saliva. The saliva digests the host tissues and this juice is then sucked up. They stay attached to the host for two or three days and the continual injection of saliva causes an intense reaction. The area around the bite becomes red and swollen and a sore often develops. The mites often attach themselves in places where the clothing is tight, for example, around the waist, under the belt, and around the ankles under socks or gaiters. The condition is sometimes called 'scrub itch'. It was common among soldiers in Niugini during the last war.

The first Trombiculidae to be described were collected from Niugini and described in 1905. The explorer D'Albertis in his book *New Guinea: what I did and what I saw* (published 1880) describes what was almost certainly an attack of scrub itch as follows:

'For some days past all on board (the boat) have complained of a violent itching in different parts of the body without finding any cause for it, but today an almost microscopical insect was discovered, which either attacked us in the forests or has been carried on board in the skins of the Gowras (Gouria) and other birds which mostly live on land. It was useless to wash ourselves either with hot water and soap, or acid, or eau-de-cologne, but today we had recourse to a new expedient, that of washing ourselves with petroleum'.

Anyone who has read D'Albertis' book will have little sympathy for his suffering. Serves him right!

Mr H. Womersley has written a book giving details of all the trombiculid mites known from South East Asia and Niugini.

Scrub typhus is carried by certain trombiculid mites of the genus *Trombicula.* It has been reported from coastal areas of Niugini and the Islands. It is caused by a rickettsia. (Rickettsias are micro-organisms that appear to be related to bacteria and viruses.) Scrub typhus is a disease of wild rats that live in the forest. Infections in man can be serious unless certain drugs are given.

Tasol

We would like to end this book with a message. We are very conscious of the debt we owe to those who have written books and articles about the parasites of man, and especially those who have written about the parasites of man in Niugini. Without them this book could not have been written.

One of us owes a debt, in particular, to Joe Bearup, who taught him something about human parasitology. He is now 'lapun tru' and lives in Sydney. He went to the Highlands with Dr Heydon in the early 1930s and he tells a story of his time there, that is worth recording.

Bearup and Heydon had to collect faeces from the Highlanders in order to find out what parasites they had. Most of the talk was in sign language. It must have been funny to watch. Joe had false teeth and he would take them out and clack them together to make a bit of a show. After a while, people heard about this, and he had to repeat this performance every day for a great crowd that came to watch it. He says that some of them had walked for several days just to see it. You probably cannot guess the price of admission to this display!

We said we had a message. We have several. We would like you to consider the photograph on the opposite page. It is a photo of a sign that used to stand outside the public toilet at Pagwi, on the Sepik River. If everyone followed the advice it offers, gut parasites would cease to be a problem.

The parasites of man cause a great deal of suffering and sickness, but they will be overcome. As this happens, we will be faced with new problems. In many ways

man is his own worst enemy. As malaria and other sicknesses are conquered, this country will have a great increase in population. It has already started. There may not be enough land to grow all the food that is needed. Birth control must be used to make sure that the population does not get too big. Niugini can become a rich country because many crops grow well here, but riches are nothing if they are shared among too many people.

Many of the 'developed' countries of the world are today facing great problems that they have made themselves: problems of pollution, overcrowding, drug addiction, traffic, sexual maladjustment, violence and an increase in mental illness. Our most important problems are concerned with the health of people; and parasitic diseases make up an important part of them. As we face these problems we must also think to the future and try and work a society which does not suffer from the ills of 'developed' countries.

APPENDIX I

Table of Measurements

(i) Length

Metric System

1000 millimicrons	=	1 micron (μ)
1000 microns	=	1 millimeter (mm)
10 millimeters	=	1 centimeter (cm)
100 centimeters	=	1 meter (m)
1000 meters	=	1 kilometer (km)

Imperial or British System

12 inches	=	1 foot
3 feet	=	1 yard
1760 yards	=	1 mile

Conversion

1 centimeter	=	0.39 inches
2.54 centimeters	=	1 inch
1 yard	=	0.914 meters

(ii) Volume

Metric System

1000 milliliters (ml)	=	1 liter (l)
1000 cubic centimeters (cc)	=	1 liter (l)

Imperial or British System

16 fluid ounces (ozs)	=	1 pint
2 pints	=	1 quart
4 quarts	=	1 gallon
231 cubic inches	=	1 gallon

Conversion

4.546 liters	=	1 gallon
35.2 fluid ounces	=	1 liter
28.41 milliliters	=	1 fluid ounce

(iii) Weight or Mass

Metric System

1000 micrograms (μg)	=	1 milligram (mg)
1000 milligrams	=	1 gram (g)
1000 grams	=	1 kilogram(kg)

Imperial or British System

16 ounces (oz)	=	1 pound (lb)
14 pounds	=	1 stone
2240 pounds	=	1 ton

Conversion

1 pound	=	453.6 grams
1 kilogram	=	2.205 pounds

(*iv*) *Temperature*

Degrees °F Fahrenheit	Degrees °C Centigrade
32	0
50	10
60	15.6
70	21.1
80	26.7
90	32.2
95	35
96	35.6
97	36.1
98	36.7
99	37.2
100	37.8
101	38.3
102	38.9
103	39.4
104	40
105	40.6
106	41.1
212	100

APPENDIX II

Malaria Data for Niugini

All data from surveys done after control programs started is preceded by an asterisk thus*.

Each line gives the place and time of the survey. Place names and areas, are arranged alphabetically under present Administrative districts. District boundaries are shown in fig. 3.30. Spleen Rates are given thus: SR 4%. All other figures are for the incidence of parasites. The following abbreviations are used. C = children; A = adults, OPR= Overall Parasite Rate (all ages); IPR= Infant Parasite Rate (children less than one year old). Age groups are shown thus <2y=less than 2 years; 2-5y=children between 2 and 5 years old; >40y = adults over 40 years etc. CD = Census Division 1970.

Where a range is given, the highest and lowest figures for a project are given.

SOUTH MAINLAND REGION
CENTRAL DISTRICT
GULF DISTRICT
MILNE BAY DISTRICT
NORTHERN DISTRICT
WESTERN DISTRICT

CENTRAL DISTRICT

Babiko	1935 SR.C 64%
Bioto	1935 SR.C 100%
Chirria	1935 SR.C 60%
Delena	1935 SR.C 68%
Goilala	1936 SR.C 0%
	1952 SR.A 17%
Koitaki	1942 SR 88%
Mekio Villages	1935 SR 2-9y 92-100%
Angabunga River Area	
Inauive	1935 SR.C 100%
Inawi	1935 SR.C 100%
Beipa	1935 SR.C 100%
Aipeana	1935 SR.C 100%
Amoamo	1935 SR.C 100%
Rarai	1935 SR.C 100%
Inauauni	1935 SR.C 92%

Mekio Villages

Bebeo	1935 SR.C 100%
Oriopetana	1935 SR.C 100%
Iesubaibua	1935 SR.C 100%
Inauai-a	1935 SR.C 100%
Eboa	1935 SR.C 100%
Inauabui	1935 SR.C 100%

Mountain Tribes

Kuni Tribe	1935 SR.C 0%
Fugi Tribe	1935 SR.C 2% Children recently to coast.

Ambo Tribe

Ikuiwei 1935 SR.C 5% Children recently to coast.

Kone 1935 SR.C 0%

Wane 1935 SR.C 0%

Goilala 1935 SR.C 0%

Motu Motu No. 1.

(Uratai) 150 miles west of Port Moresby on coast.

Aug. 1944 PR <2y 69%; 2-5y 62%; 6-10y 40%; 11-15y 32%; >16y 22%

Aug. 1944 SR <2y 59%; 2-5y 81%; 6-10y 84%; 11-15y 65%; >16y 44%

*Jan. 1945 PR <2y 68%; 2-5y 55%; 6-10y 42%; 11-15y 25%; >16y 16%

*Jan. 1945 SR <2y 76%; 2-5y 98%; 6-10y 96%; 11-15y 73%; >16y 38%

Motu Motu No. 2.

Aug. 1944 <2y 60%; 2-5y 64%; 6-10y 43%; 11-15y 26%; >16y 14%

Aug. 1944 SR <2y 65%; 2-5y 83%; 6-10y 86%; 11-15y 70%; >16y 39%

*Jan. 1945 <2y 91%; 2-5y 74%; 6-10y 45%; 11-15y 49%; >16y 32%

*Jan. 1945 SR <2y 97%; 2-5y 100%; 6-10y 98%; 11-15y 91%; >16y 59%

Mou 1935 SR 52%

Pinupaka 1935 SR 30%

Port Moresby to Redscar Villages 1935 SR 26-33%

Rapa 1935 SR 44%

Roro Villages 1935 SR 44-100%

South Coast Villages 1935 SR 12-50%

Tapini 1954 SR.C 17%

GULF DISTRICT

Isa Peapi 1944 Sept. <2y 60%; 2-5y 64%; 6-10y 43%; 10-15y 26%; >16y 14%

Isa Peapi 1944 Sept. SR 2-10y 79%

Korovagi (Purari Delta) 1947; 2-9y 33%; SR 2-9y 52%; OPR 26%; SR 38%

Lala Pipi 1945 SR <1y 63%; 2-5y 90%; 6-10y 88%; 11-15y 65%; SR.A 41%

Uritai Villages 1944 Aug. <2y 69%; 2-5y 62%; 6-10y 40%; 11-15y 32%; >16y 22%

Uritai Village 1944 Aug. SR 2-10y 88%

MILNE BAY DISTRICT

East Cape 1939 SR 50%

D'Entrecasteaux Islands

Boitalu 1953 SR 2-10y 6%

Glibila 1953 SR 2-10y 100%; 2-10y 33%

Gna-Legola 1953 SR 2-10y 91%; 2-10y 50%

Gumilabaga 1953 SR 2-10y 0%

Ialaka 1953 SR 2-10y 4%

Ialungwa-Mwaimwala (date not known) SR 2-10y 0%

Kaulagu 1953 SR 2-10y 20%; 2-10y 70%

Kaibola 1953 SR 2-10y 0%; 2-10y 7%
Kaisiga & Bulakwa 1952 SR 2-10y 78%; 2-10y 33%
Kaisiga & Bulakwa 1953 SR 2-10y 100%; 2-10y 60%
Kaduwaga 1953 SR 2-10y 19%; 2-10y 0%
Kumageya 1953 SR 2-10y 14%; 2-10y 2%
Koma-Osiwasiwa 1953 SR 2-10y 38%; 2-10y 6%
Luya 1953 SR 2-10y 5%
Munuwata Is. 1953 SR 2-10y 100%; 2-10y 14%
Nuatuba 1953 SR 2-10y 81%; 2-10y 43%
Oiuvai & Oua-Dasi 1953 SR 2-10y 9%
Oluvleni-Okaiboma 1953 SR 2-10y 14%; 2-10y 7%
Obowaba-Gayobara 1953 SR 2-10y 13%
Sinaketa 1953 SR 2-10y 76%; 2-10y 18%
Simsim 1953 SR 2-10y 100%; 2-10y 0%
Teiava 1952 SR 2-10y 64%
Teiava 1953 SR 2-10y 89%; 2-10y 67%
Tauwema 1953 SR 2-10y 40%; 2-10y 0%
Vakuta Is. 1953 SR 2-10y 85%; 2-10y 39%
Dobu 1938-39 SR 80%
Fyfe Bay 1942 SR 76%

Ferguson Island

Ferguson Island Moratau 1938-39 SR 48-87%
Ferguson Island Moratau 1957 OPR 40%
Ferguson Island Moratau 1959 OPR 40%; IPR 56%
*Ferguson Island Moratau 1963 OPR 8%; after 5 years spraying
*Ferguson Island Moratau 1964 OPR 6%; after 6 years spraying

Goodenough Island

Goodenough Island Moratau 1938-39 SR 64-93%
Goodenough Island 1957 OPR 78%
*Goodenough Island 1958 IPR 59% after 1 year spraying
*Goodenough Island 1962 OPR 26%
*Goodenough Island 1964 OPR 14%
Normanby Island (DJAU) 1938-39 SR 60-94%
*Normanby Island 1963 OPR 57%; 1965 OPR 20%; 1967 OPR 8%
*Loboda 1963 OPR 70%; 1965 OPR 44%
*Maideba 1963 OPR 75%; 1965 OPR 20%; 1967 OPR 15%
*Meudawa 1963 OPR 38%; 1965 OPR 8%; 1967 OPR 2%
*Sapoloa 1963 OPR 27%; 1965 OPR 6%
*Sisiana 1963 OPR 64%; 1965 OPR 0%; 1967 OPR 11%
*Sawatupawa 1963 OPR 63%; 1965 OPR 36%; 1967 OPR 20%
*Sipupu 1963 OPR 47%; 1965 OPR 8%; 1967 OPR 5%
Misima and Calvados Chain of Islands 1962 OPR 14%; *1964 OPR 5%; *1965 OPR 2%
Misima Is. 1962 0-1y 64%; 1-4y 33%; 5-14y 20%; > 15y 19%
*Misima Is. 1963 0-1y 12%; 1-4y 15%; 5-14y 10%; > 15y 0%
*Misima Is. 1969 May 0-1y 0%; 1-4y 1%; 5-14y 1%; > 15y 0%
*Bolou 1969 SR 0%
*Bwagoiwa 1969 SR 1.9%
*Eaus 1969 SR 2.9%
*Liak 1969 SR 0%
*Nimoa 1969 SR 3.1%
*Siagara 1969 SR 1.6%

*Rossel Island 1969 1.4%
Rogeia Island 1938 SR 48%

Mainland

Gama Dodo (date not known) 0-14y 68%; 0-14y SR 100%; >15y 26%
Raba Raba Area 1938-1939 SR 73-90%
Trobriand Islands 1947 2-9y 25%; 2-9y SR 63%; OPR 13%; SR all ages 21%
Kiriwina Villages 1938 SR 4-30%
Kiriwina 1952 OPR 7%
Kiriwina 1953 OPR 21%
Kiriwina 1962 OPR 6%
*Kiriwina 1963 OPR 2%
*Kiriwina 1964 OPR 3%
*Kiriwina 1965 OPR 0.2%
*Kiriwina 1966 OPR 1.6%
Kanataria 1952-53 SR 0-1y 10%; 2-10y 42%; 11-20y 26%; 21-30y 34%; >30y 12%
Kanataria 1952-53 OPR 7%; 0-1y 14%; 2-10y 17%; 11-20y 7%; 21-30y 4%; >30y 3%
Kantaria 1954 OPR 7%; IPR 14%; SR 27%
Kailewena 1954 2-9y 60%; SR 2-9y 100%; OPR 18%; SR all ages 80%
Kitava 1954 2-9y 2%; SR 2-9y 14%; OPR 2%; SR all ages 2%
Losuia Govt. Station 1952-53 OPR 6%; SR all ages 37%
Losuia 1954 IPR 0%; OPR 6%; SR all ages 37%
Losuia 1962 April OPR 9%; IPR 0%; 2-5y 12%; A 9%
Losuia 1963 March OPR 8%
Molosaida 1952-53, SR 0-1y 0%; 2-10y 17%; 11-20y 17%; 21-30y 5%; >31y 2%
Molosaida OPR 2%; 0-1y 11%; 2-10y 6%; 11-20y 0%; 21-30y 0%; >31y 0%

NORTHERN DISTRICT

Buma 1942 SR 68%
Inonda 1944 Feb. < 2y 100%; 2-5y 100%; 6-10y 97%; 11-15y 100%; >16y63%; SR 2-10y 95%
Inonda 1944 Oct. 0-10y 71%
Omo 1956 OPR 1%
Wanagela 1963 OPR 38%
*Wanagela 1964 OPR 20%
*Wanagela 1965 OPR 10%

WESTERN DISTRICT

Fly Delta 1956 Nov. <2y 37%; 2-4y 47%; 5-9y 29%; 10-14y 20%; 15-19y 16%; >20 12%
Fly Delta 1956 Nov. SR<2y 80%; 2-4y 74%; 5-9y 81%; 10-14y 69%; 15-19y 71%; >20y 36%
Kiunga 1956 Nov. <2y 67%; 2-4y 50%; 5-9y 50%; 10-14y 27%; 15-19y 12%; >20y 17%
Kiunga 1956 SR <2y 88%; 2-4y 100%; 5-9y 95%; 10-14y 100%; 15-19y 87%; >20y 73%
Lake Murray 1956 Nov. 5-9y 54%; 10-14y 29%; 15-19y 7%; >20y 7%
Lake Murray 1956 Nov. SR<2y 100%; 2-4y 100%; 5-9y 94%; 10-14y 86%; 15-19y 67%; >20y 49%
Middle Fly 1956 Nov. <2y 42%; 2-4y 72%; 5-9y 64%; 10-14y 64%; 15-19y 41% >20y 35%
Middle Fly 1956 Nov. SR <2y 67%; 2-4y 100%; 5-9y 90%; 10-14y 63%; 15-19y 25%; >20y 37%

HIGHLANDS REGION
CHIMBU DISTRICT
EASTERN HIGHLANDS DISTRICT
WESTERN HIGHLANDS DISTRICT
SOUTHERN HIGHLANDS DISTRICT

CHIMBU DISTRICT
Bandi 1946 Oct. C 1%; SRC 7%; A 1%; SRA 5%
Bangl-Kerowagi 1946—47 C 10%; SRC 42%; A 5%; SRA 18%
Bau-Undugu 1946 Nov.C 13%; SRC 12%; A 2%; SRA 5%
Bomai 1962 Sept. OPR 37%
*Bomai 1968 OPR 0%
Chimbu Hospital 1955-56 12.6% admissions were for malaria.
Chauve 1946-47 Jan. C. 0.3%; SRC 1.2%; A 0%; SRA 0.8%
Chauve 1947 Nov. C 0.3%; SRC 1%; A 0%; SRA 1%
*Chauve 1968 early OPR 2%
*Chauve 1968 late OPR 1%
Dinga Tribe 1947 Dec.
Kangendi C.0% SRC 3%; A 0%; SRA 5%
Bugmigl C 0%; SRC 8%; A 0%; SRA 1%
Dogbun-Koraku 1964 Sept. OPR 4%
Dom 1964 Nov. OPR 7%
Dom-Wikauma 1965 Jan. OPR 7%
*Dom-Wikauma 1968 OPR 0.05%
Duma 1947 Jan C 0%; SRC 0%; A 0%; SRA 0%
Ega Lutheran Mission 1947 Jan. OPR 0%; SRC 8%
Elimbari 1968 OPR 0.3%
Endugwa Tribe 1. 1946 Oct. C 20%; SRC 68%; A 16%; SRA 11%
Endugwa Tribe 2. 1946 Oct. C 0%; SRC 30%; A 4%; SRA 8%
Endugwa Tribe 3. 1946 Oct. C 1%; SRC 7%; A 1%; SRA 3%
Endugwa Tribes 1947 Dec.
Tokiaku C 0%; SRC 26%; A 0%; SRA 12%
Tangeku C 0%; SRC 12%; A 6%; SRA 5%
Eku C 5%; SRC 10%; A 4%; SRA 10%
Endugwa 1956 OPR 25%; SR 46%; 4800 to 5500 ft above sea level
Gembogl 1963 OPR 2%
Gembogl 1964 July OPR 5%; pre-operational data.
*Gembogl 1964 OPR 3%; after 1 spray round.
*Gembogl July OPR 1%; after 2 spray rounds.
*Gembogl 1966 July OPR 1%; after 3 spray rounds.
Gogmei 1946-47 C 0%; SRC 0.3%; A 0.4%; SRA 0.8%
Gogmei 1947 Nov. C 0%; SRC 0.3%; A 0.4%; SRA 1%
Gomla Tribe 'A' 1947 Jan. C 0%; SRC 2%; A 0%; SRA 2%
Gomla Tribe 'B' 1947 Jan. C 0%; SRC 0%; A 0%; SRA 1%
Kamagani 1946 Nov. C 0%; SRC 3%; A 0%, SRA 4%
Karamui 1962 Oct. OPR 17%
Karamui—Daribi CD 1965 OPR 19% pre-operational data.
*Karamui Daribi 1968 OPR 6%
Kerowagi 1945 Oct. C 6%; SRC 19%; SRA 19%
Kerowagi 1947 Apr. Mission School C 6%; SRC 22%
Kerowagi 1963 OPR 10%; pre-operational data
*Kerowagi 1965 OPR 4% after 2 spray rounds.
Kebai 1947 Jan. C. 0%; SRC 0%; A 0%; SRA 0%
Koge 1956 OPR 4%; SR 22%. 6400-7000 ft above sea level

Koliagu 1946 Oct. C. 3%; SRC 22%; A 0%; SRA 5%
Kombagu 1946 Nov. C. 8%; SRC 13%; A 6%; SRA 8%
Kombugu 1946 Nov. C. 5%; SRC 35%; A 4%; SRA 14%
Komanaku Tribe 1947 Dec.
 Okondia C 4%; SRC 21%; A 2%; SRA 6%
 Awagawi C 0%; SRC 0%; A 0%; SRA 1%
 Endugukani C 0%; SRC 0%; A 0%; SRA 4%
Kowandagu Tribe 1947 Feb.
 Kombri 'A' C 0%; SRC 0%; A 0%; SRA 0%
 Kombri 'B' C 0%; SRC 0%; A 2%; SRA 3%
 Kigiu C 0%; SRC 0%; A 0%; SRA 0%
 Dogbun C 0%; SRC 0%; A 0%; SRA 1%
 Girai C 0%; SRC 6%; A 0%; SRA 0%
Ku 1963 Sept. OPR 13%
Ku 1963 Oct. OPR 4%
Ku 1964 Feb. OPR 1%
Ku 1964 Apr. OPR 2%
Ku 1964 June OPR 5%
Ku 1964 Sept. OPR 3%
Kumai 1946 Oct. C 3%; SRC 21%; A 1%; SRA 6%
Kumuga 1946 Oct. C 9%; SRC 15%; A 4%; SRA 7%
Kundiawa 1946 C 0.9%; SRC 11.3%; A 1.2%; SRA 4.6%
Kundiawa 1947 Feb. C 0%; SRC 0.3%; A 0.4%; SRA 1%
Kundiawa 1947 Oct. School Children 2%; SR 18%
Kundiawa 1947 Nov. Station C. 0.5%; SRC 11%; A 1%; SRA 5%
Kundiawa 1947 Dec. OPR 1-2%; SR 5-11%
Kup 1946-47 C 3%; SRC 16.3%; A 1.8%; SRA 5.3%
Kup 1946 Oct. Mission C 3%; SRC 17%
Kup 1946 Oct. Road Gang A 5%; SRA 18%
Kup 1947 C 1%; SRC 16%; A 2%; SRA 5%
Kup 1963 OPR 1%
*Kup 1968 OPR 0%
Lower Marigl 1964 Nov. OPR 8%
*Lower Marigl 1968 OPR 0%
Mainamo Tribe 1947 Jan. C 2%; SRC 5%; A 0%; SRA 0%
Minagu 1946 Nov. C 2%; SRC 2%; A 0%; SRA 3%
Mingende 1946-47 C 5.2%; SRC 15%; A 3.6%; SRA 6.8%
Mingende 1947 Nov. Mission School C 9%; SRC 13%
Mingende 1947 Nov. C 7%; SRC 16%; A 4%; SRA 7%
Mintima 1964 Oct. OPR 3%
Mitnande 1963 OPR 2%
Mog 1963 Oct. OPR 10%
Mugwaaku 1946 Oct C 10%; SRC 25%; A 2%; SRA 13%
Nambaiyufa 1968 OPR 0.2%
Nigl Kande 1968 OPR 0.08%
Pendagu 1946 Nov. C 3%; SRC 15%; A 2%; SRA 1%
Sambugu 1946 Nov. C 11%; SRC 17%; A 6%; SRA 7%
Salt Nomani 1969 OPR 0%
Sina Sina 1965 OPR 6%
*Sina Sina 1968 OPR 0.05%
Teina 1947 April Mission School C 22%; SRC 15%
Upper Chimbu 1956 OPR 0%, in residents who have never been away
Waiye—Yongamugl 1963 OPR 7%
Waiye—Yongamugl 1965 OPR 31% pre-operational data.

*Waiye—Yongamugl 1968 OPR 0.08%
West Koronigl 1963 OPR 10%
*West Koronigl 1968 OPR 0.04%
Yongamugl Tribe 1947 Feb.
Korumba C 0%; SRC 0%; A 0%; SRA 1%
Kayangi C 0%; SRC 0%; A 0%; SRA 1%
Auru C 0%; SRC 0%; A 2%; SRA 0%
Kumun C 0%; SRC 0%; A 0%; SRA 0%
Damagu C 0%; SRC 0%; A 0%; SRA 0%

1947 Dec.
Wauku C 8%; SRC 25%; A 4%; SRA 14%
Waimgi C 0%; SRC 15%; A 2%; SRA 3%
Kondugu C 0%; SRC 26%; A 0%; SRA 4%
Damagu C 0%; SRC 13%; A 0%; SRA 4%
Guglbugl C 0%; SRC 0%; A 8%; SRA 0%
Kindiku C 0%; SRC 0%; A 0%; SRA 3%
Sigagu C 2%; SRC 5%; A 0%; SRA 1%
Yaugrimaugnem 1964 Nov. C 16%; SRC 23%; A 14%; SRA 10%
Yundei 1946 Oct. OPR 1%; SR villages 1%

EASTERN HIGHLANDS DISTRICT
Agarabi CD 1963 May IPR 5%; 1-4y 5%; 5-14y 10%; >15y 7%; OPR 7-10%
*Agarabi CD 1965 Aug. IPR 0%; 1-4y 0%; 5-14y 1%; >15y 1%; OPR 1% after 1 spray round
*Agarabi CD 1965 Sept. OPR 1-3%
*Agarabi CD 1966 Dec. OPR 1-3%
*Agarabi CD 1967 April OPR 0%
*Agarabi CD 1968 Jan. OPR 1%
Aiyura Farm 1945 Oct. OPR 2% SR 0%
Akuna 1950 Aug. C 0%; SRC 0%; A 0%; SRA 0%
Amamonta 1950 Aug. C 0%; SRC 0%; A 0%; SRA 0%
Anona 1945 OPR 3%; SR 0%
Apumaga 1950 Aug. C 0%; SRC 0%; A 0%; SRA 0%
Arau 1950 Aug. C 0%; SRC 14%; A 0%; SRA 0.8%
Asaro Feb. 1949 OPR 0%; SR 0.002%
Asaro June 1949 IPR 0%; OPR 0%
Asaro Feb. 1957 SR 4%
Asapuia 1945 Oct. OPR 3%; SR 10%
Bena Bena 1949 June C 0%; A 0%
Bena Bena 1949 July OPR 0%; SR 0%
Bena Bena 1949 Nov. OPR 0%; SR 0%
Bena Bena 1956 Nov. SR 2-9y 22%
Bena Bena 1968 Aug. IPR 1-2%; OPR 1-4%; Pre-operational data.
Bilimoia 1945 Oct. OPR 3%; SR 0%
Biobe 1956 SR 2-9y 25%
Chauve to Goroka area 1949 April-May C 0%; SRC 0%; A 0%; SRA .0025%
Dampit and Mirir 1950 Aug. People said to have died out from malaria epidemics. They trade with Markham Valley and get infected there.
Dirtiwata (Dunantina) 1949 Sept. OPR 0%; SR 0%
Dirtiwata (Dunantina) 1949 Oct. IPR 0%; SRC 0%; OPR 0%; SRA 0%
Dirtiwata (Dunantina) 1956 SR 2-9y 0%
Goroka 1949 April-May OPR 0.2%; SR 1.3%; C 0.7%; SRC 1%
Hengenofi 1949 Oct. SRC 0.05%; SRA 0%

Hengenofi 1949 Nov. Hengenofi to Ramu Survey OPR 0%; SR all ages 0%
Hengenofi 1956 SR 2-9y 1%
Hengenofi 1966 June IPR 6%; OPR 8%
Hengenofi 1967 IPR 6%; OPR 8%
*Hengenofi 1968 Nov.-Dec. IPR 2-3%; OPR 1-2% after 2 spray rounds.
Hogabe 1956 SR 2-9y 10%
Ikana 1950 Aug. C 0%; SRC 0%; A 0%; SRA 0%
Iturva CD 1964 IPR 5%; OPR 18%
*Iturva 1966 Oct. IPR 4%; OPR 3%
*Iturva 1967 May IPR 1%; OPR 0%
*Iturva 1968 May-June IPR 0-1%; OPR 0-1%
Kainantu 1934 C 0%; SRC 0%
Kainantu 1944 OPR 10%
Kainantu Villages 1945 Oct. OPR 0.8%; SR 0.8%
Kainantu Station 1945 Oct. OPR 3%; SR 5-8%

It was suggested that malaria was spread from infected police and government workers from Kainantu Station area. All other villages examined in Kainantu this survey, were negative for spleen and parasite rates.

Kainantu 1963 OPR 8%; pre-operational data
*Kainantu 1964 May IPR 5%; OPR 1-14%
*Kainantu 1965 OPR 2%
*Kainantu 1966 OPR 1%
*Kainantu 1968 Sept. IPR 1%; OPR 1%; after 5 spray rounds
*Kainantu 1969 Jan.-March IPR 0-1%; OPR 0-1%; after 6 spray rounds
Kainoa 1945 Oct. OPR 1%; SR 0%
Kamaliki 1956 OPR 5%
Koranga 1950 Aug. C 10%; SRC 25%; A 2%; SRA 32%
Kotuni 1949 July OPR 0%; SR 0.9%
Lupa 1956 SR 2-9y 21%
Omaura 1950 Aug. C 0%; SRC 0%; A 0%; SRA 0%
Onamuna 1945 Oct. OPR 4%; SR 0%
Puntabasi 1950 Aug. C 0%; SRC 0%; A 0%; SRA 2.5%
Saiora 1950 Aug. C 0%; SRC 0%; A 0%; SRA 0%
Sassaura 1950 Aug. C 0%; SRC 0%; A 0%; SRA 0%
Tombendaka 1950 Aug. C 0%; SRC 1.2%; A 0%; SRA 7.6%
Upper Asaro CD 18 1969 May IPR 1.5%; OPR 3%; pre-operational data
Uninufin Tenu 1945 Oct. OPR 3%; SR 0%
Vioka 1950 Aug. C 0%; SRC 0%; A 0%; SRA 0.7%
Watabung CD 37 1968 IPR 2%; OPR 2%; pre-operational data
*Watabung CD 37 1969 IPR 1%; OPR 1%; after 1 spray round
Wopepa 1945 Oct. OPR 3%; SR 0%
Wompur 1950 Aug. C 40% SRC 78%; A 7%; SRA 61%

WESTERN HIGHLANDS DISTRICT

Aringa & Demal 1954 May 0-10y 42%; A 19%
Aringa & Demal 1955 Jan. 2-10y 6%; 11-20y 13%; >20y 7% dry season
Aringa & Demal 1955 Feb. 2-10y 10%; SR 2-10y 77.4%; >10y 6%.

Awagi Group 1947 April

Degau 1. C 5%; SRC 5%; A 2%; SRA 2%
Degau 2. C 0%; SRC 0%; A 2%; SRA 4%
Siku C 3%; SRC 36%; A 4%; SRA 8%
Kamanagu C 17%; SRC 55%; A 6%; SRA 26%
Awagi C 5%; SRC 45%; A 10%; SRA 24%
Kerowagi Mission C 6%; SRC 22%

Baiyer River 1950 A 21%; SRA 57.6%
Baiyer River 1963 OPR 7%
*Baiyer River 1964 IPR 0%; 1-4y 0%; 5-14y 1%; >15y 0.6%.
*Baiyer River 1965 OPR 0.6%; After one year spraying.
*Baiyer River 1967 OPR 0.7%.

Bangi 1947 April

Kumbura C 9%; SRC 39%; A 10%; SRA 11%
Sagaga C 0%; SRC 40%; A 8%; SRA 10%
Kuraka C 8%; SRC 17%; A 6%; SRA 16%
Bangi—Kerowagi 1947 C 3%; SRC 42%; A 5%; SRA 18%.
Banz 1945 Oct. OPR 4%; SRC 28%; SRA 37%
Banz 1947
Moluka A 4%; SRA 19%
Andaban A 10%; SRA 44%
Kondagma A 3%; SRA 13%
Kumnimba A 6%; SRA 3%
Kundigul A 5%; SRA 13%
Nansawanku C 8%; SRC 28%; A 6%; SRA 12%
*Banz 1951 Oct. C 0.8%; SRC 16%; A 0%; SRA 9%
Banz 1965 IPR 0%; 1-4y 1%; 5-14y 1%; >15y 1%; OPR 1%

East and West Kambia

1963 IPR 20%; OPR 15%
* 1964 OPR 9% after 1 spray round
* 1964 OPR 4%
* 1968 May IPR 33% OPR 12%
Kerowil 1947 SRC 62%; SRA 37% wet season
Kerowil 1947 April
Kerowil Baiman C 19%; SRC 80%; A 6%; SRA 42%
Kerowil Kumaganeng C 3%; SRC 60%; A 2%; SRA 35%
Kerowil 1951 SRC 17%; SRA 3%; dry season
Kerowil 1955 2-10y 39%; 11-20y 29%; 21-30y 22%; 31-40y 6%; >41y 13%
Kerowil 1955 SR 2-10y 93%; 11-20y 9%; 21-30y 82%; 31-40y 76%; >41y 89%
Kerowil and Pugamil 1959 0-2y 5%; 2-4y 29%; 5-9y 11%; 10-14y 5%; 15-19y 4%; >20y 5%
Kerowil 1963 IPR 1%; 2-5y 20%; 6-14y 11%; A 4%; OPR 6%
Kon Dambi 1947 C 27%; SRC 44%; A 6%; SRA 17%
Kon Dambi Yaniga 1. 1947 C 20%; SRC 20%; A 6%; SRA 11%
Kon Dambi Yaniga 2. 1947 C 4%; SRC 0%; A 6%; SRA 4%
Kon Dambi Kono buka 1947 C 11%; SRC 23%; A 4%; SRA 17%.
Kon Dambi Berabuka C 17%; SRC 23%; A 4%; SRA 9%
Kon Dambi Kondiga C 33%; SRC 59%; A 5%; SRA 3%
Kon Dambi 1954 2-10y 24%; 11-20y 38%; 21-30y 10%; 31-40y 11%; >41y 10%
Kon Dambi 1954 SR 2-10y 63%; 11-20y 55%; 21-30y 50%; 31-40y 70%; >41y 39%
Korn Farm 1945 A 13%; SRA 21%
Kogerugi 1955 July OPR 48%
Kudjip 1947
Kudjip Kudjiil C 18%; SRC 61%; A 10%; SRA 24%
Kudjip Tungarg C 32%; SRC 48%; A 12%; SRA 12%
Kudjip Kurukupa C 16%; SRC 27%; A 8%; SRA 6%
Kurumuk 1947

Kurumuk Meiamga C 18%; SRC 37%; A 0%; SRA 8%
Karumuk Deimanga C 15%; SRC 37%; A 4%; SRA 10%
Kurumuk Kitehu C 17%; SRC 47%; A 6%; SRA 5%
Laiagum 1956 C 0%; SRC 0%; A 0%; SRA 2%
Lower Minj People 1954 2-10y 23%; 11-20y 26%; 21-30y 4% >31y 4%
Lower Minj People 1954 SR 2-10y 65%; 11-20y 81%; 21-30y 74%; >31y 64%
Lower Nebilyer CD 4% 1967 Dec. IPR 0.5%; OPR 3%
Minj 1946-47 C 17%; SRC 36%; A 7%; SRA 10%
Minj 1947 0-10y 17%; >11y 7%; 2-10y SR 17%
1954 OPR 7%; 2-10y SR 23%
Minj 1954-55 OPR 29%; 2-10y SR 6%
Minj 1956 0-2y 2%; 2-10y 4%; 11-20y 3%; >20y 3%
Minj 1960 OPR 4%; pre-operational data
Minj 1961 OPR 7%
Minj 1963 OPR 7%
*Minj 1964 OPR 5%
Mogay 1934-35 SRC 3.3%; SRA 0%
Mul CD 1966 IPR 0.5%; OPR 2%
Mt. Hagen 1945 C 6%; SRC 1.5%; A 6%; SRA 8.8%
Mt. Hagen 1964 OPR 15%
Mt. Hagen 1965 IPR 0%; OPR 18%
*Mt. Hagen 1968 IPR 1%; OPR 4%
Nondugl 1947 SRC 62%; SRA 20% wet season
Nondugl Bamina C 15%; SRC 53%; A 6%; SRA 19%
Nondugl Nodugl 1. C 20%; SRC 70%; A 8%; SRA 35%
Nondugl Nondugl 2. C 8%; SRC 60%; A 6%; SRA 16%
Nondugl 1951 SRC 9%; SRA 8% dry season.
North Wall Whagi 1955 OPR 18%
North Wall Whagi 1961 OPR 6%
North Wall Whagi 1963 OPR 7%
*North Wall Whagi 1965 OPR 0.4% after 2 years spraying
South Wall Whagi 1960 IPR 4%; OPR 38%
*South Wall Whagi 1968 IPR 0.1%; OPR 0%
Tomba and Kudjip 1960 1-4y 7%; 5-14y 5%; >15y 2%; OPR 4%
Tomba and Kudjip 1961 1-4y 5%; 5-14y 3%; >15y 9%; OPR 7%
*Tomba and Kudjip 1964 <1y 11%; 1-4y 12%; 5-14y 5%; >15y 0%; OPR 5%
Upper Minj River 1954 May 2-10y 5%; 11-20y 8%; 21-30y 0%; >31y 3%
Upper Minj River 1954 May SR 2-10y 6%; 11-20y 12%; 21-30y 8%; >31y 8%
Wabag 1956 C 0%; SRC 0%; A 0%; SRA 1.3%
Wapinamunda 1948 SRA 0.5%

SOUTHERN HIGHLANDS DISTRICT

Batri 1956 OPR 1%; SR all ages 15%
Erave 1956 C 1.9%; SRC 43.7% OPR 3%; SRA all ages 61%
Erave 1963 OPR 75%
Lamurupi 1956 OPR 3%; SR all ages 65%
Ialabu 1963 OPR 0.2%
Kagua 1956 OPR 2%; SR all ages 10%
Kompian-Wabag 1963 OPR 10%
Koiari 1956 OPR 0%; SR all ages 47%
Lake Kutubu 1963 OPR 13%
Tari 1956 C 1.3%; SRC 10.9% A 0%; SRA 0.6%
Tari 1963 OPR 16%

NORTH MAINLAND REGION
EAST SEPIK DISTRICT
MADANG DISTRICT
MOROBE DISTRICT
WEST SEPIK DISTRICT
WEST IRIAN

EAST SEPIK DISTRICT

Berui 1956 < 2y 80%; 2-4y 100%; 5-9 y 54%; 10-14y 41%; 15-19y 60%; > 20y 72%
Berui 1956 SR < 2y 100%; 2-4y 100%; 5-9y 83%; 10-14y 76%; 15-19y 75%; > 20y 44%
Bonohoi 1962 IPR 71%; OPR 33%
*Bumbita 1965 April OPR 39%; after 3 spray rounds
*Bumbita 1966 April-May OPR 17%; after 5 spray rounds
*Bumbita 1966 Sept. Oct. OPR 63% after 6 spray rounds
*Bumbita 1967 Aug.-Sept. OPR 22% after 7 spray rounds
*Bumbita 1968 Oct.-Dec. OPR 37% after 8 spray rounds
Bumbita-Urim 1962 IPR 45%; OPR 27%
Brandi 1956 5-9y 36%; 10-14y 31%; 15-19y 34%; > 20y 50%
Brandi 1956 SR 5-9y 83%; 10-14y 91%; 15-19y 78%; > 20y 83%
Deikikir 1963 OPR 43% pre-operational data
*Deikikir 1964 OPR 16% after 1 year spraying
*Deikikir 1965 OPR 3% after 2 years spraying
Hawain 1945 < 2y 92%; 2-12y 59%; SRC 92%; A 31%; SRA 67%
Australian Soldiers 5% SR 10%
Japanese Soldiers 47% SR 8%
Kunjingini 1956 < 2y 87%; 2-4y 100%; 5-9y 77%; 10-14y 70%; 15-19y 60%; > 20y 63%
Kunjingini 1956 SR > 2y 94%; 2-4y 100%; 5-9y 92%; 10-14y 91%; 15-19y 82%; > 20y 47%
Maprik & Soragum 1956 < 2y 85%; 2-4y 89%; 5-9y 80%; 10-14y 64%; 15-19y 53%; > 20y 50%
Maprik & Soragum 1956 SR < 2y 90%; 2-4y 100%; 5-9y 99%; 10-14y 79%; 15-19y 77%; > 20y 62%
Maprik 1957 IPR 95%; 1-2y 93%; 2-3y 95%; 3-4y 96%; 5-9y 82%; 10-14y 62%; 15-19y 48%; 20-29y 34%
Maprik 1957 Sept. DDT area < 1y 72%; 1-2y 83%; 2-4y 95%; pre-operational data
*Maprik 1958 Oct. DDT area < 1y 32%; 1-2y 58%; 2-4y 91%. after 1 year spraying
*Maprik 1959 Sept. DDT area IPR 8%; DLD area 18%; DLD & Daraprin area 27%
*Maprik 1960 Oct. DDT area IPR 5%; DLD area 14%; DLD & Daraprin area 4%
*Maprik 1960 June IPR 0%; 1-2y 20%; 2-3y 27%; 3-4y 31%; 5-9y 21%; 10-14y 19%; 15-19y 13%; 20-29y 19%
*Maprik 1961 Feb.-March IPR 26%; 1-2y 28%; 2-3y 39%; 3-4y 40%; 5-9y 38%; 10-14y 30%; 15-19y 21%; 20-29y 19%; 30-39y 16%; > 40y 15%
*Maprik 1962 IPR 12%; 1-4y 25%; 5-14y 22%; > 15y 12%
*Maprik 1963 OPR 18%
*Maprik 1964 May-June OPR 10%; after 14 spray rounds
*Maprik 1965 Oct.-Nov. OPR 14%; after 16 years spray rounds
*Maprik 1966 OPR 12%
*Maprik 1967 June-July OPR 32% after 19 spray rounds
*Maprik 1967 Oct.-Dec. OPR 22% after 20 spray rounds
*Maprik 1968 June-July OPR 68% after 21 spray rounds
*Maprik 1968 Dec. IPR 15%; 1-4y 14%; 5-14y 20%; > 14y 12%; OPR 16%
Pagwi 1962 IPR 36-54%; of 3 indicator villages pre-operational data

Pawgi 1963 1-4y 41%; pre-operational data
*Pawgi 1964 1-4y 1%; after 1 year of spraying
*Pagwi 1965 Feb.-March 1-4y 1%; OPR 15%; after 5 spray rounds
*Pagwi 1965 July-Sept. OPR 8%; after 6 spray rounds
*Pagwi 1966 Jan.-March OPR 11%; after 7 spray rounds
*Pagwi 1967 Oct.-Nov. OPR 38%; after 9 spray rounds
Sepik District 1936-37. SRC 43%
Wam 1962 IPR 87%; OPR 70%
Wam 1966 July OPR 18%; after 1 spray round
Wam 1967 Nov. OPR 24%; after 2 spray rounds
Wewak 1964 Mar. OPR 17%; after 3 spray rounds
Wewak 1966 Aug. OPR 4%; after 7 spray rounds
Wewak 1966 Nov. OPR 17%; after 8 spray rounds
Wewak 1967 Sept. OPR 16%; after 9 spray rounds
Wingei 1961 Dec. <1y 11%; 1-2y 27%; 2-5y 44%; 5-10y 39%; 10-20y 20%; >15y 15%
Yangoru 1963 OPR 10%; after 2 years spraying
Yangoru 1964 OPR 4-19%; After 3 years spraying
Yangoru 1965 April OPR 2%; after 5 spray rounds
Yangoru 1966 April OPR 7%; after 7 spray rounds

MADANG DISTRICT
Bogadjin 1898 <2y 80%; 2-5y 42%; >5y 0%
Bongu 1898 <2y 100%; 2-5y 46%; 5-10y 24%; >10y 0%
Bundi 1969 Aug. SR 2-9y 30%; dry season
Brahman 1969 Nov. SR 2-9y 78%; wet season
Korinogobo 1969 Aug. SR 2-9y 18%; dry season
Madang 1936-37 SR 33%
Manum Is. 1964 OPR 25%
*Manum Is. 1966 OPR 1%
Utu 1967<2y 65%; 3-4y 79%; 5-9y 74%; 11-15y 47%; >16y 16%; OPR 43%
Upper Ramu 1935 SRC 1%; SRA 0%
Wau-Kausi 1969 Aug. SR 2-9y 91% dry season

MOROBE DISTRICT
Angau Labour Camp 1944 A 11%
Anari-Onga CD Kaiapit 1965 IPR 38%; 2-4y 45%; 5-14y 19%; >15y 6%; OPR 14%
Atzeras CD 18 villages IPR 29%; 2-4y 57%; 5-14y 24%; >15y 10%; OPR 19%
Amari (Kaiapit) 1968 IPR 70%; OPR 45%
*Amari 1968 IPR 26%; OPR 17%; after 2 cycles of spraying
*Amari 1969 IPR 13%; 2-4y 11%; 5-14y 4%; A 2%; OPR 4%
Busamba 1947 OPR 42%; SR 50%; 2-9y 82%; SR 2-9y 74%
Butibum 1944 May <2y 100%; 2-5y 93%; 6-10y 84%; 11-15y 18%
Butibum 1944 July <2y 92%; 2-5y 89%; 6-10y 33%; 11-15y 37%; SR 2-10y 59%
Butibum 1944 Nov. <2y 95%; 2-5y 75%; 6-10y 39%; 11-15y 33%
Chivasing 1944 2-5y 82%; 6-10y 67%; >16y 27%; SR 2-10y 86%
Chivasing 1944 A 20%; C 61%; OPR 42%
Gabsonkek 1944 A 33%; C 64%; OPR 39%
Kiaipit (Amari) 1947 OPR 47%; SRA 47%; 2-9y 79%; SR 2-9y 83%
Morobe District 1936-37 SR 25%
North Markham area—several villages 1944 C 60%; A 25%; OPR 41%
Patep No. 2 1947 OPR 14%; SR 26%; 2-9y 29%; SR 2-9y 15%
*Seassi Island 1963 OPR 7%; 2nd year of spraying

*Seassi Island 1964 OPR 4%; 3rd year of spraying
*Tolokiwa Island 1962 IPR 25%; 2-4y 43%; 5-14 y10%; after 4 spray rounds
*Tolokiwa Island 1962 IPR 23%; 2-4y 41%; 5-14y 23%; after 5 spray rounds
Wankum 1964 Aug. OPR 40%

WEST SEPIK DISTRICT

Aitape 1944-45 Coastal Natives between Raihu & Nigia rivers
 < 2y 67%; 2-5y 72%; 6-10y 62%; 11-15y 30%; > 16y 17%; SR 2-10y 100%
*Aitape 1965 July OPR 10%; < 5y 49% after 7 spray rounds
*Aitape 1966 Jan. OPR 70%; after 8 spray rounds
*Imonda 1963 OPR 63%; after 2 years spraying
*Imonda 1964 OPR 83%; after 3 years spraying
*Imonda 1965 OPR 39%; after 4 years spraying

WEST IRIAN

Biak 1962 OPR 2%
Biak 1968 OPR 23%
Djajapura (Hollandia) 1935-39 SR 2-9y 75%
Djajapura 1962 OPR 15%
Djajapura 1968 OPR 49%
Fakfak 1967 OPR 15%
Fakfak 1968 OPR 34%
Iffar 1935-39, SR 2-9y 100%
Manokwari 1935-39, SR 2-9y 88%
Merauke 1967 OPR 6%
Merauke 1968 OPR 7%
Nimboran Valley Control Area
 1954 0-6m 79%; 6-12m 70%; 1-5y 96%; 6-14y 99%; 15-44y 93% > 45y 83%; OPR 87%
 1954 SR 0-6m 57%; 6-12m 97%; 1-5y 96%; 1-5y 96%; 6-14y 97%; 15-44y 88%; > 45y 70%; all ages 84%
 1955 July 0-6m 52%; 6-12m 88%; 1-5y 91%; 6-14y 73%; 15-44y 42%; > 45y 30%; OPR 64%
 1955 July SR 0-6m 64%; 6-12m 97%; 1-5y 87%; 6-14y 71%; 15-44y 39%; > 45y 23%; all ages 64%
Pioneer Bivak 1935-39 SR 2-9y 80%
Sabron 1935-39 SR 2-9y 100%
Sarmi 1962 OPR 30%
Sarmi 1968 OPR 49%
Serdei 1935–39 SR 2-9y 82%
Sentani 1953 OPR 46%; pre-operational data
Sentani 1955 OPR 32%; 1 year after spraying
Sentani 1956 OPR 20%
Sentani 1957 OPR 14%
Sentani 1958 OPR 12%
Sentani 1959 OPR 6%
Sentani Area 1954 0-6m 28%; 6-12m 70%; 1-5y 79%; 6-14y 49%; 15-44y 28%; > 45y 32%; OPR 47%
Sentani Area 1954 SR 0-6m 46%; 6-12m 67%; 1-5y 92%; 6-14y 99%; 15-44y 86%; > 45y 82%; all ages 80%
*Sentani Area 1955, 0-6m 0%; 6-12m 7%; 1-5y 43%; 6-14y 37%; 15-44y 14%; > 45y 13%; OPR 20%; after 1 years spraying
*Sentani Area 1955, SR 0-6m 54%; 6-12m 53%; 1-5y 67%; 6-14y 99%; 15-44y 86%; > 45y 86%; all ages 74%; After 1 year spraying

Sentani Lake Area 1954 0-6m 12%; 6-12m 7%; 1-5y 20%; 6-14y 22%; 15-44y 13%; >45y 16%; OPR 16%
Sentani Lake Area 1954 SR 0-6m 39%; 6-12m 56%; 1-5y 38%; 6-14y 52%; 15-44y 57%; >45y 50%; all ages 48%
*Sentani Lake Area 1955 0-6m 1%; 6-12m 4%; 1-5y 8%; 6-14y 10%; 15-44y 7%; >45y 5%; OPR 6%
Sentani Lake Area 1955 SR 0-6m 52%; 6-12m 43%; 1-5y 23%; 6-14y 46%; 15-44y 66%; >45y 64%; all ages 49%

ISLANDS REGION
EAST NEW BRITAIN DISTRICT
BOUGAINVILLE DISTRICT
MANUS DISTRICT
NEW IRELAND DISTRICT
WEST NEW BRITAIN DISTRICT

EAST NEW BRITAIN
Bainings Villages 1923 SR 3-11y 54%
Bainings Villages 1936-37 SR 52%
Bainings Mountains 1962 Peak PR 2-4y 25%
*Bainings Coastal 1967 Feb. CD 25; East Nakanai OPR 18%; after 13 spray rounds
*Bainings Coastal 1967 Sept. East Nakanai OPR 5% after 14 spray rounds
*Bainings Coastal 1968 April. East Nakanai OPR 3%; after 15 spray rounds
*Bainings Coastal 1968 Dec. East Nakanai OPR 14%; after 16 spray rounds.
Bainings Central Nakanai 1961 Mar. < 1y 76%; 1-4y 76%; 5-14y 62%; > 15y 40%; OPR 56%
*Bainings Central Nakanai 1962 Aug. < 1y 20%; 1-4y 28%; 5-14y 15%; > 15y 6%; OPR 13%; after 1 spray round
*Bainings Central Nakanai 1963 Aug. < 1y 14%; 1-4y 15%; 5-14y 9%; > 15y 3%; OPR 7%; after 2 spray rounds

Bainings Inland CD 18
*1966 July OPR 0.4% after 14th spray round
*1967 Jan. OPR 1% after 15th spray round
*1967 June OPR 1% after 16th spray round
*1968 Jan. OPR 6%; after 17th spray round
*1968 Aug. OPR 5%; after 18th spray round
*1969 Jan. OPR 5%; after 19th spray round
Bianings Inland CD 18, 1969 Share Islands OPR 12%; after 16th & 17th spray round
Bitapaka Hospital 1962 OPR 5%
Butawen Hospital 1962 OPR 3%

Duke of York Islands CD 17
1923 0-2y 27%; 3-6y 50%; 7-11y 53%; SR > 17y 10%
SR 0-2y 24%; SR 3-6y 57%; SR 7-11y 60%; SR 12-17y 56%
1932-33 SRC 50-80%
1935-42 OPR 58%; 1936-37 SR 56%
* 1966 Feb. OPR 0.7% after 12 spray rounds
*1966 Nov. OPR 0.5% after 13 spray rounds
*1967 May OPR 11% after 14 spray rounds
*1967 Aug. OPR 2% after 15 spray rounds
*1967 Dec. OPR 5% after 16 spray rounds
*1968 April OPR 6% after 17 spray rounds
*1968 Aug. OPR 4% after 18 spray rounds

*1969 Jan. OPR 8% after 19 spray rounds
*1969 Aug. OPR 8% after 20 spray rounds

EAST NEW BRITAIN

Gazelle Peninsula 1936-37 SR 23%
1960 OPR 15%
*1961 OPR 9%
*1962 OPR 9%
*1963 OPR 6%
*1964 OPR 0.5%
*1965 OPR 0.5%
*1966 Sept. OPR 4%; 13 spray rounds
*1967 Jan.-Apr. OPR 2-5%; 14 spray rounds
*1967 Sept.-Oct. OPR 0.3-3%; 15 spray rounds
*1968 Mar.-July OPR 0.3-3%; 16 spray rounds
*1968 Sept.-1969 Jan. OPR 1.2-2%; 17 spray rounds
*1969 Feb.-May OPR 1-7%; 18 spray rounds
* 1969 June-Oct. OPR 1%; 19 & 20 spray rounds
South Gazelle CD 24
*1966 Dec.- March 1967 OPR 0.5-3%; 16 spray rounds
*1967 Aug.-Oct. 1967 OPR 0.3-3%; 17 spray rounds
*1967 Nov.-May 1968 OPR 0-0.3%; 18 spray rounds
*1968 May-July 1968 OPR 0.3%-1% 19th spray round
*1968 Oct.-May 1969 OPR 1-3%; 20th spray round
*1969 April OPR 1%; 21st spray round
Gaulim Hospital 1962 OPR 4%
Hospital Patients 1921 SR 84%
*Kokopo 1967 March-Sept. OPR 0.5-11%; 13 spray rounds
*Kokopo 1967 Oct.-Apr. 1968 OPR 1-4%; 14 spray rounds
*Kokopo 1968 Apr.-Aug. OPR 0.1-16%; 15 spray rounds
*Kokopo 1968 July-Oct. OPR 6-9%; 16 spray rounds
*Kokopo 1969 Mar.-July OPR 5-8%; 17 spray rounds
*Labour Compound Rabaul 1961 Aug.-Sept. OPR 10%
*Malaguna Hospital 1962 OPR 11%
Napapar Health Centre OPR 6%
New Britain North Coast 1935-42 SR 13%
Pomio CD 6, 7, 8, 9.
*1967 Sept. OPR 0.6% 10th spray round
*1967 July-Dec. OPR 0-2.4% 11th spray round
*1967 Dec. April 1968 OPR 0-4.2% 12th spray round
*1968 June-Dec. OPR 6% 13th spray round
*1969 Feb.-July OPR 2-26% 14th spray round
Paparava Health Centre 1962 OPR 2%
Watom Is. CD 24.
*1967 Aug. OPR 6% 13th spray round
*1967 Sept. OPR 1% 14th spray round
*1968 March OPR 0% 15th spray round
*1969 March OPR 3% 16th spray round
Rabaul villages 1923 (within 5 miles of town)
SR 0-2y 15%; 3-6y 29%; 7-11y 40%; 12-17y 44%; >17y 19%
PR 0-2y 24%; 3-6y 32%; 7-11y 37%; 12-17y 37%; >17y 7%
*Tapipipi Health Centre 1961 OPR 29%
*Tapipipi Health Centre 1962 OPR 7%
*Tapo 1962 OPR 6%

Tolai Inland 1950 SR <2y 85% SR all ages 54%
Tolai Inland 1966 SR <2y 8% SR all ages 10%
Toma Village 1923 SR 0-2y 0%; 3-6y 0%; 7-11y 1%; 12-17y 100%; >17y 0%
Vunapope Hospital 1962 OPR 3%

BOUGAINVILLE DISTRICT
Buka 1925-26 SR 75%
Bougainville SR 50%
Buka Island 1936-37 SR 25%
*Buka Island 1960 <1y 16%; 1-2y 36%; 2-4y 33%; 5-9y 32%; mothers 6%
*Buka Island 1962 Peak PR 1-2y 36%
Buka Passage, Wakunai and Kieta 1960 OPR 22%; pre-operational survey
*Buka Passage. Wakunai and Kieta 1962 OPR 3% after 3 years spraying
*Buka Passage, Wakunai and Kieta 1963 OPR 2% after 4 years spraying
*Buka Passage, Wakunai and Kieta 1964 OPR 1% after 5 years spraying
Kieta 1936-37 SR 63%
Lemankua 1954 SR 56%
*Lemankua 1966 SR 33%
Nissan Is. 1960 May <1y 20%; 1-4y 33%; 5-14y 22%; A 13% pre-operational survey
*Nissan Is. 1962 May <1y 2%; 1-4y 18%; 5-14y 9%; A 6%
*Nissan Is. 1962 Aug. <1y 1%; 1-4y 12%; 5-14y 7%; A 2% after 3 years spraying
*Nissan Is. 1962 Peak PR 2-4y 34%
New Britain & New Ireland & Bougainville
*off shore Islands 1962 OPR 7% after 2 years spraying
*off shore Islands 1963 OPR 6% after 3 years spraying
*off shore Islands 1964 OPR 2% after 4 years spraying
*off shore Islands 1965 OPR 1% after 5 years spraying

MANUS DISTRICT
Admiralty Islands 1936-37 SR 53%
Admiralty Islands 1962 OPR 17% pre-operational survey
*Admiralty Islands 1963 OPR 8% after 2 years spraying
*Admiralty Islands 1965 OPR 4%; after 5 years of spraying
*Admiralty Islands 1966 OPR 2%; after 6 years of spraying
Mokerang Village 1944 May <2y 100%; 2-5y 45%; 6-10y 27%; 11-15y 8%; >16y 14%
Mokerang Village 1944 May SR <2y 33%; 2-5y 100%; 6-10y 100%
M' Buke 1966 SR 23%
Rambutyu Island 1944 May <2y 72%; 2-5y 58%; 6-10y 40%; 11-15y 38%; >16y 16%
Rambutyu Island 1944 May SR <2y 96%; 2-5y 97%; 6-10y 92%

NEW IRELAND DISTRICT
Kavieng Dist. 1936-37 SR 34%
Kavieng and Namatanai 1958 OPR 20%
*Kavieng and Namatanai 1961 OPR 7% after 3 years spraying
*Kavieng and Namatanai 1962 OPR 7% SR all ages 21%
*Kavieng and Namatanai 1963 OPR 5%
*Kavieng and Namatanai 1964 OPR 2.4%
*Kavieng and Namatanai 1965 OPR 1-2% after 7 years spraying
*Kavieng Hospital 1962 Jan.-March OPR 10%
*Kavieng Hospital 1962 Apr.-June OPR 18%
*Kavieng Hospital 1962 July-Sept. OPR 20%
*Kavieng Hospital 1962 Oct.-Dec. OPR 3%

*Kavieng Hospital 1962 OPR 14% average after 4 years spraying
*Kavieng Hospital 1963 OPR 26% average after 5 years spraying
Namatanai District. 1936-37 SR 32%
Namatanai District 1959 SR 33%
*Namatanai District 1966 SR 7-13%
New Hanover 1959 SR 45%
*New Hanover 1966 SR 38% after 8 years spraying
New Ireland District 1931-32 SRC 30%
*New Ireland District 1962 Peak PR 2-4y group 40%
Tigak 1954 SR 71%
WEST NEW BRITAIN
Talasea 1936-37 SR 40%
Talasea 1961 OPR 45% pre-operational data
*Talasea 1962 OPR 12% after one year spraying
*Talasea 1963 OPR 7%
*Talasea 1964 OPR 8%
*Talasea 1965 OPR 3%
*Talasea 1966 OPR 3% after 6 years spraying

BIBLIOGRAPHY

Works concerned with the parasites of Man in Niugini and West Irian.

Anon, 1943 'Malaria in the South West Pacific' *Australian Military Forces* pp. 1-16

Anon, 1944 'Malaria in the South-West Pacific' *Nature*, Lond, *153*:160

Anon, 1944 '*Plasmodium ovale* in New Guinea' *Bulletin U.S. Army Medical Dept. 81*:35

Anon, 1954 'Annotated bibliography of filariasis and elephantiasis Part I' *Technical Papers of the South Pacific Commission.* No. 65

Assem, J. van den 1959 'A window-trap hut experiment to test the effects of dieldrin under local conditions in the Merauke area (Netherlands New Guinea) *Tropical and Geographical Medicine 11*:32-43

Assem, H. van den 1961 'Mosquitoes collected in the Hollandia area, Netherlands New Guinea, with notes on the ecology of larvae' *Tijdschrift voor Entomologie. 104*: 17-30

Assem, J. van den & **Dijk,** W. J. O. M. van 'Distribution of anopheline mosquitoes in Netherlands New Guinea' *Trop. Geogr. Med. 10*:249-225

Assem, J. van den & **Metselaar,** D. 'Host plants and breeding places of *Mansonia* (*Mansonioides*) *uniformis* in Netherlands New Guinea' *Trop. Geogr. Med. 10;*51-55

Atherton, D. O. & **Lemerle,** T.H. 1943 '*An. Punctulatus* var. *moluccensis* adult habits and behaviour' *A.M.F. Hygiene Service Records*

Atherton, D. O. 1944 'Notes on the habits of *An. punctulatus* in the Ramu Valley' *A.M.F. Hygiene Service Records*

Atherton, D. O. 1944 'Notes on the malarial vectors in the Ramu Valley' Report of the Allied Malaria Control Conf. 27th Meeting, Feb. 9th 1944. *A.M.F. Hygiene Service Records*

Avery, J. L. 1946 'Parasite infections among natives of Samarai District, Papua, New Guinea' *Journal of Parasitology. 32*:25

Backhouse, T. C. 1929 'Sarcoptic skin disease in natives of the Territory of New Guinea' *Transactions of the Royal Society of Tropical Medicine and Hygiene. 23*(2)*;*173-8

Backhouse, T. C. 1953 'Filariasis in Melanesia' *Proceedings VII Pacific Science Congress New Zealand* (*1949*) *7*:239-251

Backhouse, T. C., and **Heydon,** G. A. M. 1950 'Filariasis in Melanesia: observations at Rabaul relating to incidence and vectors' *Trans. R. Soc. Trop. Med. Hyg. 44;* 291-306

Bailey, K. V. 1966 'Some aspects of anaemia, haemoglobin levels and iron metabolism in the N.G. highlands' *Medical Journal of Australia. 1*: 386-392

Bang, F. B., **Hairston,** N. G., **Maier,** J. & **Roberts,** F. H. S. 1947 'DDT spraying inside houses as a means of malaria control in New Guinea' *Trans. R. Soc. Trop. Med. Hyg. 40*: 809-822

Barnes, R. & **McKay,** J. 1962 'Report of a case of a flagellate causing a urinary infection' *Med. J. Aust. 2*: 913-914

Bearup, A. J. & **Lawrence,** J. J. 1947 'Parasitological Report' in *Report of the New Guinea Nutrition Survey Expedition; 1947* ed. by E. H. Hipsley and F. W. Clements Govt. Printer, Sydney

Bearup, A. J. & **Lawrence,** J. J. 1950 'A parasitological survey of five New Guinea villages' *Med. J. Aust. 1*: 724-732

Bern, H. A. & **Hansen,** M. F. 1950 'Parasitic infections among natives of the North Markham Area, New Guinea' *J. Parasit. 36.* 103-6

Bick, G. H. 1951 'The ecology of the mosquito larvae of New Guinea' *Pacific Science. 5*: 392-431

Black, R. H. 1954 'Malaria Experiments in the Trobriand Islands' *Papua & New Guinea Villager*, May 1954 : 37-40

Black, R. H. 1954 'Malaria in the Trobriand Islands *Tech. Pap. S. Pacif. Commn.* No. 61

Black, R. H. 1954 'A malaria survey of the people living on the Minj River in the Western Highlands of New Guinea' *Med. J. Aust. 2*: 782-787

Black, R. H. 1954 'Enlargement of the liver due to malaria in natives of the South West Pacific area' *Trans. R. Soc. Trop. Med. Hyg. 48*: 226-233

Black, R. H. 1955 'Comparison of the liver and spleen sizes in a community in Netherlands New Guinea suffering from malaria of a high degree of endemicity' *Trans. R. Soc. Trop. Med. Hyg. 49*: 62-63

Black, R. H. 1955 'Malaria control and research in Netherlands New Guinea' *Tech. Pap. S. Pacif. Commn.* No. 80

Black, R. H. 1955 'Malaria in the South West Pacific' *Tech. Pap. S. Pacif. Commn.* No. 81: 1-56

Black, R. H. 1955 'Rural malaria control in the S. W. Pacific' *S. P. C. Bull.* April 1955

Black, R. H. 1955 'The geographical distribution of malaria in the South West Pacific' *Australian Geographer. 6* (4): 32-35

Black, R. H. 1955 'Observations on the behaviour of *Anopheles farauti* (Laveran). An important malaria vector in T.P.N.G. *Med. J. Aust. 1*: 949-

Black, R. H. 1956 'The present status of malaria control in the S. W. Pacific Area' *Med. J. Aust.* 1956: *1*: 498-501

Black, R. H. 1956 'The epidemiology of malaria in the South West Pacific; changes associated with increasing European contact' *Oceania 27*: 136-142

Black, R. H., **Dew**, B. **Hennessy**, W. B. & **McMillan**, B. 1966 'Studies on depot Antimalarials I. The effect of a single injection of the depot Antimalarial C1-501 (Camolar) on relapsing vivax malaria acquired in N. G. ' *Med. J. Aust.* Sept. *24*: 588-592

Black, R. H., **Hennessy**, W. B., **McMillan**, B., & **Dew**, B. 1966 'Studies on Depot Antimalarials 2. The effect of a single injection of the Depot Antimalarial C1-564 on relapsing vivax malaria acquired in N. G.' *Med. J. Aust.* Oct. *22*: 808-811

Black, R. H. 1967 'Correspondence—the fruits of malaria control' *Papua New Guinea Medical Journal. 10*: (4): 142

Bonne Wepster, J. & **Swellengrebel**, N. H. 1953 *The anopheline mosquitoes of the Indo-Australian Region* deBussy, Amsterdam

Booth, K., **Alpers**, J. H. & **Lloyd**, A. 1969. 'Blood dyscrasia in Papua: pyrimethamine as a possible cause' *Papua New Guin. Med. J. 12* (2): 35-41

Breinl, A. 1915 'On the occurrence and prevalence of diseases in British New Guinea' *Annals of Tropical Medicine and Parasitology. 9*: 285-334

Breinl, A. & **Priestley**, A. 1916 'Malaria contracted in New Guinea by members of the expeditionary force and its treatment' *Med. J. Aust. 1*: 91

Brug, S. L. 1937 'De vooruitzichten van Filaria-bestrijding in Ned-Indië' *Geneeskundig Tijdschrift voor Nederlandsch Indië. 77*: 3202-3206

Brug, S. L. & **de Rook**, H. 1933 'Filariasis in Nederlandsch-Indië IV' *Geneesk. Tijdschr. Ned.-Indië 73*: 264-279

Burchett, P. M. 1966 'Amoebiasis in the New Guinea Western Highlands *Med. J. Aust. 2*: 1079-81

Burrows, R. B. 1945 'A survey of intestinal parasites in natives of Dutch New Guinea' *American Journal of Hygiene. 42*: 262-5

Campbell, C. H. & **Arthur**, R. K. 1964 'A study of 2000 admissions to the medical ward of the Port Moresby General Hospital' *Med. J. Aust. 1*: 989-992

Christian, S. H. 1949-51 'Series of reports on malaria, surveys in Eastern Highlands, Chimbu District, and Western Highlands' Registry, Dept. Public Health, Konedobu

Cilento, R. W. 1924 'Malaria, with special reference to Australia and its dependencies' Commonwealth of Australia Dept. of Health Tropical Division. *Service Publications.* Govt. Printer, Melbourne

Cilento, R. W. 1942 *Tropical diseases in Australia—a handbook* W. R. Smith & Paterson, Brisbane,

Cilento, R. W. 1964 'Filariasis with special reference to Australia and its dependencies' Comm. Aust. Dept. Hlth. Trop. Div. *Serv. Publs.* pp. 78

Cilento, R. W. & **Backhouse,** T. C. 1927 'Paragonimiasis: its first recorded occurrence in the Territory of New Guinea' *Med. J. Aust. 1*: 79

Clements, F. W. 1936 'A Papuan field experiment of malaria treatment' *Med. J. Aust. 1*: 235-237

Clements, F. W. 1936 'A medical survey in Papua: report of the first expedition by the School of Public Health and Tropical Medicine to Papua, 1935' *Med. J. Aust. 1*: 451-464

Cooke, R. A. & **Rodrigue,** R. B. 1964 'Amoebic balanitis' *Med. J. Aust. 1*: 114

Cotter, H., **Kariks,** J., & **Walsh,** R. J. 1958 'Haemoglobin values of some native inhabitants of Bougainville, New Guinea' *Med. J. Aust. 1*: 29

Cotter, H. & **Walsh,** R. J. 1958 'Haemoglobin values in some Pacific native groups' *Med. J. Aust. 2*, 603

Couvee, L. M. J. & **Rijpstra,** A. C. 1961 'The prevalence of *Balantidium coli* in the central highlands of Western New Guinea' *Trop. Geogr. Med. 13*: 284-286

Crichlow, N. 1921 'Blackwater fever in the Solomon Islands' *Journal of Tropical Medicine and Hygiene. 24*: 231-

Crichlow, N. 1921 'Further cases of blackwater fever in the British Solomon Islands' *J. Trop. Med. Hyg. 24*: 255

Deland, C. M. 1951 'Filariasis in Melanesia' *Trans. R. Soc. Trop. Med. Hyg. 44*: 160

Dempwolff, 1904 'Bericht über eine Malaria—Expedition nach Deutsche-New-Guinea' *Zeitschrift fur Hygiene und Infektionskrankheiten. 47*: 81-132

Desowitz, R. S. & **Saave,** J. J. 1965 'The application of the haemagglutination test to a study of the immunity to malaria in protected and unprotected population groups in Australian New Guinea' *Bulletin of the World Health Organisation. 32*: 149-159

Desowitz, R. S., **Saave,** J. J. & **Sawada,** T. 1966 'Studies on the immuno epidemiology of parasitic infections in New Guinea Part 1. Population studies on the relationship of a skin test to microfilaraemia' *Ann. Trop. Med. Parasit. 60*: 257-264

Detzner, H. 1921 'Medizinische und hygienische Streifilichter aus dem innern von New Guinea' *Archiv for Schiffs-u. Tropenhygiene. 25*: 67

Dijk, W. J. O. M. van 1958 'Mass chemoprophylaxis with chloroquine additional to DDT indoor spraying: report on a pilot project in the Demta area Netherlands New Guinea' *Trop. Geogr. Med. 10*: 379-384

Dijk, W. J. O. M. van 1958 'Transmission of *Wuchereria bancrofti* in Netherlands New Guinea' *Trop. Geogr. Med. 10*: 21-33

Dijk, W. J. O. M. van 1961 'Mass treatment of *Wuchereria bancrofti* filariasis with diethylcarbamazine: a trial at Inanwatan, Western New Guinea' *Trop. Geogr. Med. 13*: 143-159

Dijk, W. J. O. M. van 1965 'The role of *Culex fatigans* in the transmission of *Wuchereria bancrofti* in New Guinea' *Trop Geogr. Med. 17* (4): 317-324

Dowell, M. F., **Booth,** P. B. & **Walsh,** R. J. 1967 'Blood groups and haemoglobin values amongst the Ewage and Orakaiva people of the Northern District of Papua' *Archaeology and Physical Anthropology in Oceania. 31*: 47-56

Egerton, W. S. 1954 'An account of the state of health and problems of medical interest in Goilala, Papua' *Med. J. Aust. 2*: 444

Fairley, N. H. 1945 'Chemotherapeutic suppression and prophylaxis in malaria' *Trans. R. Soc. Trop. Med. Hyg. 38.* (5): 311-366

Fairley, N. H. 1946 'Malaria in the South West Pacific with special reference to its chemotherapeutic control' *Med. J. Aust. 2*: 145-162

Fairley, N. H. et al. 1946 'Researches on paludrine (M4888) in malaria' *Trans. R. Soc. Trop. Med. Hyg. 40*: 105-162

Fairley, N. H. et. al. 1946 'Atebrin susceptibility of the Aitape-Wewak strains of *P. falciparum* and *P. vivax*' *Trans. R. Soc. Trop. Med. Hyg. 40*: 229-273

Fairley, N. H. 1949 'The efficacy of paludrine (proguanil) as a theraputic agent' *Trans. R. Soc. Trop. Med. Hyg. 42*: 623-626

Ford, E. 1943 'Malaria in the South West Pacific' *A.M.F. Land H.Q. Publication*

Ford, E. 1950 'The malaria problem in Australia and the Australian Pacific Territories' *Med. J. Aust. 1*: 749-

Fulleborn, F. 1911 'Untersuchungen über Microfilarien in der Südsee' *Arch. Schiffs—u. Tropenhyg. 15*: 358-

Fulleborn, F. 1912 'Über Microfilarien des Menschen im deutschen Südsee—Gebiet und dereu 'Turnus' nebst Bemerkungen über die klinischen Manifestationen der dortigen Filariasis' *Arch. Schiffs—u. Tropenhyg. 16*: 533-547

Gal, K. 1962 'Tropical eosinophilia in a Papuan child' *Papua New Guin. Med. J. 6* (1): 23-4

Giblin, W. E. 1954 'Daraprim in the Wahgi Valley of New Guinea' *Med. J. Aust. 1*: 9-10

Gunther, C. E. M. 1956 'The problems of malaria in New Guinea' *Med. J. Aust. 1*: 493-8

Gunther, J. T. 1955 'Epidemiology of malaria in Papua and New Guinea' *Papua New Guin. Med. J. 1*:1-

Hagen, B. 1899 *Unter den Papua's Beobachtungen und Studien über Land und Leute, Thier-und Pflanzenwelt in Kaiser-Wilhelmsland.* Wiesbaden Kriedels Verlag

Hairston, G. G., **Bang,** F. B., & **Maier,** J. 1947 'Malaria in the natives of New Guinea' *Trans. R. Soc. Trop. Med. Hyg. 40*: 795-807

Heydon, G. A. M. 1923 'Malaria at Rabaul' *Med. J. Aust. 2*: 625-633

Heydon, G. A. M. 1927 'The differences between the infective larvae of the hookworms of man' *Med. J. Aust. 1*: 531-

Heydon, G. A. M. 1940 'Parasitological and immunological observations in natives of Central New Guinea' *Far East. Assoc. of Trop. Med. Comptes-Rendus du Dixieme Congress Hanoi*, 1938, Tome II: 621-634

Hoeven, J. A. van der & **Rijpstra,** A. C. 1957 'Intestinal parasites in the central mountain district of Netherlands New Guinea: an important focus of *Balantidium coli*' *Documenta de Medicina Geographica et Tropica. 9*: 225-228

Hoeven, J. A. van der, & **Rijpstra,** A. C. 1962 'Animal parasites prevalent in the human intestine in central south New Guinea' *Trop. Geogr. Med. 14*: 350-354

Hopla, C. F. 1946 'Studies on filariasis in Papua New Guinea' *Mosquito News 6;* 189-192

Horsfall, L. E. & **Porter,** D. A. 1947 'Biologies of two malaria mosquitoes in New Guinea' *Annals of the Entomological Society of America. 39*: 549-560

Ivinskis, V., **Kooptzoff,** O., **Walsh,** R. J., & **Dunn,** D. 1956 'A medical and anthropological study of the Chimbu natives in the Central Highlands of New Guinea' *Oceania. 27*: 143-157

Iyengar, M. O. T. 1954 'Distribution of filariasis in the South Pacific Region' *Tech. Pap. S. Pacif. Commn.* No. 66. pp. 52

Iyengar, M. O. T. 1956 'Annotated bibliography of filariasis and elephantiasis Part 2, studies on mosquitoes of the South Pacific Region' *Tech. Pap. S. Pacif. Commn.* No. 88, p. 1-114

Iyengar, M. O. T. 1959 'Annotated bibliography on filariasis and elephantiasis Part 3: symptomatology, aetiology, pathology and diagnosis of filariasis due to *Wuchereria bancrofti* and *W. malayi*' *Tech. Pap. S. Pacif. Commn.* No. 109, pp. 1-25

Iyengar, M. O. T. 1959 'Annotated bibliography of filariasis and elephantiasis Part 4: treatment' *Tech. Pap. S. Pacif. Commn.* No. 124

Iyengar, M. O. T. 1960 'Annotated bibliography of filariasis and elephantiasis Part 5: prophylaxis and control of filariasis due to *Wuchereria bancrofti* and *W. malayi*'

Tech. Pap. S. Pacif. Commn. No. 129 pp. 1-102

Iyengar, M. O. T. 1960 'A review of the mosquito fauna of the South Pacific' *Tech. Pap. S. Pacif. Commn.* No. 130

Iyengar, M. O. T. 1960 'Summary data on filariasis in the South Pacific' *Tech. Pap. S. Pacif. Commn.* No. 132, pp. 1-92

Iyengar, M. O. T. 1965 'Epidemiology of filariasis in the South Pacific' *Tech. Pap. S. Pacif. Commn.* No. 148, pp. 183

Iyengar, M. O. T., **Rook,** H. de, & **Dijk,** W. J. O. M. van 1959 'Interruption of transmission of *Anopheles* bourne filariasis by indoor residual spraying in Netherlands New Guinea' *Trop. Geogr. Med. 11*: 287-290

James, C. S. 1939 'Malarial nephritis (nephrosis) in the Solomon Islands and Mandated Territory of New Guinea' *Med. J. Aust. 1*: 759

Kariks, J. 1958 'Tropical pulmonary eosinophilia' *Med. J. Aust. 2*: 773-5

Kariks, J., **Kooptzoff,** O., **Cotter,** H., **Walsh,** R. J. 1958 'A study of the heights, weights, haemoglobin values and blood groups of the natives of the Goilala sub-district, Papua' *Oceania, 29*: 117

Kersten, H. E. 1915 'Zur Frage des Bevölkerungsrückganges in Neupommern (Deutsch Neuguinea)' *Arch. Schiffs—u. Tropenhyg. 19*: 561

Kidson, C., **Curtain,** C., & **Gorman,** J. G. 1966 'Childhood anaemias in Melanesia: a haematological survey among children of Papua-New Guinea' *Med. J. Aust. 2*: 880-883

Koch, R. 1900a 'Zusammenfassende Darstellung der Ergebnisse der Malaria-Expedition' *Deutsche Medizinische Wochenschrift.* Leipzig. *25*: 781

Koch, R. 1900b 'Dritter Bericht über die Tätigkeit der Malaria-Expedition. Untersuchungen in Deutsch Neu Guinea während der Monate Januar and Februar 1900' *Deutsche. Med. Wchnschr. 26*: 281-284

Koch, R. 1900c 'Vierter Bericht über die Tätigkeit der Malaria-Expedition, die Monate März und April 1900 umfassend' *Deutsche. Med. Wchnschr. 26*: 397-398

Koch, R. 1900d 'Fünfter Bericht über die Tätigkeit der Malaria Expedition. Untersuchungen in Neū Guinea während der Zeit vom 28 April bis zum 15 Juni 1900' *Deutsche. Med. Wchnschr. 26*: 541-542

Kopp, K. 1913 'Zur Frage des Bevölkerungsrückganges in Neupommern' *Arch. Schiffs —u. Tropenhyg. 17*: 727-750

Laird, M. 1946 'Observations on *Anopheles punctulatus* Donitz, 1901, and *Anopheles farauti* Laveran, 1902, at Palmalmal and Manguinuna, New Britain, during July and August, 1945' *Trans. R. Soc. N.Z. 76*: 148

Laird, M. 1952 'Notes on the mosquitoes of Nissan Is., Territory of New Guinea' *Pacif. Sci. 6*:151-156

Lambert, S. M. 1920 'Hookworm disease in Papua' *Australian Hookworm Campaign Reports, Survey No. 7*

Lambert, S. M. 1921 'Hookworm disease in the Territory of New Guinea' *Australian Hookworm Survey Report No. 16*

Laurence, B. R. 1968 'Elephantiasis and Polynesian origins' *Nature*, Lond. *219* (5154): 561-3

Lee, D. J. 1944 'An atlas of the mosquito larvae of the Australasian Region' *A.M.F.* Melbourne pp. 1-119

Lee, D. J. 1946 'Records of *Anopheles* (Diptera:Culicidae) from high altitudes of New Guinea' *Australian Journal of Science. 8*: 165

Lee, D. J. and **Woodhill,** A. R. 1944 'The anopheline mosquitoes of the Australasian Region' *University of Sydney Zoology Dept. Monograph No. 2.* pp. 1-209

Levine, N. D. & **Harper,** P. A. 1947 'Malaria and other insect-borne diseases in the South Pacific campaign 1942-1945 Part IV. Parasitological observations on malaria in natives and troops, and on filariasis in natives' *American Journal of*

Tropical Medicine. 27 (Suppl.): 119-128

McMillan, B. 1960 'The importance of *Culex fatigans* as a vector of *Wuchereria bancrofti* in New Guinea' *Trop. Geogr. Med.* 1960 183-185

McMillan, B. 1963 'Filariasis Research Studies Progress Report' Public Health Department, Konedobu.

McMillan, B. & **Kelly,** A. 1967 'Ovale malaria in Eastern New Guinea' *Trop. Geogr. Med. 19*: 172

Mackerras, I. M. 1947 'The Australian anophelines as vectors of malaria' *Med. J. Aust. 1*:1-

Mackerras, I. M. & **Aberdeen,** J. E. C. 1946 'A malaria survey at Wewak, New Guinea' *Med. J. Aust. 2*: 763-771

Maclennan, R., **Kooptzoff,** O. & **Walsh,** R. J. 1960 'A survey of the blood groups and haemoglobin values of natives in Mount Hagen area' *Oceania, 30*: 313-320

Marsden, P. D., **Connor,** D. H., **Voller,** A., **Kelly,** A., **Schofield,** F. D. & **Hutt,** M. S. R. 1967 'Splenomegaly in New Guinea' W.H.O. duplicated report W.H.O./Mal/67. 593

Mense, C. 1899 'Berichte über das Schwarzwasserfieber in Indien und New Guinea' *Arch. Schiffs-u. Tropenhyg. 3*: 166

Metselaar, D. 1955 'Some observations on *Anopheles karwari*'. James. *Documenta Med. Geogr. Trop. 7*: 193-196

Metselaar, D. 1956 'A pilot project of residual-insecticide spraying to control malaria transmitted by the *Anopheles punctulatus* group in the Netherlands New Guinea' *Am. J. Trop. Med. Hyg. 5*: 977-987

Metselaar, D. 1957 'A pilot project of residual insecticide spraying in Netherlands New Guinea; contribution to the knowledge of holo-endemic malaria' Thesis, Leiden University. Utrecht pp. 1-128

Metselaar, D. 1959 'Two malaria surveys in the central mountains of Netherlands New Guinea' *Am. J. Trop. Med. Hyg. 8*: 364-367

Metselaar, D. 1960 'Relative increase in the prevalence of *Plasmodium falciparum* some years after the beginning of a house spraying campaign in Netherlands New Guinea' *Trans. R. Soc. Trop. Med. Hyg. 54*: 523-528

Metselaar, D. 1961 'Malaria in Netherlands New Guinea' *Trans. R. Soc. Trop. Med. Hyg. 55*: 103 (Correspondence)

Metselaar, D. & **Thiel,** P. H. van 1959 'Classification of Malaria' *Trop. Geogr. Med. 11*: 157-161.

Morahan, R. J. 1956 'Medical Survey of Omu, a Papuan Highland Village 17th November, 1956 to 24th November, 1956' *Papua New Guin. Med. J. 2*: 48-52

Mylius, R. E. & **Tenseldam,** R. E. J. 1962 'Venereal infection by *Entamoeba histolytica* in a New Guinea Native Couple' *Trop. Geogr. Med. 14* (1): 20-26

Nevermann, H. 1934 Admiralitats—Inseln. '*In Thilenius, Ergebnisse de Südse Expedition 1908-1910.*' II Ethrographie A. Melanesia, Band 3

Nieuwenhuis, A. W. 1930 'Die Entstehung der Polynesier und ihrer Kultur. Der Einfluss endemischer Malaria in Ozeanien' *Acta Leidensia. edita cura et sumptibus Scholae Medicinae Tropicae. 5*: 128

Penington, R. G. 1931 'Blackwater fever with hyperpyrexia in a New Guinea Native' *Trans. R. Soc. Trop. Med. Hyg. 25*: 137

Penn, G. H. 1949 'The pupae of the mosquitoes of New Guinea' *Pacif. Sci. 3*: 3-85

Peters, W. 1957 'A malaria survey in the Western District of Papua' *Papua New Guin. Med. J. 2*: 25-38

Peters, W. 1959 'Malaria control in Papua and New Guinea' *Papua New Guin. Med. J. 3*: 66-75

Peters, W. 1960 'Body temperature and holoendemic malaria in New Guinea' *Med. J. Aust. 2*: 658-660

Peters, W. 1960 'Studies on the epidemiology of malaria in New Guinea Part I.

Holoendemic malaria—the clinical picture' *Trans. R. Soc. Trop. Med. Hyg. 54*: 242-9

Peters, W. 1960 'Studies on the epidemiology of malaria in New Guinea. Part III. Holoendemic malaria—analysis of data and relation to control' *Trans. R. Soc. Trop. Med. Hyg. 54*: 254-60

Peters, W. 1962 'A critical survey of the results of malaria eradication and control programmes in the South West Pacific' *Ann. Trop. Med. Parasit. 56*: 1: 20-32

Peters, W. 1965 'Investigations of malaria and mosquitoes in New Guinea.' Thesis submitted for degree of M.D. in the Faculty of Medicine University of London. 1965.

Peters, W. & **Christian,** S. H. 1960 'Studies on the epidemiology of malaria in New Guinea. Part IV. Unstable Highland malaria, the clinical picture' *Trans. R. Soc. Trop. Med. Hyg. 54*: 529-536

Peters, W. & **Christian,** S. H. 'Studies on the epidemiology of malaria in New Guinea. Part V. Unstable Highland malaria—the entomological picture' *Trans. R. Soc. Trop. Med. Hyg. 54: 537-542*

Peters, W. & **Christian, S. H.** 1960 'Studies on the epidemiology of malaria in New Guinea. Part VI. Unstable Highland malaria—analysis of data and possibilities for eradication of malaria' *Trans. R. Soc. Trop. Med. Hyg. 54*: 542-548

Peters, W. & **Christian,** S. H. 1963 'The bionomics, ecology and distribution of some of the mosquitoes (Diptera: Culicidae) in the Territory of Papua and New Guinea' *Acta Tropica. 20* (1):1-

Peters, W., **Christian,** S. H. & **Jameson,** J. L. 1958 'Malaria in the Highlands of Papua and New Guinea' *Med. J. Aust. 2*: 409-416

Peters, W., & **Standfast,** H. A. 1957 'Report on a malaria survey in the Sepik District' *Med. J. Aust. 1*: 85-88

Peters, W. & **Standfast,** H. A. 1958 'The susceptibility of mosquitoes of the *Anopheles punctulatus* group to chlorinated hydrocarbon insecticides' *Aust. J. Sci. 20* (7): 218-9

Peters, W. & **Standfast,** H. A. 1960 'Studies on the epidemiology of malaria in New Guinea. Part II. Holoendemic malaria—the entomological picture' *Trans. R. Soc. Trop. Med. Hyg. 54* (3): 249-254

Pitney, W. R. 1961 'Report: haematological survey in Papua and New Guinea' Dept. Public Health, Port Moresby

Powell, K. C., & **Booth,** K. 1969 'Anaemia in the Sepik District, New Guinea' *Papua New Guin. Med. J. 12*: 3-6

Rieckmann, K. H. 1967 'A new repository antimalarial agent C1-564, used in a field trial in New Guinea' *Trans. R. Soc. Trop. Med. Hyg. 61*: 189-198

Rieckmann, K. H., **Kooptzoff,** O., & **Walsh,** R. J. 1961 'Blood groups and haemoglobin values in the Telefomin area, New Guinea' *Oceania, 31*: 296

Rodrigue, R. B. 1962 'Intestinal obstruction caused by *Ascaris lumbricoides*' *Papua New Guin. Med. J. 6*: 22-

Rook, H. de 1935 'Health of Netherlands New Guinea' *Report of the Second International Pacific Health Conference held at Sydney. 3rd-6th Sept. 1935.* Canberra Govt. Printer

Rook, H. de 1938 *In Gezondheidstoestand Nieuw Guinea 3;* 835-899. ed. by W. C. Klein, de Bussey, Amsterdam.

Rook, H. de 1957 'Filariasis in the Island of Pam' *Documenta Med. Geogr. Trop. 9*: 197-212

Rook, H. de 1957 'An investigation on filariasis in the Berau Region, North West New Guinea' *Tech. Pap. Pacif. Commn.* No. 105 pp. 1-19

Rook, H. de and **Dijk,** W. J. O. M. van 1959 'Changing concept of *Wuchereria bancrofti* transmission in Netherlands New Guinea' *Trop. Geogr. Med. 11*: 57-60

Saave, J. J. & **Desowitz,** R. S. 1966 'Studies on filariasis in New Guinea.: observations on periodicity' (Demonstration) *Medical Journal of Malaya 20*: 335-336

Saave, J. J. 1967 'Malaria Service' *Annual Report Department of Public Health T.P.N.G.* 1966-7

Sangster, C. B. 1946 'Hookworm disease in Australian soldiers with reports of cases' *Med. J. Aust. 1*: 385-393

Schellong, O. 1901 'Die Neu Guinea—Malaria einst und jetzt. Zugleich ein Beitrag zu der Geschichte der Gründung der Kolonie Kaiser—Wilhelmsland' *Arch. Schiffs—u. Tropenhyg. 5*: 303

Schofield, F. D. 1962 'Differences in palpable liver and spleen rates between men and women of the Sepik District, New Guinea. The effects of pregnancy and lactation' *Trans. R. Soc. Trop. Med. Hyg.* 56: 60-69

Schofield, F. D., **Parkinson,** A. D., & **Kelly,** A. 1964 'Changes in haemoglobin values and hepatosplenomegaly produced by control of holoendemic malaria' *British Medical Journal. 1*: 587-591

Scragg, R. F. R. 1957 'Depopulation in New Ireland (Territory of Papua & New Guinea) Port Moresby' (Reprint of Thesis)

Scragg, R. F. R. 1968 'Specialists and spraymen' *Papua New Guin. Med. J. 11*: 43-48

Scragg, R. F. R. 1969 'Mortality changes in Rural New Guinea' *Papua New Guin. Med. J. 12*: 73-83

Seligmann, C. G. 1901 'Filariasis in British New Guinea' *Journal of Pathology and Bacteriology. 7*: 308

Slooff, R. 1961 'Field observations on the biting activity of *Anopheles koliensis* Owen' *Trop. Geogr. Med. 13*: 67-76

Slooff, R. 1962 'Susceptibility to DDT of six species of New Guinea anophelines' *Trop. Geogr. Med. 14*: 80-86

Slooff, R. 1964 'Observations on the effect of residual DDT house spraying on behaviour and mortality in species of the *Anopheles punctulatus* group' Thesis, Leiden University, Holland

Spencer, M. 1961 'Malaria control: pre-operational entomological report April, 1960' *Papua New Guin. Med. J. 5*: 21-23

Spencer, M. 1964 'Blood preferences of *Anopheles farauti*' *Papua New Guin. Med. J. 7*: 19-22

Spencer, M. 1967 'Anopheline attack on mother and infant pairs, Fergusson Island' *Papua New Guin. Med. J. 10* (3): 75

Spencer, M. 1969 'Sporozoite rate in *Anopheles farauti* Laveran related to types of catch and seasonal conditions' *Proceedings Linnean Society of New South Wales. 94*: 100-104

Spencer, T. E. T. 1960 'Report on pre-operational malaria status on Buka Island, Bougainville District, January 1960' Registry. Dept. Public Health, Konedobu T.P.N.G.

Spencer, T. E. T. 1962 'The malaria pattern in the Territory of Papua and New Guinea' *Med. J. Aust. 2*: 873-877

Spencer, T. E. T. 1967 'Haemoglobin levels in the D'Entrecasteaux Islands in relation to malaria and nutrition II' *Med. J. Aust. 1*: 855-6

Spencer, T. & **Spencer,** M. 1960 'Malaria assessment methods' *Papua New Guin. Med. J. 4*: 55-61

Spencer, T. E. T., **Spencer,** M., **Jameson,** M. T., **Tommerup,** J. W. J. 1956 'Special patrol report—malaria in the Mount Hagen area' *Papua New Guin. Med. J. 1*: 110-113

Sweeney, A. W. 1968 'Variations in density of *Anopheles farauti* Laveran in the Carteret Islands' *Papua New Guin. Med. J. 11*: 11-18

Taylor, F. H. 1946 'Spiders, ticks and mites, including the species harmful to man in Australia and New Guinea' Commonwealth of Australia Dept. of Health School of Public Health and Tropical Medicine, Sydney, *Service Publications No. 6*: pp. 367

Toffaletti, J. P. & **King,** W. V. 1947 'Some records of mosquito dissections in northern New Guinea' *Journal of the National Malaria Society. 6*: 32-36

Van Thiel, P. H. & **Metselaar,** D. 1954 'A pilot project of residual insecticide spraying to control malaria transmission by the *Anopheles punctulatus* group in Netherlands New Guinea' *Bull. WHO 11*: 521-524

Van Thiel, P. H. & **Metselaar,** D. 1955 'A pilot project of residual spraying as a means of controlling malaria transmitted by anophelines of the *punctulatus* group in Netherlands New Guinea' *Documenta Med. Geogr. Trop. 7*: 164-181

Venkatachalam, P. S. 1962 'A study of the diet, nutrition and health of the people of the Chimbu area (New Guinea Highlands)' *Public Health Monograph No. 4*, Port Moresby.

Vines, A. P. 1965 'Methodology of the Papua and New Guinea, Epidemiological Survey' *Papua New Guin. Med. J. 8*: 35-

Vines, A. P. 'The epidemiological sample survey of the Highlands Mainland and Islands Regions of the Territory of Papua and New Guinea.' M.D. Thesis, University of Sydney. Govt. Printer. Pt. Moresby 1970.

Vines, A. P. & **Kelly,** A. 1966 'Highlands regional survey of intestinal parasites' *Med. J. Aust. 2*: 635-40

Vines, A. P., **Muirden,** H. J., & **Kelly,** A. 1967 'Haematological indices in a random sample survey of the highlands region of T.P.N.G.' *Med. J. Aust., 1*: 814-818

Voretzsch 1942 'Über die Entwicklung der Gesundheitsverhältnisse insbesondere der wichtigsten Infektionskrankheiten in Deutsch Neu Guinea und in Bismarck-Archipel in den Jahren 1922-35.' *Malaria Deutsche Tropen Medizinische Zeitschrift. 46*: 119

Waite, J. H. 1917 'Preliminary report on ankylostomiasis in Papua' *Med. J. Aust.* Sept. 15

Walsh, R. J., **Cotter,** H., **Macintosh,** N. W. G. 1959 'Haemoglobin values of Natives in the Western Highlands, New Guinea' *Med. J. Aust. 1*: 834

Walsh, R. J., **Jamieson,** J. L., **Kooptzoff,** O. 1960 'Blood groups and haemoglobin values of natives from Minj, New Guinea' *Oceania 31*: 77

Waterhouse, D. F. 1943 'Observations on the persistent lethal effects of DDT against anopheline mosquitoes' *A.M.F. Hygiene Service Records*

Waterhouse, D. F. & **Atherton,** D. O. 1947 1. 'Spray tests against adult mosquitoes.' 2. 'Spray tests with anopheline (*Anopheline punctulatus farauti*) adults' Comm. of Aust. *Council for Scientific and Industrial Research* (*C.S.I.R.O.*) *Bull* No. 219:29.

Williams, F. E. 1933 'Depopulation of the Suau District' *Territory of Papua—Anthropology Report* No. 13

Womersley, H. 1952 'The scrub typhus and scrub itch mites (Trombiculidae, Acarina) of the Asiatic-Pacific Region' *Records S. Aust Museum. X.* (2 Parts).

GLOSSARY

This glossary defines some of the words that we have used. Many other terms have been explained in the text and may be found by referring to the index. Where an * has been placed after a word, that word is also listed in the glossary.

absorption: the passage of digested food molecules from the gut into the circulatory system; the uptake of a substance, for example, by a cell

acidic: a solution with a pH* of less than 7; the opposite of alkaline*

acute sickness: sickness which starts suddenly, lasts only a short while, and usually has severe symptoms; the opposite of chronic*

adaptation: a characteristic of an organism that helps it to survive or makes it more efficient

adolescent: a person who is not yet an adult, but who is starting to become sexually mature

alkaline: a solution with a pH* or more than seven; the opposite of acidic*

allergic reaction: a reaction to a foreign substance, usually a protein. For example, asthma*, hay fever, and some kinds of skin rashes are allergic reactions

anaemia: not enough haemoglobin in the blood. It can be due to many causes, for example, blood loss due to injury or hookworms, or the destruction or red blood cells by malaria parasites

anatomy: the study of the structure of living organisms*

antibody: a substance produced by certain white blood cells which combines with foreign substances called antigens*, so that they do not harm the body

anticoagulant: a substance which stops the blood from clotting; it prevents the red blood cells from sticking together and forming a clot

antienzyme: a substance which stops certain enzymes from acting. Some helminths* produce antienzymes which prevent them from being digested by the host

antigen: any foreign protein or large carbohydrate molecule which causes an allergic reaction*. The body usually makes antibodies* which combine with antigens

asthma: a disease which makes breathing difficult; it is often due to an allergic reaction

bacteria: single celled organisms that do not have a definite nucleus; many cause disease; others live in soil, food and water

bacteriologist: a person who knows about bacteria* and the diseases they cause

blood donor: a person who gives blood to the Red Cross or to a hospital for use in blood transfusions*

blood transfusion: blood taken from one person is put into the vein of another person who is short of blood because of a sickness or an accident

breeding site: a place where mosquitoes breed and larvae* are found

carbohydrates: chemicals found in living organisms, containing carbon, hydrogen, and oxygen atoms; for example sugars, starch and cellulose

cell: most living organisms consist of many units called cells, each contains a nucleus and is surrounded by a cell membrane

chitin: a hard strong substance forming the outside 'skin' or cuticle of insects

chlorophyll: green pigment found in plants, essential for photosynthesis*

chronic sickness: sickness that lasts a long time and usually does not get better without treatment

cloaca: the last part of the gut which also receives gametes from the sex organs and urine from the kidneys, found in most vertebrates but not mammals

constipation: a person is constipated when the faeces* become hard and dry and are difficult to pass from the body

contaminated: polluted; impure, due to the contamination by some other substance, for example, faeces, bacteria, infective stage of a parasite

copulation: sexual intercourse, mating, push push (pigin)
copulatory spicule (of some nematodes): a hard spine or rod in the tail of the male which is used in copulation*
culture (of parasites): some parasites can be grown in test tubes or bottles in special mixtures of chemicals
cyst: a resting stage with a thick protective outer wall or membrane
decompose: to break down into smaller molecules; the process is often due to the action of bacteria
defaecate: to pass faeces*
diagnosis: the identification of a disease; in the case of parasitic diseases, diagnosis usually depends on finding and identifying the parasites
diarrhoea: the runs, the frequent passing of liquid faeces* (see also dysentery)
digestion: the breakdown of large food molecules into small molecules, by enzymes*, which can then be absorbed
disinfectant: a substance that kills the infective stages of bacteria and other parasites. Disinfectants are not medicines and are not given to kill parasites that are living in a host
disintegrate: to fall to pieces
dorsal: upper surface or back; opposite of ventral*
drug: a chemical given to a person to treat disease
dye: a stain or chemical that colours certain structures and makes them easier to see
dysentery: the frequent passing of faeces containing blood, often caused by *Entamoeba histolytica*
ectoparasite: a parasite that lives on the outside of its host, for example fleas, ticks and mosquitoes. (See also endoparasite)
elephantiasis: a chronic disease due to the blockage of the lymphatic vessels by *Wuchereria* worms, in which the skin becomes thickened and there is great enlargement of the legs or genital organs or, rarely, some other part
endoparasite: a parasite that lives inside its host (in the gut, blood, lungs, liver etc.) for example tapeworms, malaria and hookworms
environment: all the things that influence the survival, growth, behaviour, and reproduction of a living organism; for example, food, temperature, humidity, rainfall, pH*, parasites and predators
enzyme: special protein molecules which allow many of the chemical reactions, which occur in the body, to proceed. The enzyme is not used up, but the reaction cannot happen without it. A different enzyme is needed for each kind of reaction
eosinophil: one of the kinds of white blood cell
eosinophilia: an abnormally high number of eosinophils in the blood. Usually less than 5 per cent of all white blood cells are eosinophils, if the percentage is higher than this, eosinophilia is present; a common symptom of helminth infection
epidemic: a sudden outbreak of disease which infects many people at the same time
epidemiology: the study of disease in communities. It includes information about how many people are infected, how the disease is transmitted, and the factors which influence these things
evolution: a slow change in the characteristics of a species which occurs over many successive generations
extrinsic cycle (of malaria): the stages of malaria that live in the mosquito vector
faeces: wastes from the gut: *pekpek*
fats: chemicals often found in living organisms that are made up of glycerol and fatty acids
fertilization: the joining of an egg and a sperm to form a zygote*, which grows into a new individual
fertilizer: substance which makes plants grow better
fever: an increase in the temperature of the body

final host: a host in which the adult, sexually mature stage, of a parasite lives

fossil: the remains of a living organism (such as a hard shell or bones) that are preserved in rock, mud or sand

fumigate: to apply fumes or gases to disinfect or kill parasites in (for example) a room or house

gamete: a sex cell (egg or sperm)

globulins: large protein molecules found in the blood of vertebrates; antibodies are globulins

haemoglobin: a red coloured chemical containing iron, which makes up most of the material in red blood cells

helminths: parasitic worms and related free living worms; for example nematodes, cestodes, trematodes

hermaphrodite: an organism having both male and female sex organs in the one individual

histogram: a graph in which the quantity or number is represented by a column of a certain height

host specificity: the restriction of a parasite to one host, or to a few related hosts, for example the human malaria parasites are host-specific because they can only infect man and a few species of monkeys

humidity: the amount of water in the air

hydrocoele: a swelling due to the accumulation of fluid, usually in the testis

immunity: the ability of an animal to resist infection by parasites

incidence of infection: the per cent of people who are infected with a particular disease

incubation period: the period of time from the entry of a parasite into the body until the appearance of the first symptoms* of disease

infant parasite rate: the incidence of malaria in children less than one year old

infection: the establishment of a parasite in a host; an infection is a parasitic organism

infective stage: the stage of a parasite that is ready to infect a host

influenza: an acute infection* of the lungs, windpipe and nose

ingest: to eat

inherited characteristic: one that is passed from the parents to the offspring in the nucleus of the gametes*

insecticide: a chemical or mixture of chemicals that kills insects

insomnia: inability to fall asleep

intermediate host: a host that is infected with the immature or larval stages of a parasite; (see also final host)

invertebrate: an animal without a backbone, for example insects, worms, oysters, crayfish, snails, and jellyfish; (see also vertebrate)

larva: an immature stage that is different in form from the adult stage

longevity: length of life

lymphatic system: part of the circulatory system, a system of vessels, containing lymph, that drains into the blood; concerned with defence against infection

lymph: the fluid in the lymphatic system; almost the same as blood without any red blood cells

lymph nodes: glands containing large numbers of white blood cells, situated along lymphatic vessels

malariologist: a person who knows about and studies malaria

malnutrition: a condition which comes because a person does not eat enough of the right kinds of food

mental retardation: a person is mentally retarded when he cannot think and learn as fast as or as well as normal people of the same age

microfilaria: the larval stage of filarial worms such as *Wuchereria* which occur in the blood of the host

migrate: to move from one place to another. Most migrations are in a particular direction and not just round and about, for example, hookworm larvae migrate from the skin to the lungs and then to the gut

molecule: the smallest particle of a substance that still has the properties of that substance. Most molecules are made up of a number of different kinds of atoms

moulting: the process of shedding the skin (seen in insects and nematodes)

multinucleate: having more than one nucleus* in a cell, as in *Entamoeba* cysts

nectar: sugary fluid in flowers

nomadic: wandering

nucleus: a structure found inside most cells (not in red blood cells of man) which controls the activities of the cell and the development of the organism

nymph: an immature stage (of certain insects) which resembles the adult stage in appearance, but does not have mature sex organs or fully developed wings

organism: a living thing; for example bacteria, viruses, plants and animals are all organisms

overall parasite rate: the incidence of a disease (usually applied only to malaria) in the population; abbreviated to OPR

paralysis: when a muscle will not work, it is said to be paralysed

parasite rate: the incidence of infection*; usually used for malaria

parasitologist: a person who knows about and studies parasites

pathogenic: causing disease; not harmless

pathology: the study of the changes in the host that are produced by disease

periodicity: the microfilariae of the periodic form of *Wuchereria* show periodicity; they reappear in the peripheral blood* at regular intervals

peripheral blood: blood taken from the skin

pH: a measure of the acidity* or alkalinity* of a solution. It is measured on a scale going from 1 to 14. Solutions with a pH = 7 are neutral. If the pH is less than 7 they are acidic, and alkaline if the pH is more than seven. Pure water has a pH of 7

phagocytes: certain kinds of white blood cells which eat bacteria and other foreign particles

photosynthesis: the process whereby green plants make molecules of sugar from carbon-di-oxide and water. Light is necessary for this process

pilot project: a small project to find out the best methods to use in a malaria eradication program. It must be done before the full program starts

plague: a bacterial disease transmitted to man by fleas; rats are the reservoir host*

pneumonia: an acute inflammation or disease of the lungs which is often caused by bacteria of several different kinds

polluted: impure, dirty, or contaminated* with faeces or an infective organism

precipitate: an insoluable substance formed when two dissolved substances react together to form a new substance which comes out of solution

pregnant: with young developing inside the body

proboscis: mouthparts arranged to form a piercing or sucking organ, as in mosquitoes, and bedbugs

protein: large molecules found in living organisms, containing carbon, hydrogen, nitrogen and usually sulphur; built up of amino acids

pseudopod: the false foot of an amoeba

pupa: the stage in some insects, between larva and adult; pupa does not feed or move about much; adult develops inside the skin of the pupa

relapse: the reappearance of the symptoms of a disease (especially malaria) without reinfection by a new lot of infective organisms

research: careful scientific investigation by observations and experiments to discover new facts

reservoir host: a host that is naturally infected with a parasite, and provides a source of infection to man

residual insecticide: an insecticide that keeps on acting for a long time, for example DDT

rickettsia: small organisms like bacteria that cause various fever diseases, for example, typhus and Q fever

saliva: fluid secreted from the salivary glands into the mouth

sanitation: things done to prevent people from becoming infected with diseases; especially methods for treating faeces*

schizogony: asexual reproduction in which one nucleus splits up into several or many nuclei, as in malaria and other sporozoa

schizont: the stage in which schizogony* occurs; it gives rise to merozoites or trophozoites*

secrete: to produce a secretion or fluid, from a cell, gland or organ

sewage: faeces*, urine and other waste, often carried from each house in a sewer, or pipe, to a sewage treatment works

species: a group of living organisms of the same kind, having the same characteristics

sperm: male gametes, produced by the testis

spleen rate: (in malaria surveys) the per cent of people with enlarged spleens; usually determined in children only

sporogony: asexual reproduction which follows fertilization*, as in malaria and other sporozoa; results in the formation of sporozoites

sputum: saliva plus secretions of mucus from lungs and throat

stain: a dye or chemical which colours certain structures and makes them easier to see

suppressive drug: (in malaria) a drug that stops attacks (of malaria) by stopping the parasites from reproducing in the blood

surgery: the treatment of disease or injury by cutting the body open in an operation; diseased parts may be cut out and removed

surveillance techniques: (in malaria eradication) special methods to find infected people by examining the blood of all people with fever

susceptible host: one that can be infected with a particular parasite

symptom: a sign of disease, for example a pain, fever,* eosinophilia* etc

thorax: chest, the part of the body between the head and abdomen

toxic: poisonous

trophozoite: the feeding and growing stage of protozoa

urethra: tube which carries urine from the bladder to the exterior; in the human male it also carries sperms to the end of the penis

vagina: part of the female sex organs which receives the penis of the male during copulation*

vector: an organism that carries an infection from one host to another; it may be an intermediate host of the parasite concerned

venereal disease: a disease spread by sexual intercourse or copulation*

venom: poisonous substance from snakes, spiders etc

ventral: the belly side of an animal; the opposite of dorsal* or upper surface

vertebrate: an animal with a backbone; all fish, reptiles, birds and mammals are vertebrates.

virologist; a person who studies viruses*

viruses: infective organisms of very small size that can only reproduce when living inside cells of a host; for example small pox, warts, encephalitis, and influenza

vitamins: chemicals occurring in food in small amounts that are needed by animals; lack of a vitamin causes a vitamin deficiency disease

vomiting: the movement of stomach contents out of the mouth; *draut* (pigin)

water table: the level of water in the ground; or the level to which the water rises in a well

yaws: a disease causing ulcers, or sores, in the skin

zygote: a new organism which starts from the joining of an egg and a sperm*

INDEX